41 MICRO-COUNTRIES

How Economic Competitive Disadvantages, Climate Change Disasters, and Identity Divides Threaten the World's Smallest Countries

Samuel Humes IV

41 MICRO-COUNTRIES: HOW ECONOMIC COMPETITIVE DISADVANTAGES, CLIMATE CHANGE DISASTERS, AND IDENTITY DIVIDES THREATEN THE WORLD'S SMALLEST COUNTRIES

1405 SW 6th Avenue • Ocala, Florida 34471 • Phone 352-622-1825 • Fax 352-622-1875
Website: www.atlantic-pub.com • Email: sales@atlantic-pub.com
SAN Number: 268-1250

Library of Congress Cataloging-in-Publication Data

Names: Humes, Samuel, author.

Title: 41 micro countries : how economic competitive disadvantages, climate change disasters, and economic divides threaten the world's smallest countries / by Samuel Humes. Other titles: Forty-one micro countries

Description: Ocala, Florida : Atlantic Publishing Group, Inc, [2019] | Includes bibliographical references and index. | Summary: "The United Nations recognizes the 41 independent countries with less than one million inhabitants (40 were admitted as members; diplomatic missions were accepted with the Vatican). Many scholars and statesmen were initially skeptical of the survival of the small countries established outside mainland Europe in the wave of colonization following World War II. Nevertheless, they have survived the initial decades. But will these countries survive the economic, environmental, and ethnic challenges that will confront them in the next few decades?"-- Provided by publisher.

Identifiers: LCCN 2019037114 | ISBN 9781620236727 (paperback) | ISBN 9781620236734 (ebook) Subjects: LCSH: States, Small. Classification: LCC JC365 .H86 2019 | DDC 330.9--dc23

LC record available at https://lccn.loc.gov/2019037114

PROJECT MANAGERS: Meaghan Summers, Katie Cline, and Crystal Edwards
INTERIOR LAYOUT AND JACKET DESIGN: Nicole Sturk

To **Lynne De Lay,** who, for more than four decades, has encouraged my writing and tolerated my eccentricities.

TABLE OF CONTENTS

MAPS, TABLES, AND DIAGRAMS

PREFACE

"In order to survive, newborn states need to possess a set of viable internal organs, including a functional executive, a defense force, a revenue system, and a diplomatic service. If they possess none of these things, they lack the means to maintain an autonomous existence, and they perish before they can breathe and flourish."

—Norman Davies

WHAT?

What so fascinated me about micro-countries that led me to write this book? Three issues stand out. First, what were the events and issues that shaped the present micro-countries—from the precolonial through the postcolonial eras? Second, what have been the major concerns impeding their development? Third, how has the development of global and regional organizations assisted micro-countries? My aim has been to assess the many sociological, economic, and political factors affecting their viability in this interdependent world and what may continue to impact economically disadvantages, environmental disasters, and ethnic and other identity divides.

The post-World War II rapid increase of newly independent countries raised the least populous, least resourced, most isolated ones are especially worthy of attention to the innumerable challenges confronting managing their independence in this increasingly interdependent world. As my research and writing progressed, my focus extended to include the emergence of global and regional organizations and the evolution of more representative, responsible, and respectful governance.

The book's aim is to enlighten interest in small countries and their threats. The challenge confronting those under 1 million population is especially appealing, because their independence marked the end of empires as we knew them (or, more insightfully, the transition of political empires to economic ones). The tripling of "independent" countries (the addition of 35 new micro-countries and the redefinition of the criteria for independence) and the emergence of potent international institutions have fundamentally changed how countries manage their governance.

This study begins by describing how the nuances that affected how European post-medieval history affected the exploration of, expansion into, exploitation of Africa, Asia, America, and Oceania—and how post-World War II circumstances affected dependencies, becoming independent. The book focuses on the triple threat to the viability, even the existence of a few micro-countries: the economic competitive disadvantages, climate-change related disasters, and ethnic and other "identity" divides that foster dissension, delay, and deceit. The extensive use of tables enables comparisons to be made between micro-countries. The maps facilitate readers hoping to locate each of the micro-countries; they also illustrate the proximity of some micro-countries to one another and mainland and the distances that separate others.

The extent to which the survival of so many new small countries has been facilitated by the emergence and support of international institutions, including global ones such as the United Nations (UN) and its affiliates; those led by former colonial powers such as the (British) Commonwealth of Nations and the Organisation International de la Francophonie (OIF), and of a host of regional organizations of governments such as the European Union (EU), the Organization of American States (OAS), the African Union (AU), the Association of Southeast Asian Nations (ASEAN). I wish I could have given as much attention to the emergence of regional organizations of governments as I have to micro-countries. Their development is a political innovation comparable to the emergence of the federal concept at the end of the 1700s.

A global overview of the setting, the challenges, and the governance issues are presented in the first part of the book (Part I). In discussing the setting, the 41 micro-countries are identified and the world-shaping events that determined the history of these countries are highlighted. In examining the challenges the countries face, the threats to the micro-countries' continued viability and even their existence are uncovered (e.g., the populist/nationalist trade war driven economic disadvantages, climate change propelled environmental disasters, and ethnic and other identity-driven political divides). In describing governance issues, how global international institutions and the increasing presence and efforts of the regional organizations of governments has strengthened micro-country governance and development is revealed. Will these efforts strengthen their ability to meet the challenge of the three major threats to their viability and existence?

Brief biographies of each of the 41 micro-countries are presented in Part II of the book. They highlight their pre- and post-independence histories and most potent emerging challenges that will confront them in the coming decades, such as increased economic disadvantages, more and worsening environmental disasters, and the continued impact of ethnic and other identity divisions that are frustrating governance. They have survived the challenges of the initial decades of independence! How many will survive those of the mid-2000s?

"Looking Ahead," the final section in the book, concludes the text by presenting how the further development of regional organizations of governments may improve their participation in world governance as well as their own governance. Consideration of the crises confronting micro-countries provides a lens for appreciating the threats to most countries.

Will micro-countries survive the economic, environmental, and identity-divide challenges confronting them in the next few decades? Five forewords, written by scholar-practitioners, analyze the issues confronting small countries today.

WHY?

Why write about the world's 41 least populous independent countries? Why was I so curious and concerned about them? Together, these 41 micro-countries, each with fewer than 1 million people, account for less than

13 million of the world's 7-plus billion people, yet they constitute about one-fifth of the world's independent countries! How many have read anything recently about Palau, Comoros, San Marino, or Suriname?

Appreciating the issues confronting the very smallest countries provides an insight into the issues confronting all smaller countries. I was prompted by curiosity as well as concern! How does a Lilliput survive in a world still run by (21st-century version) empires? Can they continue to survive in a world in which many countries are becoming more inwardly oriented, affecting their skepticism of foreign aid and free trade?

This book aims to further an interest in and understanding of micro–countries, all but a few of which became independent within the last few decades. Micro-countries provide a unique perspective on the challenges of globalization, including increasing inequality among countries and people, global governance, and small government challenges. The prospect of writing about their challenges and prospects was irresistible.

My interest in small governments propelled a lifelong odyssey. In elementary school, I designed governments for my toy soldier island country. Even before my father died—before I became a teenager—my interest in governance was kindled by listening to my father's insights, talking to his friends in the courthouse where my father's office was and on the campaign trail.

My odyssey continued through Williams College, Wharton Graduate School of the University of Pennsylvania, and the University of Leiden. Three books comparing local governments throughout the world followed: *The Structure of Local Government Throughout the World* (which was my University of Leiden Faculty of Law doctoral dissertation, sponsored by the International Union of Local Authorities at the request of the United Nations Educational, Social and Cultural Organization); *The Structure of Local Government, A Comparative Survey of 81 Countries*; and *Local Governance and National Power, A Worldwide Comparison of Tradition and Change in Local Government.*

Directing the first regional council of local governments, serving as chief administrator of a municipality of 600,000 residents, preparing the plan for reorganizing local governments in Nigeria, serving on the commission that recommended the reorganization on the government of Kenya, and writing two books describing the millennium—plus transition from tribalism to nationalism—one about Nigeria, the other about Belgium.

I resisted the temptation to write a study of the amazing development of the European Union; too many well-informed and qualified scholars, including my friend Anthony Teasdale, have already written such studies. The time is coming when a comparative survey of regional organizations of government/confederations will be written by a younger scholar. Such a study would deserve the attention of the political studies community.

WHO?

My parents, teachers, career mentors, and scholars provided the foundation for my efforts as well as those who read and provided insight regarding this manuscript. The tutelage of my parents created a desire to learn. My father, a state judge whose death came far too early, succeeded in exciting in me an interest in politics and governance that has lasted. My mother challenged and helped me pursue my varied interests.

My teachers at Williamsport public schools, The Hill School, Williams College, the University of Pennsylvania, and the University of Leiden gave me the confidence to be curious. My colleagues, staff, members of governing bodies, and others with whom I associated mentored me in the nuances of governing. Later, my professional faculty colleagues and students taught me to persist in asking provocative questions. Many of these are no longer available to thank for helping me develop my confidence, curiosity, and creativity.

James MacGregor Burns, a Pulitzer Prize winner and my Williams College Honors thesis mentor reminded me that "writing about the past is a collective enterprise." C.H. Polak, my Leiden 'promoter' reminded me that giants provide the shoulders upon which enterprising scholars stand. Professors Burns and Polak gave me the confidence to write for publication.

Williams College provides a nurturing haven for authors. The library's extensive collection, its global scope, its open stacks, and the helpfulness of its staff were indispensable. Magnus Bernardsson introduced me to the field of micro-countries. Sam Crane, James Mahon, and professor-emeritus John Hyde provided advice. Roger Bolton, professor of economic emeritus at Williams, did a thorough and thoughtful job of editing an early version of the manuscript, suggesting basic improvements. Richard Hespos and Susan Stetson Clarke applied their eagle eyes to proofreading. Adam Hall creatively and resourcefully crafted several tables and located the maps to illustrate the themes in the text.

A.J. Mediratta, a Greylock Capital senior partner, provided critical insights and facilitated meetings with government leaders. Liam Locario, another Greylock Capital partner, also provided assistance. Greylock Capital's support was critical to the success of this project.

My Atlantic Publishing project manager and editors, Meaghan Summers, Katie Cline, and Crystal Edwards not only coordinated the process but also made valuable suggestions that extended from the title to editing out superfluous and misleading words; for that, I was lucky and am very grateful. It was a genuine pleasure to work with them. Nicole Sturk designed the cover and many of the interior pages.

Four scholars, whose expertise and insights I highly regard, wrote forewords introducing this book and highlighting themes asserted in the text. They are as follows: DeLisle Worrel, former governor of the Central Bank of Barbados; Douglas Yoder, deputy director of the Miami-Dade Department Water and Sewers and member of the Florida Water Resources Advisory Commission (and a former doctoral student and then academic colleague of mine); Michael Peyrefitte, the attorney-general of Belize and former speaker of its House of Representatives; and Anthony Teasdale, director-general of European Parliament Research Centre and co-author of *The Penguin Comparison to European Union*. I am profoundly grateful that they accepted my invitation and wrote such provocative messages.

The experience of living and working in three European countries, three African countries, and three Asian countries, as well as Canada, 10 U.S. states, and visiting 88 other countries has fostered my appreciation of the diversity of cultures and governance around the world. While all places face threats, the specifics and severity of these challenges differ. I have appreciated the frank and informative conversations with persons of most of the 41 micro-countries (those quoted are listed under "Interviews" in the bibliography). These insights have been critical to my understanding of micro-countries their distinctive histories, their governance, and their challenges.

This survey has been an extended family affair in several respects. All the Humes (by birth or marriage) not only tolerated but also encouraged my absorption with the project. My older son, Samuel Hamilton, read and critiqued the final draft. My younger son Hans not only urged me to write a book on the world's smallest countries and shares with me an interest in developing countries but also wrote the chapter on the impact of COVID-19, as the founder and managing partner of Greylock Capital, which specializes in restructuring the debt of countries.

Lynne De Lay, my wife and confidant of 44 years, has always encouraged my plans and projects and tolerated most of my eccentricities. Without her support, encouraging me to undertake this project, accompanying me on working visits to micro-countries, and editing the manuscript, I would not, and could not, have completed this survey.

The withering of colonial empires and the advent of micro-countries has been a notable feature of the past several decades. The maturing of micro-countries has been accompanied by the formation of the United Nations and affiliated global organizations and the development their assistance and aid to developing countries as shelter countries. The emergence of regional organizations of governments has increased the momentum leading our world community to where we are and where we may be going. Let us hope that the challenges will be met in a way that is more effectively representative—that is, a more responsive, responsible, and respectful global governance system. I hope I have stimulated your curiosity and engaged your concern about the world's smaller countries as well as the 41 micro-countries.

FOREWORD

THREATS CHALLENGING MICRO-COUNTRIES

A TEMPLATE AND PACESETTER FOR REGIONAL
ORGANIZATIONS: HOW MICRO-COUNTRIES CAN TEACH US

*Anthony Teasdale, Director General of the European Parliament Research Centre
and co-author of* The Penguin Companion to the European Union

Ever since writing his undergraduate thesis under James MacGregor Burns
in the early 1950s, Dr. Samuel Humes has devoted a large part of his life to
the study of how power is exercised in multilevel political systems—ranging from the federal politics of the United States and Belgium to the structure and operation of local government worldwide. In this latest book, Dr.
Humes turns his hand to analyzing the experience of the smallest countries
in the global community, picking out the 41 micro-states that have fewer
than 1 million inhabitants, a majority of them ex-colonies that became
independent following the Second World War.

Although the patterns of politics and power that Dr. Humes chart vary
enormously—encompassing countries as diverse as Bhutan, Liechtenstein,
Djibouti, the Bahamas, Cape Verde, Tuvalu, and the Vatican—they often have certain common characteristics. Notable among these is access to
and reliance on bigger, "sheltering" states that protect them and help them
survive in a potentially hostile environment, offsetting some of the vulnerability that comes of their size. Such mentor countries may be friendly
neighbors or ex-colonial powers on whom they are economically or polit-

ically dependent—or they may be wider groups of countries aspiring to some form of collective strength through regional integration.

The latter phenomenon is linked to the broader importance of international law in making the proliferation of micro-states easier in the post-war world. The idea that greater self-determination might be viable among very small countries has been underpinned by the notion of equality between states and a move away from the more brutal realpolitik and "might is right" principles that previously prevailed.

Membership of a wide range of international political and economic organizations, operating within a rules-based system based on mutual respect—ranging from the United Nations to the Alliance of Small Island States—has provided structures in which micro-states can now find their place. Membership of 13 regional organizations of governments has proved especially important in this regard. Whether future decades will offer as hospitable a climate for micro-states to flourish as the last 70 years remains to be seen.

One of the interesting patterns identified by Dr. Humes is that micro-countries often perform better economically than their larger neighbors. The reasons for this are almost as varied as the countries themselves and can include simpler and more efficient governmental structures, distinctive appeals as tourist destinations, higher per-capita aid receipts, the hosting of foreign military bases, and/or their use as tax havens. Micro-countries are also characteristically somewhat more democratic than other countries worldwide: three-quarters qualify as democracies in some form.

Dr. Humes' definition means that 10 countries in Europe are considered to be micro-countries, and, interestingly, all now operate within the broad framework of the European Union (EU), even if only three of them are actually member countries (Cyprus, Luxembourg, and Malta). In different ways, it is the EU that offers them a strong regional framework of stability in which to develop and prosper. Two of the remaining seven states are very closely linked to the EU (Liechtenstein and Iceland); three are administered in part through individual EU member states and use the European Union's currency, the euro (Andorra, Monaco, and San Marino); one aspires to join the EU and already uses the euro (Montenegro); and another non-EU-member euro (the Vatican).

The EU is the most advanced example of regional integration in the world, with common institutions adopting and applying a wide range of joint policies, underpinned by supranational law. It is unlikely that any other regional grouping will ever achieve the same level of interlinking and collective collaboration as that currently seen in Europe. One of its principal features—and, arguably, its source of success—is the existence of an executive body, in the form of the European Commission, which sees its role as defending equally the interests of members big and small. Likewise, in the Council of Ministers, there is no sense that the rotating presidency of a country like Luxembourg or Malta is less legitimate than that of Germany or France, even if their voting power is obviously much smaller. The EU has been a model for the many regional organizations that have developed throughout the world and may continue to serve as a model as they grow in the ability to promote joint action in common programs and advocating regional interests.

Dr. Humes has done a great service by drawing attention to the rich tapestry of micro-states that exist around the world. Their varying experiences and distinctive contribution to the global community merit greater study, and this book goes a long way to correct an important missing gap in political analysis. The differences and similarities sketched here are fascinating and offer a powerful portrait of a significant but little-understood component of today's global order.

WHY SMALL ECONOMIES MUST BE MANAGED DIFFERENTLY

Dr. DeLisle Worrell, former governor of the Central Bank of Barbados

We, who live and work in countries with fewer than 1 million people, find it enlightening to entertain the views of thoughtful commentators who take an intimate interest in our economies and societies from a fresh perspective. Dr. Humes is intrigued by our survival, a sentiment shared by many who live in municipalities that are larger and endowed with more resources than ours. The observation that micro-countries such as Bermuda and Iceland have standards of living among the highest in the world has deceived the mainstream of the economics profession into the belief that fiscal and exchange-rate policies suitable for countries including the United States, Brazil, and Nigeria are equally suitable for Fiji, the Comoros, and Barbados. The experience of small states suggests otherwise. Small states may do well economically but only if they use the tools appropriate to their circumstances, especially where practice departs from orthodoxy.

Contemporary economics may have lost sight of the fact that small economies are necessarily outward looking. Just as the subsistence of a family depends entirely on its own resources, the small economy that shuts itself away from international commerce deprives its population of the amenities of civilized life. Small state economies flourish to the extent that they produce goods and services that can be sold competitively at ruling prices on international markets. The proceeds they use to purchase from abroad the requirements for a modern way of life. The more successful they are in adding value to the things they sell to the world, the more prosperous the economy.

The reality of the small open economy is not well understood by those who share the common belief that devaluation of a country's currency stimulates the economy to faster growth. That cannot be the case, no matter how many purportedly authoritative studies are published in support of this fallacy. With devaluation, domestic residents will have more local currency for every U.S. dollar, but they earn no more U.S. dollars and, therefore, will not be able to buy anything they need for improving their livelihood. This is because those are all imported from abroad.

The International Monetary Fund and most mainstream economists also recommend that small countries set up independent central banks, or joint regional ones, with a mandate to control inflation, just as large countries are reputed to do. However, the prices of most things small states consume are set abroad; the only things affected by central bank policy are government services and other services that, by their very nature, cannot be traded. So long as the central bank does not add to expenditures by lending to government agencies in local currency, the rate of inflation will remain at the international level, determined by the prices of imports.

The drive to increase domestic saving as a means of speeding economic growth is another fallacy of mainstream thinking about small economies. Growth potential is augmented not by savings per se, but when savings are invested to increase the capacity to produce competitive exports and services. But, as with everything else, most materials used in investment have to be bought abroad. For these, local currency savings cannot be used. It follows that there is an upper limit to the domestic savings ratio, set by the proportion of investments that has to be imported. It is often the case in small economies that the domestic savings rate will *fall* when major investment projects are underway, because the imports needed for investment projects are so substantial.

Many small countries have prospered by explicitly recognizing that international commerce is their economic lifeblood. Their governments have maintained surpluses of tax revenues over operating expenses, combining these savings with prudent foreign borrowing to fund investments in social and economic infrastructure. When required, these governments were able to borrow U.S. dollars to make good temporary shortages due to circumstances beyond their control.

The contrast between the countries that have pursued disciplined fiscal policies and those that have allowed operational expenses to run ahead of revenues is reflected in overall economic performance and the citizens' quality of life. When a government borrows to fund current expenditures, it impoverishes the country, because there is a burden of new debt to be serviced with no new capacity to generate income to service that debt. In addition, when a government that issues its own currency borrows from the central bank, the excess spending results in additional imports, which have to be paid for by drawing on the central bank's foreign reserves. If such borrowing persists, the foreign reserves are eventually depleted.

The central bank's foreign reserves are what allow governments to protect the purchasing power of the local currency in U.S. dollars. Since all local consumption and investment use imports, that matters. When the central bank no longer has sufficient foreign reserves to supply any temporary foreign currency shortage, the government loses control of the exchange rate, and the value of the local currency plummets. In small states, devaluation is the evidence of fiscal extravagance. The situation can be corrected only when government reverts to the Golden Rule: that operating expenses must be contained below government revenues, allowing for fiscal savings.

Good fiscal policy provides the basis for economic success in small countries. Tax policy should reinforce competitive strengths of exports and services, providing incentives for investment in capacity and productivity. Current expenditures should be devoted to providing public services effectively, within the limits of tax revenues not generally regarded as overly burdened. Government should maintain a small surplus on the current account, which—along with foreign borrowing to provide much-needed foreign currency and some local borrowing—will be used to build infrastructure and upgrade technological capabilities. Micro-countries have demonstrated that they can outperform larger countries. Those consistently implementing prudent fiscal policies will survive and prosper.

NATURAL DANGERS FOR MICRO-COUNTRIES: PREPARING FOR ENVIRONMENTAL DISASTERS

Dr. Douglas Yoder, Deputy Director of the Dade County (FL) Water and Sewer Department and member of Florida Water Resources Advisory Commission

In 2017, multiple powerful hurricanes made their way through the Caribbean basin, causing extreme damage to many island countries and Florida, lingering long enough that its 90-mile-per-hour winds caused substantial flooding, the massive destruction of structures and infrastructure, and the almost total loss of electric power. Two years later, much of the damage in the Caribbean remains in a condition of precarious recovery. The differential is, in part, geographical: islands are inherently more difficult to service logistically. Part is the economic base, the political resolve, and the foreign support to prepare for and repair the damage.

Areas prone to various types of recurring natural disasters prepare for those recurrences to the extent that their resources allow. Climate change represents a new and dynamic set of conditions that will challenge communities and nations in different ways for the foreseeable future. Miami has been identified as the most vulnerable metropolis in the world to the risk of sea level rise in terms of the value of property that is exposed to damage and loss.

As the local population has begun to experience the consequences of sea level rise in the form of "clear sky" street flooding at extreme high tides, southeast Florida communities have banded together to assess vulnerability and assess adaptation strategies to mitigate risks that may be realized over the next 50 years. This expression of local political will, not necessarily matched at the state and federal levels at this time, has resulted in substantial local investment to use sophisticated modeling and analytical tools to evaluate future conditions such as increasing storm surges, increasing frequency of chronic flooding, and increasing salt water intrusion into the fresh water supply.

Water and sewer infrastructure, much of which is vulnerable to storm surge damage and flooding, is now being designed to withstand conditions that include the projected amount of sea level rise that is likely to occur over the life of the asset. The incremental cost of mitigating this future risk as infrastructure is being replaced or new infrastructure is being built to meet new demands is relatively modest and manageable in the case of utility infrastructure, but it is also important for communities to understand that all types of infrastructure will be impacted. This could mean elevating roads and other transportation infrastructure, elevating buildings, and re-thinking drainage systems. Secondary concerns that may be directly related to the effectiveness of public responses to climate-related risks include the willingness of insurers to continue offering insurance products and the willingness of investors to support and invest in regional economies.

Over the longer term, of course, there may be a point at which mitigating risk to sea level rise is not technically or economically feasible. Many island micro-countries, along with low-lying coastal areas such as south Florida, face this possibility. Retreat is a very complex strategy to plan and manage. Environmental disasters have led to thousands of residents moving to other locations, many of whom have not returned. For less developed economies in nations with very limited boundaries, this can easily become a refugee problem of significant proportions.

The international system for handling transnational refugees is inadequate. Providing adequate international aid where the disaster occurred after a natural disaster is challenging. Protecting threatened land and people and failing that moving large numbers of people out of harm's way and into some alternative country on a permanent basis. Political destabilization is driven by cultural and economic limitations. As Dr. Humes reports, the very international institutions that need to facilitate such efforts are struggling to remain viable; nationalism challenges a world order of trade, peacekeeping, and disaster response that once held promise for addressing such issues. This has been very evident in the long-standing negotiations to address climate change itself through control of carbon emissions that may, at least, limit the longer-term consequences that could totally disrupt the natural world in addition to the political world.

Many micro-countries operate in fragile circumstances. In some cases, historic ties with former suzerain or neighboring countries provide a degree of economic and political stability that can be particularly stabilizing following natural disasters and similar challenges. In other instances, former colonies maintain such associations with their former empires. In the end, the ability of the micro-countries to plan for and survive, in some fashion, the challenges presented by climate change will depend on successful international collaboration at some institutional level. This may happen more readily where an alliance appeals to various players (e.g., Dr. Humes' reference of China) seeking to gain political or economic advantage in the world. That leaves situations where a country's finances, location, and attractions provide insufficient incentives for an alliance.

In the absence of international will to plan for and develop strategies to address threats, the world order will surely struggle to respond to these natural disasters. Climate change adds another layer of risk to the already risk-burdened circumstances of many micro-countries, particularly flat and low islands that are already challenged by the rising sea. Few threatened seacoast cities have property values sufficient to justify investing billions on seawalls. Will the world respond by allowing only the survival of the fittest?

THE CASE OF MICRO-COUNTRY BELIZE: AN ANGLO ENCLAVE IN LATIN AMERICA

Michael Peyrefitte, Attorney General of Belize

Belize is micro-country geographically situated on the Caribbean coast of Central America. It is an English-speaking country surrounded by an almost wholly Spanish-speaking Central American sub-continent. Belize covers 8,867 square miles and has a population of about 380,000 persons. In addition to be being a small state, Belize is also a small, developing state, and, as such, faces special development challenges resulting from its limited resources, remote location, and geographic isolation from other countries that share the same culture and language as well as small population.

As a small state, size is a constant impediment to sustained economic growth. Belize's economy rests on a narrow agricultural base of just a handful of commodities such as sugar, bananas, citrus, and marine products. The corresponding small population means that there is low domestic demand for goods and services produced. This in turn leads to overdependence on external trade. As a consequence, Belize is exposed to market shocks that affect income, employment, and expenditure—exogenous shocks, including volatile world market prices for commodities.

Additionally, small island developing countries like Belize are highly vulnerable to climate change and natural disasters. Belize has experienced a frequency in the occurrence of hurricanes and droughts and increased severity of impact. The recurrent financial, climate, and disaster shocks reduce a government's fiscal space (i.e., increase a government's financial flexibility). The government is, thus, challenged to respond to shocks and disasters and instead become dependent on external aid or loans. In Belize, the government also faces the challenge of providing public services and infrastructure for a population that is scattered across the country, including in remote areas.

Among the few advantages of small states is that size is conducive to democratic decision-making and consensus building. Belize is a longstanding democracy; free and fair elections are a normal part of the political cycle. In addition, a small state like Belize has a smaller administrative bureaucracy. Another advantage of small statehood is that it enables better strategic planning and management of resources. This is due, in part, to the fact that resources are limited as well as indispensable. Small states have also gained experience in adapting to changing circumstances swiftly. Belize is a pioneer in, for example, its eco-based sustainable management approach to fisheries.

Sugar is an important foreign exchange earner for Belize, but changing market conditions in the EU, the main importer of Belize sugar, necessitated critical reforms in Belize's sugar industry in order to maintain competitiveness. Since small states have open economies and are very dependent on trade, they must adjust to constantly changing terms of trade.

To overcome these challenges and to maximize advantages, small states have had to join forces and build regional organizations to amplify their voices, advocate issues of common interest, and work on joint projects. The fact that Belize does not share the same culture and language as its immediate Central American neighbors limits its inter-country cooperation with them, especially since it has a long-simmering boundary dispute with Guatemala.

The Caribbean Community and Common Market (CARICOM) is an integrative organization of 15 small states through which we are able to expand economic and trading opportunity by establishment of a single market and to pool resources for common institutions such as the Caribbean Court of Justice. Belize also belongs to the Central American Integrative System (SICA), the Community of Latin American and Caribbean States (LIMUN), and the Organization of American States (OAS). In addition, Belize is member of several multi-continental groupings of small states, such as the Alliance of Island States (AOIS) in which we advocate on climate change, and the Forum of Small States in which we coordinate on issues before the United Nations.

PART ONE
THE 41 MICRO-COUNTRIES, A GLOBAL PERSPECTIVE

"Successful statehood, in fact, is a rare blessing. It requires health and vigorous good fortune, benevolent neighbors, and a sense of purpose to aid growth and reach maturity. All the best-known polities in history have passed through this test of infancy, and many have lived to a grand old age. Those which have failed the test have perished without making their mark. In the chronicles of bodies politic, as in human condition in general, this has been the way of the world since time immemorial."

—Norman Davies

"One peculiar feature of our age is the acceleration of the pace of change to an unprecedented degree as a result of 'the annihilation of distance' through the extraordinary recent advance of technology. History is now being made so fast that it is constantly taking us by surprise."

—Arnold Toynbee

Table 1.1 Dependencies, Micro-, Mini-, and Mega-Countries ranked by 2016 population

Dependencies Population: over 1,000	Pop.	'Micro-Countries' Population: less than 1 million	Pop.	'Mini-Countries' Population: between 1 and 5 million	Pop.	'Mega-Countries' Population: over 50 million	Pop.
Tokelau NZ	1,276	Vatican	842	Timor-Leste	1,211,245	South Korea	50,503,933
Niue NZ	1,612	Nauru	9,450	Mauritius	1,277,459	Myanmar	54,363,426
Falkland Is UK	2,912	Tuvalu	10,869	eSwati (Eswatini)	1,304,063	South Africa	54,978,907
Saint Helena UK	3,956	Palau	21,265	Estonia	1,309,104	Tanzania	55,155,473
Montserrat UK	5,154	Monaco	30,535	Trinidad & To	1,364,973	Italy	59.801,004
St.Pierre & Miqu F	6,301	San Marino	33,020	Bahrain	1,396,829	France	64,668,129
Wallis & Futuna F	13,112	Liechtenstein	37,624	Gabon	1,763,142	U.K.	65,111,143
Anguilla UK	14,763	St.Kitts & Nevis	51, 936	Guinea-Bissau	1,888,429	Thailand	68,146,609
Cook Is NZ	20,948	Marshall Is	72,192	Latvia	1,955,742	Turkey	79,622,062
CarNetherland N*	25,328	Dominica	73,607	Gambia	2,054,986	DR Congo	79,722,624
Virgin Is UK	30,659	Andorra	85,580	Slovenia	2,069,362	Iran	80,043,146
Gibraltar UK	32,373	Seychelles	92,430	Macedonia	2,081,012	Germany	80,682,351
Turks & Caicos UK	34,904	Antigua & Barbu	92,436	Lesotho	2,160,309	Egypt	93,383,574
Sint Maarten Fr	39,538	St,Vincent & Gr	102,627	Qatar	2,291,368	Viet Nam	94,444,200
Faeroe Is D	48,239	Micronesia	105,218	Botswana	2,303,820	Ethiopia	101,853,268
N. Marianas US	55,389	Tonga	106,501	Namibia	2,513,981	Philippines	102,250,133
America Samoa US	55,602	Kiribati	106,711	Jamaica	2,803,362	Japan	126,323,715
Greenland D	56,196	Grenada	107,327	Lithuania	2,850,030	Mexico	128,632,004
Cayman Is UK	60,764	Saint Lucia	163,,992	Albania	2,903,700	Russia	143,439,832
Bermuda UK	61,662	SaoTome & Prin	194,006	Mongolia	3,006,444	Bangladesh	162,910,864
Isle of Man UK	88,421	Samoa	197, 773	Armenia	3,026,048	Nigeria	186,987,563
Aruba N*	104,263	Vanuatu	272,264	Uruguay	3,444,071	Pakistan	192,826,502
U.S. Virgin Is US	106,415	Barbados	290,604	Bosnia & Here	3,802,134	Brazil	209,567,920
Curaçao N*	158,635	Bahamas	324,627	Georgia	3,979,781	Indonesia	260,581,100
Channel Is K	164,466	Iceland	331,918	Panama	3,990,406	United States	324,118,787
Guam US	172,094	Belize	347,369	Kuwait	4,007,146	India	1,326,801,576
N. Caledonia F	266,431	Maldives	393,353	Mauritania	4,166,463.	China	1,367,485,388
Fr. Polynesia F	285,735	Malta	413,968	Croatia	4,225,001		
W. Sahara M	584,206	Brunei	429,646	New Zealand	4,565,185		
Macao C	597,126	Cape Verde	545, 993	Liberia	4,615,222		
Puerto Rico US	3,680,772	Luxembourg	570,252	Oman	4,654,471		
Palestine Gaza I	4,797,239	Suriname	579,633	Ireland	4,713,993		
Hong Kong C	7,141,106	Solomon Is	622, 469	Congo, Rep of	4,740,992		
		Montenegro	647,673	Costa Rica	4,857,218		
		Guyana	735,222	C. African Rep	4,998,493		
		Equator. Guinea	740,743				
		Bhutan	741,919				
		Comoros	780, 971				
		Djibouti	828,324				
		Fiji	990,389				
		# Cyprus/	189,197				

Source: Data from the 2016 UNs Population Division and *The World Almanac and Book of Facts (2016)*. Data assembled by author.

\# The UN considers the Republic of Cyprus includes the Turkish Republic of Northern Cyprus (TRNC) and lists it with a population of 1,189,197; but 275,265, of these reside in area controlled by the TRNC.

^ The population of China does not include Hong Kong, Macao, and Taiwan.

(N*) The Netherlands does not consider these polities to be territories.

Left column letters name governing power: C=China, D=Denmark, F= France, I=Israel, M=Morocco, N=Netherlands, NZ=New Zealand, UK=United Kingdom, and US=United States.

The UN considers six *de facto* autonomous regions as parts of member countries: Abkhazia and South Ossetia (two separatist regions [formerly part of Georgia SSR] are still considered part of Georgia); Transnistria and Nagana-Karabakt (two separatist regions formerly part of Moldova SSR); Somaliland (a separatist region split off from Somalia still considered part of Somalia); and the Turkish Republic of North Cyprus (TRNC), located in the north-eastern part of Cyprus which Turkey occupies and is not considered as part of Cyprus. Taiwan—with a population of 2,415,126—is no longer recognized by the United Nations. Seventy-seven countries recognized Taiwan in 1972; in 2018, the number was down to 17.

1. INTRODUCING THE 41 MICRO-COUNTRIES

"We live in a world of nearly two hundred states. Each flaunts symbols of sovereignty . . . These states, big and small, are in principle equal members of a global community . . . Throughout history, most people have lived in political units that did not pretend to represent a single people. Making a state conform to a nation is a recent phenomenon, neither fully carried out nor universally desired . . . What, then, is an empire . . . Empires are large political units, extended over space, polities that maintain distinction and hierarchy as they incorporate new people. The nation-state, in contrast . . . proclaims the commonality of its people—even if reality is more complicated—while the empire-state declares the non-equivalence of multiple populations . . . The concept of empire presumes that different peoples within the polity will be governed differently."

—Jane Burbank and Frederick Cooper

Map 1.1 Contemporary World Map with the 41 Micro-countries Marked

Source: World Map by M. Ruskin Co. LLC

EUROPE

Luxembourg
• Liechtenstein
   • San Marino
   •   • Montenegro
Vatican
   • Malta   • Cyprus

ASIA

   • Bhutan

AFRICA

   • Djibouti

   • Marshall
     Islands

   • Fed. States
• Brunei   • Palau   of Micronesia   Kiribati

• Maldives

   Seychelles
      •

ea •
•
ipe

   • Comoros

   • Solomon
     Islands   • Tuvalu

   • Vanuatu

International Date Line

      • Nauru

      • Samoa
   • Fiji   • Tonga

N
W   E
S

MICRO-COUNTRIES: COUNTRIES WITH POPULATIONS LESS THAN 1 MILLION

Of the 41 micro-countries[1]—countries with fewer than 1 million inhabitants whose independence is generally recognized—40 were United Nations (UN) members in 2016; the Vatican, not a UN member, is the 41st micro-country. Since the UN's founding in 1946, admittance to its membership affirms a country's independence. The UN lists 193 independent and dependent countries, which are as follows:

Population	Independent/UN members	Independent but not UN members	Dependencies with over 1,000 population total
Under 1 million: 40	Vatican: 1	30	71
Over 1 million: 153	Taiwan and Kosovo: 2	Hong Kong, Palestine, and Puerto Rica: 3	158
Total: 193	3	33	229

Ten of the 41 countries are European. The six in mainland western Europe have long been quasi-independent. The other micro countries—six African, three fringe South Asian, 11 in the extended Caribbean American, and 11 in island Oceanian—are tropical.

Of the 41 micro-countries, 28 are islands. Distance and ethnic and other "identity" schisms undermined the openness of communities to merge with a polity with which they shared little sense of community. Of the 34 micro-countries not on the European mainland, 21 are former British dependencies. This fact reflects not only the extensiveness of the British Empire but also its decolonizing strategy, which stressed the importance of granting independence to small entities. The French and Dutch prefer to keep the countries within their sovereign umbrella as local governments or "autonomous kingdoms."

---

1. The author prefers the terms "country" and "micro-country" rather than "nation" or "state" for an independent polity. The word "nation" is avoided because The Shorter Oxford English Dictionary defines "nation," first, as "a distinct race or people, characterized by common descent, language, or history" and, second, as a "separate political state." The word "state" is avoided because it is also used for a component part of a federal country, such as the United States, Germany, Brazil, and India.

The mid-2020s world contains over 7 billion people (compared to about 1.7 billion in 1900); China and India each account for more than 1.3 billion of the world's total population, and the United States, Indonesia, and Brazil account for over 200 million, meaning these five countries contain almost 47 percent of the world's population. In contrast, the micro-countries in terms of population are Lilliputians. They range in populations similar to the Vatican (with fewer than 1,000 people) and Nauru and Tuvalu (with fewer than 11,000 people) to Fiji, Djibouti, and the Republic of Cyprus (with just under 1 million inhabitants). Monaco, the Vatican, the Maldives, and Malta rank with Singapore as the most densely populated countries in the world. Iceland, on the other hand, is the world's fourth least-populated country.

Gross Domestic Product (GDP) and Gross Domestic Product per capita (GDPpc) in the mid-2020s are compared in Table 1.2. Six European countries had the highest GDPpcs: micro-countries Monaco, Liechtenstein, and Luxembourg, as well as Sweden, Norway, and Ireland. Micro-country Montenegro in Europe, four micro-countries in Africa, Bhutan in Asia, and seven micro-countries in Oceania had among the lowest per-capita GDPs.

WILL THE NEW MICRO-COUNTRIES SURVIVE THE MID-2000s?

Skeptics questioned whether the newborn small countries had infrastructures, industries, and institutions sufficiently developed to support governments capable of managing external as well as internal affairs. This survey of the 41 micro-countries focuses on the following: (1) how the new micro-countries gained their independence and survived their first decades and (2) whether they will survive the economic disadvantages, environmental disasters, and ethnic and other divides that threaten them in the mid-2000s. Will changing the way countries' governments relate to one another help them overcome these challenges?

Table 1.2 41 Micro-countries: Population, Budget, GDP, GDPpc, Growth, Industries, and Tourism

Country	Population Nearest 1,000	Budget Nearest $1,000,000	GDP Nearest $1,000,000	GDPpc Nearest $1,000	Growth over previous year %	Principal Industry(s)	Tourism Nearest $1,000,000
EUROPE							
Vatican	1	326	N/A	N/A	N/A	Religious and Financial	N/A
San Marino	33	719	1,802	48	N/A	Tourism and Banking	N/A
Monaco	31	1,100	6,559	168	N/A	Banking	N/A
Andorra	86	1	3,249	32	-1.60	Tourism and Banking	N/A
Liechtenstein	38	890	5,647	164	N/A	Electronics	N/A
Luxembourg	570	27,800	60,131	102	2.9	Banking	5,400
Iceland	332	7,300	15,330	61	1.8	Fish	1,400
Cyprus	924	9,400	24,047	27	-2.3	Tourism	2,800
Malta	414	4,500	9,971	27	3.5	Tourism	1,500
Montenegro	626	7,000	4,228	7	7.4	Steel	906
AFRICA							
Cape Verde	546	3,300	1,861	3.0	1	Food	418
Equator. Guinea	741	6,300	18,532	9	-3.1	Oil	N/A
São Tomé & Princ	194	135	342	1.8	4.5	Construction	13
Comoros	781	183	622	1.5	3.3	Fishing	39
Seychelles	92	473	1,445	15	2.9	Fishing and Tourism	398
Djibouti	828	648	1,456	2.0	6	Construction	13
ASIA							
Maldives	393	876	2,836	9	5	Tourism	2,700
Brunei	420	6,300	16,111	27	-.07	Oil	92
Bhutan	742	614	1,781	2.8	6.4	Cement	89
AMERICAS							
Barbados	290	1,500	4,228	20	-0.3	Tourism	947
Grenada	109	209	831	9	1.5	Food	128
St.Vincent & Gren	104	185	709	7	1.1	Tourism	360
Saint Lucia	164	222	1,336	8	-1.1	Tourism	360
Dominica	79	186	498	8	1.1	Soap	75
Antigua & Barbu	92	207	1,241	14	2.4	Tourism	330
St. Kitts & Nevis	52	222	743	17	7	Tourism	104
Bahamas	325	2,100	8,420	29	1.3	Tourism and Banking	2,300
Belize	347	500	1,624	4.7	3.4	Clothing	351
Guyana	735	2,400	2,990	4.7	3.8	Bauxite	568
Suriname	580	1,400	5,299	6	4.1	Bauxite	95
OCEANIA							
Solomon Islands	622	450	1,073	1.9	1.5	Fishing	55
Fiji	909	1,500	4,034	5	4.1	Tourism of Sugar	752
Vanuatu	272	175	800	3.0	2.9	Food and Fish	295
Tonga	107	147	440	3.7	2.3	Tourism	41
Samoa	198	258	691	4.2	1.9	Food	145
Tuvalu	11	33	38	3.0	2.2	Fishing	2
Kiribati	166	180	175	1.7	3.8	Fishing and Handcrafts	N/A
Nauru	9	52	153	9	N/A	Phosphate	N/A
Marshall Islands	72	114	189	3.5	0.5	Copra	N/A
Micronesia	105	192	333	3.1	0.1	Tourism	21
Palau	21	98	240	14	8	Tourism	112

SOURCE: United Nations, Dept. of Public Information, "List of Countries by Population (2015)"; " GDP budget, GDPpc, growth over previous tear %, principal industry, and tourism income (if listed) The World Almanac and Book of Facts (2016).

* This is the population controlled by the Republic of Cyprus, it does not include the population is listed within Turkish Republic of Northern Cyprus (TRNC)

The creation of new independent countries has been facilitated by changed criteria for independence. The support of former colonial powers, the rise of global institutions, and the emergence of regional organizations have facilitated their survival. Such support has enabled the development of infrastructure including schools, clinics, and community halls as well as roads, railways, ports, power stations, water and sanitation facilities, and services spurring the new countries' economies. What is their future—and the future for all small countries—in their own governance their participation in world governance?

To what extent have accommodations in international arrangements assisting the survival of the new micro-countries signaled the beginning of a system of world governance, one at least marginally less dominated by empires and super-powers? Formal legal equality of countries has long been a critical convention of international relations. In reality, most small countries have lacked the prosperity, presence, and prestige to function as genuinely independent members of the global community. Does their growing experience with global international organizations and with regional organizations of countries provide reason to hope they may gain international standing, if only marginally, in the future?

In the late 1400s, imperialism extended worldwide. By 1900, 10 empires ruled 95 percent of the world's people whereas the Chinese and British Empires ruled 60 percent of the world's people. New political, economic, and social bonds shaped world governance. The dynamics enabled the growth of transcontinental empires over five centuries, the emergence of many new countries over the past several decades, and world governance, which they continue to change. Evolutions in country governance and international relations have begun changing relations among countries and, thus, changing the way the world has been governed for the last five centuries.

Since World War II, many micro-countries have not only survived but have also gained their independence. The exigencies of securing independence and surviving are altering relations among countries and reforming world governance. Developments in the last few decades may be more fully appreciated in the perspective of the last five centuries in which their lands and people were added to European empires, exploited by European interests, and indoctrinated with European sacred and secular faiths, values, and practices.

Table 1.3: The 41 Micro-countries and Their Colonial Heritage

MICRO-COUNTRY (by Population)'	GEOGRAPHY Island (IS), Coast (CO), Land-locked (LL)	POPULATION (nearest 1,000)	AUSTRALIA	AUSTRIA	BRITAIN	FRANCE	GERMANY	ITALY	JAPAN	NETHERLANDS	PORTUGAL	SPAIN	USA	Other former colonial powers which granted independence	Independence DATE	UN member DATE
Vatican	LL	1						IT							1871*	NA
Tuvalu	IS	10			BR										1978	2000
Nauru	IS	9	AU				G		J						1968	1999
Palau	IS	21					G		J			S	US		1994	1994
San Marino	LL	33						IT							1862*	1992
Liechtenstein	LL	38		A										SWISS	1806*	1990
Monaco	CO	31				FR									1861*	1993
Marshall Is	IS	72											US		1991	1991
St. Kitts & Nevis	Is	52			BR				J			S			1983	1983
Dominica	IS	74			BR										1978	1978
Andorra	LL	92				FR'									1993	1993
Antigua &Barbu	IS	82			BR										1981	1981
Seychelles	IS	92			BR	FR									1976	1976
FS Micronesia	IS	105					G		J			S	US		1986	1991
Kiribati	IS	107			BR								US		1979	1999
Tonga	IS	107			BR										1970	1999
Grenada	IS	103			BR	FR									1974	1974
St Vincent&Gr	IS	103			BR										1979	1980
Saint Lucia	IS	164			BR										1978	1979
Samoa	IS	193												NEW ZEALAND	1962	1976
SaoTome&Princi	IS	194									P				1975	1975
Vanuatu	IS	273			BR	FR									1980	1981
Barbados	IS	291			BR										1966	1966
Iceland	IS	332												DENMARK	1944	1946
Belize	CO	347			BR										1981	1981
Maldives	IS	393			BR										1965	1965
Bahamas	IS	326			BR										1973	1975
Brunei	CO	430			BR										1984	1984
Malta	IS	414			BR	FR									1964	1964
Cape Verde	IS	546									P				1975	1976
Luxembourg	LL	543				FR	G					S		BELGIUM	1866	1945
Suriname	CO	580			BR					N					1975	1975
Solomon Is	IS	622			BR				J						1978	1978
Montenegro	CO	678												SERBIA	2006	2006
Comoros	IS	780				FR									1975	1975
Bhutan	LL	742			BR									INDIA	1971	1971
Equatorial Gu	IS CO	741									P	S			1968	1968
Guyana	CO	735			BR					N					1966	1966
Cyprus	IS	847			BR										1960	1968
Fiji	IS	909			BR										1970	1970
Djibouti	CO	827				FR									1977	1977

A few European countries (those with an *) gained independence incrementally on multiple dates: the date shown here is the earliest claimed,

Thanks, in part, to the UN and their affiliate bodies—and fast-developing regional organizations of countries—micro-countries are developing a more active role in "global governance." Review of the experience of recent decades may provide a perspective on how many of the newly created micro-countries will survive the political, economic, and environmental challenges of the mid-2000s.

Reporting the challenges confronting the new micro-countries provides important insight on the challenges confronting many of the smaller and less advantaged countries.

2. FROM DEVELOPING TO DISMEMBERING GLOBAL EMPIRES (late 1400s–late 1900s)

"The blocking of the land paths proved a godsend. Driven by new incentives to go to sea, Europeans would discover waterways to everywhere. The science of cartography first flourished on the sea. There the needs of working mariners shifted the interest of geographers and map-makers . . . Christian geography had become . . . more interested on everyplace than in anyplace, more concerned with faith than with facts."

—Daniel J. Boorstin

"Thus circumstanced, the two nations of Castile and Portugal were naturally led to turn their eyes to the great ocean which washed their western borders, and to seek . . . new domains, and, if possible . . . undiscovered track towards the opulent regions of the East."

—William H. Prescott

"Never certainly did any nation since the world began, assume anything like so much responsibility."

—J. R. Seeley in *The Expansion of England*

MAP 2.1 World Map (1554) by Sebastian Munster

Ptolomy, an Egyptian geographer (c. 90–168AD) presented a world limited to Europe and adjacent parts of Africa and Asia. Sebastian Munster included a reproduction of it in his atlas, *Altera Generalis Tab, Secundum Ptol*, published in 1554. (This map is in the possession of an individual who has given permission for its use.)

EMPOWERED EUROPEAN MONARCHIES EXPLORE, EXPAND INTO, AND EXPLOIT OVERSEAS

A few European kingdoms emerged from the chaos, conflict, and commerce of medieval Europe to colonize much of the rest of the world. Emboldened by their European progress and prosperity, Spain and Portugal, England and France, and Holland and Denmark embarked upon colonizing beyond Europe. The relentless momentum of European history—propelled by the Renaissance, Reformation, Counter-revolution, the Thirty Years' War and other countless other conflicts, and the aggression and avarice their leadership—exploring, expanding into, and exploiting Africa, Asia, the Americas, and Oceania.

Even before the Renaissance, an interactive series of socio-economic and political change began undermining the most significant institutions of pre-1500 Europe. The first was the all-embracive feudal system, which legitimized the reciprocal obligations of military service and land use that governed relationships from the lowest peasants to its most powerful monarchs. The other was the transnational and hierarchical papacy, which maintained its God-given primacy and power to sanctify feudal titles and rights. Pyramidal feudalistic kingdoms began to be replaced by "sovereign" ones with power concentrated within its court and a lack of allegiance to the papacy.

At the beginning of 1300s, Europe had more than 5,000 quasi-independent feudal baronies, duchies, principalities, and kingdoms. By the beginning of the 1600s, the more powerful countries had reduced the number to about 500 independent countries. By the beginning of the 1900s, they had further reduced the number to about 250. In the early 2000s, there were only a few dozen independent countries.

The larger, more aggressive monarchies, emboldened by their resources and encouraged by unabated ambitions to internally strengthen as well as geographically enlarge their empires, took steps to mold the newly acquired peoples into a unified "we." Connecting roads and waterways were constructed, a single creed fostered, a common vernacular propagated, and common laws adopted. Permanent armed services and civilian bureaucracies were developed, and schools and universities (if only for the elite) were founded. And a mythology was cultivated celebrating events and heroes with holidays, rallies, and flags—and even fighting wars against a "them"— nurtured a sense of nationhood in the enlarged country.

The strengthened European monarchies, energized by covetousness and curiosity, were tempted to explore and expand beyond Europe. During their adventures, they were able to rely on more accurate maps and charts, the advent of compasses and chronometers, the development of multi-mast and wide-bottom ships and the emergence of ambitious monarchs and adventurous mariners enabled longer voyages. Commercial links between Europe and Asia had, in fact, existed for centuries through land trade routes from eastern Mediterranean gateways, dominated by the Genoa and Venice since the 1000s, to the riches of Arabia, India, and the Indies. Visionary and vigorous rulers, determined and daring navigators, enterprising merchants, and devout clerics energized exploring and trading. The European quest for new opportunities for trade and commerce drove the early transoceanic expeditions.

The southern tips of the Americas, African, and Asian continents were gateways challenging world explorations: the Cape of Good Hope linking the Atlantic and Indian Ocean; Malacca Strait connecting the Indian Ocean with the South China Sea and the Pacific; and the Strait of Magellan linking the Atlantic and Pacific. Portuguese mariners were the first Europeans to navigate these three critical gateways. Driven by commerce and curiosity, several other European colonial powers soon followed.

EXPANDING HORIZONS ENCOURAGED EXTENDED EXPLORATIONS

Before 1500, Europeans knew little of the world beyond adjacent parts of Africa and Asia. The little that was speculated about Africa included stories about Prester John. What little that was known about Asia, ambiguously called the Indies—a term that included India, Malacca, the Spice Islands, and sometimes loosely even extended to China and Japan—was fascinating. European perceptions of Asia colored by the 1300s stories by Marco Polo and other more fanciful tales such as those extolling Prester John, books authored by Sir John Mandeville and others, and popularized by many books made possible by the advent of the printing press.

Maps inspire adventure. European adventure in the latter half of second millennium began with exploration and continued with empire-building. Annexation was followed by partial, prejudiced, and asymmetric (and uneven) integration of the economies and cultures of the non-contiguous lands and peoples. Since the early 1000s, interactive advances in commerce and trade, science and culture, and politics and government propelled world-girdling explorations from the late 1400s.

The lure of spices and treasure of the mystical East combined with Ottoman blockage of land routes to Asia drove the European search for alternate passages to the east. Portugal, even before 1492, began establishing footholds along both coasts of Africa and, later, to India and the Indies where the Islamic diaspora had brought Muslim competition to eastern

South America (namely Brazil). Spain, within a few decades of driving the Moors from the Iberian Peninsula, began dramatically extending the Spanish-Habsburg Empire inherited by Charles V.

In the mid-1400s, the Portuguese Prince Henry the Navigator (son and brother of Portugal kings and grandson by his English mother of John of Gaunt) organized expeditions going down the west coast of Africa in search of the southern tip. Ferdinand Po and Bartholomew Diaz explored the African west coast, landing in Cape Verde in 1456. The Portuguese navigator Bartholomew Diaz went east around the Cape of Good Hope in 1488, opening up seaways to the east African coast, India, and the Indies. Vasco de Gama sailed in 1497 around Africa to India, encountered Indian Ocean merchant networks run by Indians and other Asians, and projected European power into South and East Asia. The Portuguese reached India in 1498, Brazil in 1500, China in 1514, and Japan in 1543. In 1511, the Portuguese had gained control of the Strait of Malacca and sent ships to China, Japan, and the Pacific.

Spain, even before Christopher Columbus's 1492 accidental "discovery," followed Portugal's lead in exploring the West African coast. From then on, until Philip II's death in 1598, the Spanish empire was extended to the Caribbean and South America. The voyages of Columbus across the "Western Ocean" and Magellan's and Vasco de Gama's across the "Eastern Ocean" (as the Atlantic and the Pacific oceans were labeled in some atlases as late as the 1700s) opened the lands beyond the oceans to the European race for conquest, colonization, commerce, and conversion.

Spanish monarchs sponsored Columbus, Hernando Cortez, Francisco Pizarro, and Amerigo Vespucci in their voyages across the Atlantic. Martin Waldseemuller, a cosmographer at the court of the Duke of Lorraine, having been

entirely convinced by Vespucci's narratives . . . [published a new world map which combined] the "Old World" of Ptolomy, updated by the Portuguese map of Africa, with the "New World" of Vespucci. It was Vespucci's insight in this respect which contrasted so strongly with Columbus's impatience with the new western lands as mere impediments on the route to Asia, and which secured Vespucci his immortality. (Peter Whitfield, *New Found Lands in the History of Explorations*)

The claim that his voyages proved the existence of two new continents convinced Waldseemuller to name the two continents "America" on his well-known map of 1507.

Six years before Balboa's famous first sight of the Pacific, the two newly European discovered continents first received the names they would have for more than 500 years. In 1520, Ferdinand Magellan crossed from the Atlantic to the Pacific via the passage (the Straits of Magellan), which King John II of Portugal named after him, opening up the west coast of the Americas and the Pacific.

The British Navy had its roots centuries earlier in the Saxon era when, under the lead of Alfred the Great, it defeated an attempted invasion by Danes of his island kingdom in 892. David Hanna points out in *Knights of the Sea*, "The defeat of the mighty Spanish Armada in 1588; the victories over England's North Sea rival, Holland in the seventeenth century; and its string of victories over France the 18th century." According to Hanna, "this unbroken legacy of maritime success" encouraged English-sponsored adventurers such as Walter Raleigh, Francis Drake, James Cook, and John Cabot (*Giovanni Cabot* in Italian) and facilitated the development of a navy-dependent overseas empire stretching from neighboring Ireland across the Atlantic to Jamaica and Barbados, north to Hudson Bay, and clear across the world to Bengal and Malaya. The wealth generated by the acquisition of foreign resources and markets fed Britain industrial growth and imperial power.

Even before the Netherlands' revolt against Spain had succeeded in having its independence recognize in 1648, adventurers sailing for the Dutch explored new lands and claimed them for the Netherlands. In 1608, Henry Hudson sailed up the river now named the Hudson, claiming the land for the Dutch. Among the diversity of explorers following Magellan in the Pacific were Dutchmen Jacob Hogeveen and Abel Janson Tasman.

The lingering presence of the Portuguese and Spanish empires, the rise of the Dutch empire, the growth of the English Navy and rapid extension of the British Empire, and the land strength of France set the stage for battles determining the future of Asia, Africa, the Americas, and the Pacific. The Portuguese, Spanish, Dutch, French, and English/British vigorously competed for overseas trade and territories until the 20th century.

CONTIGUOUS EUROPEAN EMPIRES BECAME MULTI-CONTINENTAL AND NON-CONTIGUOUS

Aggressive rulers followed up their global explorations by expanding their contiguous continental empires overseas to non-contiguous lands, becoming multi-continental empires. In the effort to develop their dependencies economically the empires relied on the efforts of the military, merchants, and missions. Such efforts were not novel. Greece, under Alexander's and his successors' rule, spread its culture as well as its rule into west Asia and northeast Africa. The Roman Empire expanded its rule and its civilization not only into northern Europe but also into North Africa and west Asia.

The Ottoman Empire, from the 1300s until 1922, controlled Asia Minor, North Africa, and southeast Europe, blocking European land access to the lands of the East and fostering Islam and Arab ways. The Mongols, under Genghis Khan and Kublai Khan, threatened Western Europe. China, under the Ming dynasty, extended its reach over adjacent lands including Tibet, Inner Mongolia, and Xinjiang Uighur and its commercial competitiveness into the Indian Ocean. Empires such as the Egyptian and Zulu empires in Africa, the Mughal and Mongol enpires in Asia, and the Aztec and Inca empires in the Americas extended far beyond their initial borders.

Spain, a marital union of the Kingdoms of Aragon and Castile, cleared the Iberian Peninsula of Moors by 1492. England annexed Wales, Scotland, and Ireland, and, for a while, parts of France, becoming the United Kingdom Britain. A Frankish kingdom expanded to become France. A Nordic empire once embraced Norway, Sweden, and Denmark. Prussia led the unification of Germany. A Kiev-based kingdom expanded to become a Russian empire that stretched to the Pacific. Rulers, along with the military, merchants, and missions, employed a variety of tools to unify (usually incompletely) the acquired lands and peoples into their empires, including imposing sacred as well as secular overlords, fostering a single language and faith, and developing common laws and mythology. Such steps facilitated merging (if imperfectly) many small cultures into fewer mega-cultures.

The Renaissance, Reformation, Counter-Reformation, Enlightenment, and Industrial Revolution energized European expansion beyond Europe into non-adjacent lands, and sustained efforts to introduce a sense of imperial identity throughout the empire. Only later did 20th-century individualism, nationalism, and a tempering of absolute monarchy facilitate the development of representative institutions. A few empires were still in contention in the 20th century. Late in the hegemonic struggle of the Western European powers, Japan and the United States entered the Asian-Pacific colonial fracas. The march of time and thought introduced visions of new lands and riches, inspired ventures of exploration, conquest, and colonization—adding newfound areas into the known world. Spain, Portugal, France, Holland, Denmark, and England, often through trading companies, traded goods and planted settlements.

Spain and Portugal planted settlements in Africa, India, and the Americas in the 1500s. The United Provinces, through its Dutch East India Company, developed an Asian and South American presence even before it won its formal independence in 1648. Denmark colonized Caribbean islands. Until the early 1700s, the colonies of Portugal, Spain, and Holland preceded those of England. Since the 1700s, Britain acquired possessions in America, Asia, and the Pacific.[2]

During the Napoleonic Wars, Britain increased its colonies from 26 to 43 and continued expanding in the 1800s. By then, half of the 41 colonies that later became micro-countries were British. While Spain, Portugal, Holland, and France retained many possessions, they lost island possessions to Britain, whose navy "ruled the waves," enabling the sun to never set on the British Empire.[3] The 1815 defeat of Napoleon shifted the pecking order of world colonial powers. Preparation for, and the failure of, the Armada sped the build-up of the English sea power.

Britain, despite the earlier loss of part of North America in the American Revolution, which France helped the colonies win, used its naval power to enhance and protect British colonial and commercial interests. After defeating Napoleon's aggressive imperial ambitions, Britain deepened

---

2. From 1604 when James I became the English as well as the Scottish king, the term Britain was used; it became the formal title in 1707 when the two parliaments were united.
3. The sun also never set on the French Empire. Since the French have integrated several former colonies into their internal local governments system, the sun still does not set on France.

its involvement in India and South Asia, the West Indies, and the Pacific. By the beginning of the 1900s, the British Empire consisted of over "100 political units (not including some 600 Indian princely states)."

To solidify their rule over the newly acquired non-contiguous territory, the European multi-continental empires applied measures similar to those used with their contiguous single-continent empires: imposing an overseer; constructing roadways and waterways; establishing armed forces and bureaucracies; enforcing common laws; developing a common currency; and encouraging banks, founding schools, universities for the elite, and teaching prospective leaders to promote an empire-wide mythology celebrating its events and heroes attempting to arouse a uniting spirit. Such steps played a decisive part in forging empire mindset strengthening the empire.

The scramble for Africa began after the Berlin Conference of 1884-5 divided the "dark continent" south of the Sahara Desert and between the coasts. Not only did the long adventurous Portugal, Spain, Britain, and France participate, but also the more recently acquisitive Germany, Italy, and Belgium's King Leopold II's established colonies did.

The British and the French, aided by their navies, competed in Asia and the Pacific. After the Napoleonic Wars, Germany under Bismarck and Italy under Cavour developed overseas ambitions, which were successively brought to an end with Germany's defeat in World War I and Italy's in World War II. The United States, after expanding across North America in the 1800s and displacing Native Americans with European-American settlers, launched its global empire annexing independent Hawaii and then the Spanish possessions (i.e., Puerto Rico, the Philippines, and Guam) in the course of the 1898 Spanish-American War. After the defeat of Germany in World War I, its Pacific dependencies were assigned to Japan, but Germany lost them after its defeat in World War II. Around the same time, the United States separated Panama from Columbia, facilitating the construction of the Panama Canal.

The colonial powers imposed an array of suzerain relationships—including asymmetric "alliance," "protectorate," "crown colony," "indirect rule," and "direct rule,"—that defy attempts to define and assess the extent of control exercised by the colonial power. 20th century events, which stem from roots planted earlier, have continued this transformation, beginning the change of the empire-centric governance of the pre-20th century world into a more pluralistic 21st-century governance. European empires increased hegemony within expanding borders by armed forces, extending infrastructure and encouraging liturgical and language conformity. Monarchs and merchants increased their power and profits collaborated in managing exploration and exploitation.

The Vatican, Andorra, Liechtenstein, Luxembourg, Monaco, and San Marino have long exercised *de facto,* if not *de jure*, independence. The other micro-countries gained, or regained, their independence and nationalism fomented by World War II in the rising tide of anti-colonialism. Iceland became independent in 1944, remote Bhutan in 1949 (following India in 1948). In the late 1950s and '60s, the rising tide of self-governance restored independence to Cyprus and Malta in the Mediterranean, garnered independence for Barbados and Guyana in the Americas, Equatorial Guinea in Africa, the Maldives in the Indian Ocean, and Samoa and Nauru in Oceania. The number of independent micro-countries augmented in the 1970s. Montenegro separated from Serbia and admitted to the UN in 2006, and South Sudan separated from Sudan and entered the UN in 2011, enlarging membership to 193. Sovereignty and identity with the nation had become the *sine qua non* of modern polities and politics.

CULTURAL COLONIZATION:
LANGUAGE, LITURGY, AND INSTITUTIONS

European empires assigned overseers to govern built roads and later railways to facilitate transportation, and promoted their language and their version of Christianity within their contiguous lands and peoples. The process facilitated the effort to tie the conglomerate of peoples in a multi-continental empire into one extended community with a variety of constructed and cultivated bonds. Arising in different eras and different circumstances, nationalism shares many characteristics with a sense of tribalism.

These expansions and national community extensions efforts proceeded in stages: England and then the British Isles, Aragon and then Spain, Burgundy and then France, Prussia and then Germany, Piedmont and then Italy, Holland and then the Netherlands, the Kiev polity and then Russia. As internal frictions continue to indicate, the efforts to continue nation-building in adjacent lands and cultures were not as successful as earlier ones. They were even less successful in efforts to continue nation-building in non-adjacent lands and cultures on other continents.

Europe's Thirty Years' War (1618–1648)[4] marked a stage in the long ruthless combat among European powers to enhance their interests. Aggressive, expansionist, and centralizing rulers, anointed with a royal pedigree, asserted the divine right of kings and a royal *d'état, c'est moi* certitude in expanding their realms and centralizing their governance by undermining the prerogatives of secular and sacred lords and strengthening a sense of national community.

The end of the Thirty Years War marked an assertion of Germans, Italians, Hungarians, Poles, and Rumanians to be independent. The revolutions generally failed, but their aspirations continued to dominate European politics by being adopted, in one form or another, by quasi-revolutionary movements. Hobsbawn notes the following:

> If the rise of working-class parties was one major by-product of the politics of democratization, the rise of nationalism was another . . . It came to be used for all movements to whom the 'national cause' was paramount in politics: that is to say for all demanding the right of self-determination, i.e. in the last analysis to form an independent state, for some nationality defined group.

The long-established feudal system gave way to centralized monarchies that supported exploration, conquest, and colonization. These steps strengthened a sense of an extended national community, arousing patriotic allegiance for the expanded country-empire ahead of loyalty to county, duchy, and principality. In time, allegiance, loyalty, and enthusiasm for the expanded country, cultivated and constructed by strong monarchs, gradually surpassed reverence for the monarch. Paralleling the growth of a sense of nationhood was the demand for and extension of participation in governing. Once concentrated in a sovereign ruler, meaningful participation in representative governance very gradually and successively extended to aristocrats, to professionals, to property-holders, to all males, and to almost every adult.

---

4. The Thirty Years' War was a series of wars stating before 1618 and continuing after 1648.

Each European country endeavored to apply similar hegemony tactics to the non-adjacent lands and peoples in newly acquired colonial territories despite far greater differences in distance, history, ethnicity, and faith. The colonial rulers not only imposed colonial officers; their teachers taught the national heroes and events well as the language of the colonial power, and their missionaries converted to the faith practiced at the home country in its overseas territories. The effort to indoctrinate the "native" elites as quasi-Frenchmen, -Englishmen, -Dutchmen, -Spaniards, and -Portuguese with home country traditions and culture had limited success. In each country, this effort developed a more distinctive sense of how their heritage made them different. This aroused a corollary passion, a drive for independence that shook the colonial world.

A series of multi-continental wars following the Thirty Years' War provided the British Navy the opportunity to expedite the British acquisition of the possessions of the initial colonizing powers. These included the First and Second Anglo-Dutch wars (1652–4 and 1665–7), the War of Spanish Succession (1702–1713), the War of Austrian Succession (1743–1748), the Seven Years' War (1755–1763), the American War of Independence (1775–1783), and the years of intercontinental wars with the French, which ended with the Napoleonic Wars (1792–1815).[5]

As a result of these wars, the British won more than they lost to the Spanish, French, and Dutch. The Napoleonic wars signaled the rise of the British Navy as the dominant sea power, which facilitated the British Empire's efforts to construct embraced a range of overseas constitutional, commercial, and cultural relationships. In *The Influence of Sea Power upon History 1600–1783,* A.T. Mahan observed

> the history of Sea Power is largely . . . a narrative of contests between nations, of mutual rivalries, of violence frequently culminating in war.
>
> The profound influence of sea commerce upon the wealth and strength of countries was clearly seen long before the true princi-

---

5. The Thirty Years' War (1618-1648), involved many powers including the Holy Roman Empire, England, Denmark, Sweden, and the Netherlands (which had already emerged as a power before gaining its independence was formally recognized in 1648) and triggered future conflicts in Europe and throughout the world.

ples, which governed its growth and prosperity, were detected. To secure to one's people a disproportionate share of such benefits, every effort was made to exclude others.[6]

The American War of Independence and French Revolution propelled change, shifting the popular mood and setting the stage for later broad representative democratic governance. By the 1800s, the Second Industrial Revolution prompted workers to migrate from farms to factories. A professional middle class emerged. Governments developed ministries, established a civil services and a permanent military establishment, expanded public services, and extended schooling. Extending suffrage advanced "democracy" as a popular populist principle of governance. In the 19th century, countries grew larger, empires expanded, Germany and Italy unified, and the United States emerged as a superpower.

The 20th century began with the demise of the weakest and most decayed empires. Since its founding, UN membership grew from 47 to 193 countries. The post-World War II "wind of change" helped countries gain their independence—Iceland in 1944, Montenegro in 2006, and the South Sudan in 2011. Before the 20th century, colonial acquisitiveness drove the growth of empires; by the late 20th century, nationalism dismembered them.

Having won in World War II—after which the African, Asian, American, and Oceania micro-countries won their independence—Britain and its leaders (most notably Churchill) went on believing that the British world-system would continue in the postwar world. Even after the 1956 Suez Crisis, their confidence in what remained of their empire, if properly managed, would be an invaluable asset in the postwar world.

The transition from traditional rule-based empires to economic-hegemony based ones was in the future. Time was needed for recognition that holding on to the colonies was a liability, given the rising role of native political leaders. It was not yet obvious how threatening nationalism was, hence the dynamics of the cold war. The threat could not be ignored. In Asia and Africa, the British realized that, with world opinion, local oppo-

---

6. While Britain's naval capital tonnage supremacy was limited by the Washington Naval Treaty of 1932, which established ratios of 5 for Britain and the United States, 3 for Japanese, 1.75 for French, and 1.75 for Italy, the pre-World War II arms build-up negated the impact of that treaty.

sition, and the costs of maintaining authority, rendering holding onto the empire was impractical if not impossible. By the end of 1963, the British government hierarchy was no longer resisting, they were supporting its remaining dependencies in their efforts to secure independence. By the end of the 20th century, the dismembering of traditional empires was nearly completed.

The initial promotion of identification with the colonial home country by introducing the home country language, faith, and other values in colonies was circumscribed by the limited horizons of the vast bulk of people. Because the colonial power was distant from their land and disconnected from their local interests, it was distrusted. The greater the distance and disconnect, the greater the potential distrust.

As European monarchies added adjacent land and peoples and started constructing connective infrastructure and cultivating a common faith and language among their subjects, they were attempting to extend the French, British, Spanish, German Portuguese, and Dutch culture and to mold their subjects into French, British, Spanish, Germans, and Dutch citizens, respectively. When the European empires annexed non-adjacent lands and people in Africa, Asia, Americas, and Oceania, they increased their control by imposing steps like those applied earlier to annexed European lands and peoples. While the circumstances and cultures of the empires' overseas annexations differed significantly, they were sufficiently successful; the billions of people gradually adopted significant parts of their culture. The official and workplace language in South American, African, and south Asian former dependencies was English, French, Spanish, Portuguese, and/or Dutch. Christianity was the dominant faith in many of them.

After World War I, many European dependencies, asserting their own suppressed national heritage and values, became independent. After World War II, the peoples in dependencies similarly asserted their nationalism and gained their independence. Nationalization of subject lands and peoples included extending the capital's infrastructure, language, liturgy, and mythology throughout the subjected territory. Patriotism and nationalism were the heirs of the effort to develop identification with the sovereign power.

European empires attempted to extend a version of its capital-based top-down inclusive nationalization to each of its non-adjacent dependencies just as it had earlier to its adjacent ones. Instead, however, the people of non-adjacent dependencies developed their own distinctive sense of community supported by their ethnic heritage, historical narrative, commonalities of language and faith, infrastructure, heroes, and legends. These led to the development of nationalism distinct from, and in opposition to, the empire loyalty cultivated by the suzerain power. A local nationalism aroused a deeper loyalty than the colonial power's efforts to cultivate its empire-wide sense of community.

3. SOVEREIGNTY, SKEPTICISM, AND SURVIVAL

"The land on which we live has always shaped us. It has shaped the wars, the power, politics, and social development of the peoples that now inhabit nearly every part of the earth. Technology may seem to overcome the distances between us in both mental and physical space, but it is easy to forget that the land where we live, work, and raise our children is hugely important and that the choices of those that lead the seven billion inhabitants of this planet will to some degree be shaped by the rivers, mountains, deserts, lakes, and seas that constrain us all—as they always have."

—Tim Marshall

"Many of the jumbled ingredients of nationhood—beliefs, myths, institutions, customs, and loyalties . . . gained potency because they persisted . . . In such ways nations and identities are 'constructed': that is, made by people, and not determined by geography, genes or blood."

—Robert Tombs

SOVEREIGNTY: ORIGINS AND CRITERIA

A tide of anti-colonialism, sparked by nationalism and an accompanying "wind of change," swept the developing world in the last half of the 1900s. Colonies demanded that they be recognized as independent countries, as sovereignty over their land and people. The concept of sovereignty, as first articulated by Jean Bodine in his *Six Books of the Commonwealth* (1576), defined sovereignty as "absolute and perpetual power vested in a commonwealth" and the distinguishing mark of an independent country. Three centuries later, German sociologist Max Weber clarified the relation of "sovereignty" to power, politics, and political systems, stating "a state is a human community that claims the monopoly of the legitimate use of physical force within a given territory." The concept of "sovereignty" inspired, informed, and gave impetus to the era of the sovereign independent country.

Recent centuries have witnessed transformation of the governance of countries. Concomitant with industrial, infrastructure, education, suffrage, and other evolutions, rulers transformed the socio-economic and political mission of governance by binding together disparate parts into a better-ordered polity. The process included escalating the efforts constructing transport systems that interconnect countries, encouraging a common spoken vernacular, missionizing a single mode of worship, and forging a patriotic mythology with parades, gradually expanding centralized countrywide services including education, health, and safety as well as a civil and military bureaucracy, and later widening the suffrage. These efforts facilitated forming a collective community and a centralized government. To the extent rulers succeeded in this process of instilling a sense of "we-ness," they enabled the idea of the "nation"—a people purported to share a common descent, language, faith, and history organized as a separate polity and occupying a definite territory.

In Europe, the process of national integration enabled the gradual shift from the absolutism of divine right of kings to the pluralism of representative governance. Innovations in travel and communication multiplied and dramatically expanded the middle and professional classes, thereby enlarging communities of common interests throughout the world. But they also increased consciousness of local identity, loyalty, and allegiance. Increased contacts outside the local comfort zone confirmed feelings of "we versus them." World War I sped up self-determination and forced the breakup of the German, Austrian-Hungarian, and Ottoman empires. After World War II, colonial rule rapidly became a less and less acceptable form of subordination. World War II drove nationalism, forcing colonialism to give way to "self-government," defined as limited elected representativeness and a curtailed scope of responsibility, before the granting of sovereign independence.

Since the 1600s, the concept of "sovereignty" drove international relations. Expressed in national pride and aggressive power, it drove expansion and empires. Starting in the late 1900s in the Africa, Asia, the Americas, and Oceania, nationalism pushed the aggressive push for sovereign/independence. Nationalism pervaded the world, driving the movements that led to the creation of newly independent countries after World War I and even more after World War II, but many of the newly independent countries were hardly considered capable of standing by themselves as independent/sovereign states.

Churchill and De Gaulle and many other statesman and scholars expressed skepticism regarding whether the proposed newly independent countries were sufficiently developed to handle the responsibilities of independence. Small islands, separated from their neighbors by strenuous and infrequent travel and the paucity of shared values and cultural norms, do facilitate knowing and working with strangers. Small, isolated polities are disadvantaged in several ways. Their smallness makes it difficult to develop the advantages of specialization and scale, and their lack of proximity to other countries is an economic impediment to developing trade and tourism. These conditions do not encourage the countries to invest in the infrastructure and institutions that fuel progress and economic development.

Characteristics long considered to be the basic criteria of "sovereignty" have included a territory, a population, a government, and an independent existence separate from a suzerain power. The basic *raison d'être* of its government must primarily be to bring law, order, and stability to the land and its people, which requires locally based officers managing its affairs. A sufficient degree of formal and real independence enabling it to manage its internal and external affairs without dependence upon another country should be basic criteria of independence.

A striking change since World War II has been the ignoring of the long-established criteria for recognizing impendence. After World War II,

Britain's Colonial Office considered it "clearly impossible in the modern world for the present separate communities, small and isolated as most of them are, to achieve and maintain self-government on their own." Curiously, this critical assessment did not foreclose the prospects of small, underdeveloped countries gaining independence. Instead, it weakened the traditional criteria for granting independence.

For the past several decades, whether a country has been recognized as independent by the international community has no longer depende on application of the same long accepted criteria for the recognition of independence that had traditionally been used. Many micro-countries clearly do not meet the traditional "sovereignty" criteria for independence. Yet, they have been internationally recognized by most countries (although not necessarily placing diplomatic missions within their countries) and as members of the United Nations.

The difference, if any, between those whose independent status is compromised and a few who remain dependent is not clearly evident. Palau and the Marshall Islands gained their independence and were admitted to the United Nations only after entering into a "free association" agreement with their former colonial masters, provided that the United States manages the foreign and defense affairs of these countries. The Cook Islands has entered into a similar agreement with New Zealand.[7] A similar observation may be made regarding the Dutch's autonomous kingdoms.

Many former colonies gained independence according to the traditional criteria but continued to depend upon substantial aid from their former colonial powers. International recognition by other countries, especially the major powers, admittance to membership in the United Nations, the current international recognition of independence, has become the accepted yardstick of independence. The extent, however, to which many of today's independent countries exclusively manage their foreign, defense, and trade affairs vary significantly. For the smaller countries, management of international interests depends on the extent with which they depend on the good will and support of other countries, the United Nations, other global international organizations, and regional organizations of countries.

---

7. The Cook Islands remains a dependency and is not a member of the United Nations, even though it has full treaty-making capacity and maintains its own United Nations office.

SKEPTICISM: ABOUT THEIR SURVIVAL

In the postwar 1900s, when the many developing countries were gaining their independence, many skeptics continued to voice concerns: that the soon-to-be liberated colonies lacked adequate physical infrastructure, competitive industry, and developed educational, and other institutions. Scholars who focused on the handicaps that faced small developing independent countries included Demas, Clarke and Payne, Cobban, and Vital in the 1960s, Jainarain in the 1970s, Hughes in the 1980s, and Alesina, Spolaore, and Ott in the early 2000s. In 2012, DeLisle Worrall added a dimension to the way small countries are disadvantaged: "they face a foreign exchange constraint that cannot be alleviated by depreciation of the real exchange rate or other policies. These constraints affect monetary, fiscal, and exchange rate policy including fiscal sustainability, debt management, and patterns of economic growth."

Securing independence from a suzerain power demands leaders with passionate hearts, persuasive tongues, and persevering guts. It required vision, skill, and dexterity to win independence, but also courage (and fearlessness) to confront the reality of managing independence in an increasingly competitive (some would call it cutthroat) global arena. Countries with small populations, few resources, poor infrastructure, few industries, and an absence of nearby prosperous neighbors are disadvantaged, attempting to compete in a fragmented global market. The increase in newly independent, small, developing countries led many to express concern regarding their ability to manage public services, economic development, and international interests.

Sustaining micro-country economic development is only possible if powered by sources that attract foreign income—such as agricultural products, manufacturing, financial services, tourism, and foreign loans and assistance. To develop such infrastructure and encourage industry, most micro-countries must depend on investments from foreign industrial corporations, funding from international agencies such as the World Bank and the International Monetary Fund, and loans from foreign financial firms (the more prudent the better).

Imprudent fiscal management combined with economic disadvantages has escalated debt. High debt burdens undermine the prospect of securing further loans, as reflected in the credit ratings made, or not made, by firms such as Standard & Poor's, Fitch Ratings, and Moody's. Nine European micro-countries and three American ones—Bahamas, Suriname, and Saint Vincent and the Grenadines—have been rated as A (not risky) or B (slightly vulnerable). Of the African, Asian, and Pacific micro-countries, only the Seychelles, the Maldives, Fiji, and the Federated States of Micronesia have secured an A or B rating.[8]

The smaller population of micro-countries limits the number and diversity of resources, decrease the variety of export products, reduce the domestic market, and increase the need to export and import. The alternatives open to small countries face a number of issues including securing foreign investment on favorable terms from investors. Limited opportunities to develop product specialization and the economies of scale and scope inhibit what Michael E. Porter describes as "competitive advantage." Small country economies are challenged not only by their size but also, in many cases, by their insularity. European micro-countries—with their readily accessible prosperous neighbors and long inter-active histories with them—have overcome the smallness challenge. Geographically isolated economies have been additionally constrained by narrow economic bases, limited domestic markets, diseconomies of scope and scale, deficient and expensive transport, a world of trade restrictions, and bullying by foreign corporations and mega-countries.

Firms tend to do better when they are close together, close to sufficient and capable workers, resources, and customers. Interchange among a number of firms is more likely to generate new ideas and new products. Producers, without a competitive advantage, tend to survive precariously, move, or fail. Micro-countries have small internal markets and less diversified economies; few have skilled populations or strategic locations. They tend to lack sufficient resources (minerals, agriculture, fish. sand and surf, military location, etc.) needed to attract investment by larger countries and large corporations, to seek them out, and to strike bargains with them.

The lack of economic diversification compels a country to rely on one (or a few) product(s) for export for generating foreign revenue with which

---

8. A few countries, such as Monaco, Equatorial Guinea, and Brunei, have sufficient enough resources that they have not sought funding and therefore have not been rated.

to purchase import. Lacking diversification, micro-countries find it difficult to absorb economic shocks such as fluctuations in prices or yields of crops. Small, developing countries have tended to be overly dependent on a narrow range of activities—such as fishing, farming, mineral extraction, native handicraft, and tourism—producing goods and services especially subject to the variables of international trade, marketing, and pricing. Frequently, the meagerness of their potential for supporting more than poverty-level standards of living, let alone a government capable of developing and maintaining a countrywide local infrastructure and supporting a foreign affairs establishment capable of securing assistance.

Small, isolated, rural communities tend to resist innovation. The economies of some smaller countries overly rely on one traditional product such as cocoa, cotton, bananas, and fish—later oil and gas—that its owners (often foreign) can use to exert an irresistible influence on their governance. Such influence has created what are called "banana republics." In a few countries (e.g., Equatorial Guinea and Brunei), fossil fuel extraction generates so much revenue that there is little interest in the development of new industries. International firms generate few high-level jobs for the local inhabitants.

Most foreign-owned/multinational enterprises pay foreign nationals more than the local employees. Such firms contribute to GDP growth and budgets and reduce unemployment, but inadequate budgets and the number of local poverty-stricken remain major challenges. The relationship between big firms and small countries, as Louis Goodman has pointed out in *Small Nations, Giant Firms,* is multi-faceted, nuanced, and potentially hazardous. Small countries are at a disadvantage by dealing not only with larger countries but also with large companies. Most lack the financial and human capital that enables effective bargaining with governments and multinational firms.

Climate change threatens the economies of small islands countries and their very existence. Global warming increases storms' frequency and raises sea levels, flooding the habitable coasts on which many homes, firms, and farms are located. More later on, the impact of inundation and devastation catastrophies, which small independent countries are especially challenged to prepare for, cope with, and seek relief and restoration funds for.

Achieving independence is a glittering, if also illusory, goal, particularly for those whose past has been marked by humiliation and victimization. Gaining independence, especially after a bitter struggle, is a victory driven by heartfelt passion. Many polities want to be independent or, at least, less dependent. Many more want at least the rights and respect of independence. Desire for less dependence, for community and country, has increasingly inspired mankind.

The escalating 20th Century demands for self-determination and independence arose from the emerging concepts of liberty and equality, affecting governance, a trend that may not have ended. Countries must navigate internal and external conflicts between the passions of the heart and the reasoning of the brain. The heart deals with emotional and principled values. Coping with the vexing needs pressing a country, the tendency for peripheral communities (especially those in separate islands, mountains, and ethnically distinct communities) to resist the real or imagined erosion of local power and escalate the cries for local rule. Small and fractured countries are disadvantaged when competing in global markets.

SURVIVING: DESPITE THE ODDS

Several types of international institutions have helped the vulnerable new micro-countries become sufficiently viable. These include the United Nations and other global intergovernmental institutions, "shelter countries" such as the former colonial powers, and regional organizations of governments. Self-determination and democratization, the growth of supranational institutions, freer trade, and the reduction of warfare sped up the post-World War II formation of smaller countries.

Leaders of small, recently independent countries have had to choose between the benefits of a larger size and their preference for local control. While colonial powers have tended to promote joining neighboring colonies, many individual colonies have opted to remain single. Despite the warnings and the opposition, having opted to stay small, they have survived. Can they continue to survive increasing populist, parochial, protectionist global economic competition and more frequent and ferocious climate change threats?

Since World War II, polities and politics have undergone a metamorphosis. Concomitant with dismemberment of long-established empires and the multiplication of independent countries, a more complex system of global governance is emerging. The displacement of the imperial system has introduced a world of at least theoretically sovereign states. The major powers have changed the way they exert their global role; empirical domination by political rule has been replaced by economic hegemony. A host of new independent countries, at least as such recognized, have emerged. A multitude of global and regional organizations have developed that promote collective action on transnational issues.

Passions fueled by national aspirations, populist demagoguery, and skillful leaders drove the campaigns for independence. Driven by the need to compete in a world dominated by major powers required a mix of political artistry and administrative skills. The 41 micro-countries, contrary to initial skepticism, have secured international recognition of their independence and also sufficient economic viability to survive. The advent of global institutions and regional organizations and continued support from the former colonial powers played a vital role in assisting newly independent countries to maintain economic viability.

The League of Nations, created in 1919, drove efforts to collectively manage international relations. The formation of the United Nations immediately after World War II recognized the inadequacies of the League of Nations, the increasing interdependence of the countries, and the consequent need for concerted action to meet the growing array of global issues. The United Nations, and more than 20 similar specialized organizations, have played a major role in promoting programs around the world and have critically assisted micro-countries.

Shelter countries defend their *sheltered* countries, protect their trade interests, and assist their economic development. Britain and France, and, more recently, Russia, not only maintain individual shelter relations with former parts of their empires but also lead post-imperial associations. Britain heads the Commonwealth through which it works with many former colonies. Russia has its Commonwealth of Independent States (CIS). France leads the Organization internationale de la Francophonie (OIF or La Francophonie).

An alternative path to decolonizing is followed by France and the Netherlands. France converted its dependencies into French local governments—communes (municipalities), départements (counties), and regions with the same status as mainland local governments. The Netherlands replaced the status of their Caribbean colonies; three "special municipalities" became autonomous "kingdoms" within the Kingdom of the Netherlands. Neither France nor the Netherlands considers these governments to be dependencies.

More than 40 other regional organizations have been founded throughout the world. Regional organizations have proliferated since World War II. The European Union (EU), the product of courageous and imaginative political innovation since its birth in the aftermath of World War II, is the most developed regional organization. Regional organizations begin by focusing on trade. Other regional associations, while not yet as potent, have taken steps to initiate customs union and regional projects in the hopes of increasing the extent of regional collaboration. Regional organizations of governments have been a significant factor contributing to the "quasi-globalization" of international trade.

An important factor affecting the viability and vitality of the African, Asian, and American, and Oceanic micro-countries may be the extent to which more effective representation in global politics and global organizations will impact global issues, especially freer trade and steps to reduce human-enhanced global warming and consequent inundation. Further improvement of micro-country institutions of representative governance and reducing corruption will strengthen the creditability of the countries advocating their causes. Maintaining the political-economic ties and support of the former colonial power will help. Perhaps most important will be the development of key regional organizations not only to coordinate regional endeavors but also to advocate micro-country causes more effectively.

How many will survive the more menacing and interrelated economic disadvantages, environmental disasters, ethnic and identity divide challenges of the mid-2000s?

4. ECONOMIC COMPETITIVE DISADVANTAGES

"The size of a country affects the size of its economy. To the extent that larger economies and larger markets increase productivity, larger countries should be richer . . . Many of the difficulties of small states and territories are directly related to under-development in general; lack of resources, inadequate cadres, literacy, etc."

—Alberto Alesia and Enrico Spolaore

SMALLNESS AND SUSTAINABILITY

Despite gloomy forecasts regarding the political stability and economic sustainability of many postwar micro-countries (i.e., those that were founded after World War II), micro-countries have survived. While their sovereignty may have been compromised and the reality of independence infringed, they have secured membership in the United Nations membership and the accompanying international recognition. However strained their economies are, most micro-countries have kept pace with, or more than kept pace with some of their neighbors, albeit achieved with outside assistance (see Table 4.1). But the increased competitiveness of the global marketplace is imposing more significant economic disadvantages on smaller countries.

Striking is the fact that of the 187 countries, whose standard of living the International Monetary Fund (IMF) compares using per capita Purchasing power parity (PPP), only three micro-countries (Comoros, Kiribati, and the Solomon Islands) are among the 22 poorest; the others are mainly mainland African countries.[9] Micro-country PPPs tend to be sim-

---

9. PPP estimates the standard of living by using the cost of the same basket of goods in different countries to in conjunction with per capita GDP.

ilar (or better) than their continental neighbors. Pacific micro-countries possess similar per capita GDPs as Papua New Guinea and the Philippines. The per capita GDPs of African micro-countries are similar to the Cameroons and Madagascar. In the Americas, Guatemala and Honduras are actually a few thousand PPPs less than Guyana and Barbados.

In fact, several micro-countries have a higher per capita GDP (GDPpc) than their larger neighbors: the micro-countries in Europe, Equatorial Guinea, the Seychelles in Africa, Brunei in Asia, the Bahamas, Nauru and Palau in the Pacific, and three micro-countries in the Caribbean. *The Economist* concurs, "What helps is to be small. Among 56 low and middle-income countries, the top ten aid recipients per person include seven of the least populous." Smaller countries tend to have less bureaucracy, so arrangements and payments can be made more readily. A little cash may be more noticed in a small country.

The per capita GDPs of European micro-countries are not only higher than those of other micro-countries but also are higher than those of several other European countries. The few non-European ones with significantly higher GDPpc's depend upon oil (Brunei and Equatorial Guinea), tax havens (Luxembourg, Brunei, and San Marino), exceptional sun and surf (Bahamas, Maldives, and Fiji), or manufacturing (Malta and Liechtenstein.) The relative prosperity, however, allows for little cheer or complacency. More important, the average or mean GDPpc statistic can be quite misleading, because it doesn't take into account the extent that wealth inequality (many with high incomes) distorts the per capita GDP and, thus, disguises the extent of poverty.

Per capita PPP, nevertheless, misleads by comparing average incomes, not median incomes (the income below which half of the country's people struggle to live). The higher the incomes of the wealthiest few within a country, the greater the disparity between an average per capita GDP and median per capita GDP. In a few countries, the average GDPpc may be as much as 12 times higher than a median per capita GDP, thus masking the extent and depth of poverty affecting the country. The economically, ethnically, socially, politically advantaged dominate public affairs and concentrate the distribution income among their peers. The greater the disparity, the more likely a glaring contrast of slums and shanties with opulent neighborhoods and ostentatious mansions and apartments will appear.

Table 4.1 Per Capita GDP Measured in Purchasing Power Parity (PPP)Rankings

Rank	Country		Int$
1	Qatar		124,927
2	Luxembourg	X	109,192
3	Singapore		90,531
4	Brunei		76,743
5	Ireland		72,632
6	Norway		70,590
7	Kuwait		69,669
8	United Arab Emirates		68,245
9	Switzerland		61,360
10	San Marino	X	60,359
11	United States		59,495
12	Saudi Arabia		55,263
13	Netherlands		53,582
14	Iceland	X	52,150
15	Bahrain		51,846
16	Sweden		51,264
17	Germany		50,206
18	Australia		49,882
19	Taiwan		49,827
20	Denmark		49,613
21	Austria		49,247
22	Canada		48,141
23	Belgium		46,301
24	Oman		45,464
25	Finland		44,050
26	United Kingdom		43,620
27	France		43,550
28	Japan		42,659
29	Malta	X	42,532
30	South Korea		39,387
31	New Zealand		38,502
32	Spain		38,171
33	Italy		37,970
34	Cyprus	X	36,557
35	Israel		36,250
36	Czech Republic		35,223
37	Equatorial Guinea	X	34,865
38	Slovenia		34,063
39	Slovakia		32,895
40	Lithuania		31,935
41	Estonia		31,473
42	Trinidad and Tobago		31,154
43	Portugal		30,258
44	Poland		29,251
45	Hungary		28,910
46	Malaysia		26,871
47	Seychelles	X	26,712
48	Russia		27,890
49	Greece		27,776
50	Latvia		27,291
51	Saint Kitts and Nevis	X	26,849
52	Antigua and Barbuda	X	26,198
53	Turkey		26,453
54	Kazakhstan		26,071
55	Bahamas, The	X	25,080
56	Chile		24,588
57	Panama		24,262
58	Croatia		24,095
59	Romania		23,991
60	Uruguay		22,445
61	Mauritius		21,628
62	Bulgaria		21,578
63	Argentina		20,677
64	Iran		20,030
65	Mexico		19,480
66	Lebanon		19,486
67	Gabon		19,266
68	Maldives	X	19,178
69	Turkmenistan		18,680
70	Belarus		18,616
71	Botswana		18,146
72	Thailand		17,786
73	Barbados	X	17,508

Rank	Country		Int$
74	Montenegro	X	17,439
75	Azerbaijan		17,433
76	Costa Rica		17,149
77	Iraq		17,004
78	Dominican Republic		16,965
–	World [1]		16,779
80	Palau	X	16,472
81	Brazil		15,500
82	Macedonia		15,203
83	Serbia		15,164
84	Algeria		15,150
85	Grenada	X	14,779
86	Colombia		14,455
87	Suriname	X	13,876
88	Saint Lucia	X	13,579
89	South Africa		13,403
90	Peru		13,342
91	Sri Lanka		13,001
92	Egypt		12,994
93	Mongolia		12,551
94	Jordan		12,487
95	Albania		12,472
96	Venezuela		12,388
97	Indonesia		12,378
98	Dominica	X	12,035
–	Kosovo		12,003
99	Nauru	X	11,995
100	Tunisia		11,987
101	Saint Vincent and the Grenadines	X	11,623
102	Namibia		11,528
103	Bosnia and Herzegovina		11,404
104	Ecuador		11,234
105	Georgia		10,644
106	Swaziland		9,882
107	Fiji	X	9,857
108	Libya		9,792
109	Paraguay		9,785
110	Jamaica		9,212
111	Armenia		9,098
112	El Salvador		8,934
113	Bhutan	X	8,720
114	Ukraine		8,656
115	Morocco		8,612
116	Belize	X	8,341
117	Guyana		8,266
118	Philippines		8,229
119	Guatemala		8,173
120	Bolivia		7,543
121	Laos		7,367
122	India		7,174
123	Uzbekistan		6,990
124	Cape Verde	X	6,942
125	Vietnam		6,876
126	Angola		6,813
127	Congo, Rep.		6,707
128	Myanmar		6,285
129	Nigeria		5,927
130	Nicaragua		5,823
131	Samoa	X	5,737
132	Moldova		5,657
133	Tonga	X	5,805
134	Honduras		5,499
135	Pakistan		5,354
136	Timor-Leste		5,008
137	Ghana		4,805
138	Sudan		4,580
139	Bangladesh		4,561
140	Mauritania		4,474
141	Cambodia		4,010
142	Zambia		3,997
143	Lesotho		3,869
144	Côte d'Ivoire		3,857

Rank	Country		Int$
145	Tuvalu	X	3,804
146	Papua New Guinea		3,603
147	Kyrgyzstan		3,652
148	Djibouti	X	3,567
149	Kenya		3,496
150	Marshall Islands	X	3,436
151	Micronesia	X	3,392
152	Cameroon		3,359
153	Tanzania		3,263
154	São Tomé and Príncipe	X	3,206
155	Tajikistan		3,131
156	Vanuatu	X	2,780
157	Nepal		2,690
158	Senegal		2,678
159	Chad		2,433
160	Uganda		2,352
161	Yemen		2,300
162	Zimbabwe		2,277
163	Benin		2,219
164	Mali		2,169
165	Solomon Islands	X	2,145
166	Ethiopia		2,113
167	Rwanda		2,081
168	Guinea		2,039
169	Kiribati	X	1,958
170	Afghanistan		1,889
171	Burkina Faso		1,884
172	Haiti		1,810
173	Guinea-Bissau		1,806
174	Sierra Leone		1,791
175	Gambia, The		1,686
176	South Sudan		1,503
177	Togo		1,612
178	Comoros	X	1,560
179	Madagascar		1,554
180	Eritrea		1,434
181	Mozambique		1,266
182	Malawi		1,172
183	Niger		1,153
184	Liberia		867
185	Burundi		808
186	Congo, Dem. Rep.		785
187	Central African Republic		681

Source: IMF list of countries by per capita GDP (PPP). https://en.wikipedia.org

X = micro-country; IMF does not list Andorra, Liechtenstein, Monaco, the Vatican and Guyana

Shelter countries and international institutions have assisted them, the terms of assistance frequently infringing upon micro-country sovereignty, especially in international affairs. The forming of regional organizations has fostered cooperative efforts. What challenges confront the continued development of these micro-countries?

A major continuing limitation has been the lack of economies of scale and scope, which Michael Porter cites as a critical competitive advantage. Many micro-country disadvantages stem directly from lack of resources, inadequate infrastructure, and ill-funded services. The smaller the country, the fewer the taxpayers/customers among whom to spread the capital, overhead, and operational costs of roads, utilities, education, health, and other public services.

The less dense the population, or the more isolated the area, the scarcer the resources and the feebler the entrepreneurship the more inadequate the infrastructure there will be. These deficiencies impede local commerce, including transporting produce to local and overseas markets. The inadequacy of infrastructure and institutions handicaps initiatives for developing productivity and enhancing of prosperity—measured in both mean and median incomes. Micro-countries that have significantly increased their GDPs by developing new sources of revenue such as manufacturing, tax haven/banking, and tourism (see Table 4.2).

The cost of defense and foreign affairs establishments capable of defending not only its shores but also its international interests does not diminish proportionately with the size of the country. Smaller countries seldom have the sufficient critical mass in a product or sufficient revenue to support its development. Larger countries not only are more likely to generate critical masses in cultivated and fabricated products, but they also have better opportunities for lowering production and marketing costs and negotiating better trade deals. Smaller countries are simply disadvantaged.

Ethnic separatism, elite stratification, tradition-reverence, and rural parochialism enable gross disparities of income and inhibit the developing infrastructure, services, and representative institutions—as well as facilitate corruption. These conditions affect development of critical systems, which directly impact the opportunities for manufacturing, tourism, and other economic enterprise. Such systems include such basic services as telephone

Table 4.2 Micro-country increases in per capita GDP (mid-1990s to mid-2010s)

Continent and Country	Per capita GDP. mid 1990s	Per capita GDP. mid-2010s	Decrease in per capita GDP in 20 years	0-100% increase	100-200% increase	over 300% increase	Tourism major factor in economy	Banking major factor in economy	Crops/Fish Food Production major factor in economy	Specific and Other factors affecting economy
Europe										
Vatican	n/a	n/a					X	X		Catholic HQ
San Marino	20	61				X	X	X		Textiles
Monaco	25	79				X	X	X	X	Gambling
Andorra	18	37				X	X	X	X	Skiing. Timer
Liechtenstein	23	89				X	X			Electronics
Luxembourg	33	92			X			X		EU
Iceland	21	44			X		X			Fishing
Cyprus	15	31			X		X			Ship Build & Repair
Malta	12.9	33			X					Ship Build & Repair
Montenegro	NA	15								Steel
Africa										
Cape Verde	1.3	6.3				X			X	Fishing
Equat Guinea	1.5	32				X				Oil
São Tomé and Príncipe	1.0	3.2				X			X	Timber, Fishing, Textiles, Copper
Comoros	0.7	1.5			X		X		X	Fishing
Seychelles	7.0	26				X	X	X		Diving
Djibouti	1.2	3.1			X					Military
Asia										
Maldives	1.8	14				X	X	X		Fishing
Bhutan	0.7	8				X	X	X	X	Military
Brunei	13	73				X	X			Oil
The Americas										
Barbados	10	16		X				X	X	Sugar
Grenada	3.2	12				X	X	X	X	Textiles
Saint Vincent & theGrenadines	2.2	11				X	X	X	X	Cement
Saint Lucia	3.8	12				X	X	X	X	Clothing
Dominica	2.5	11				X	X	X	X	Soap, Coconut oil
Antigua & Barbuda	7.0	23				X	X	X	X	Cotton
St.Kitts&Nevis	6.0	21				X	X	X	X	Sugar
Bahamas	19	25		X			X	X	X	Off-shore of US
Belize	3.0	8			X		X		X	Garments
Guyana	2.5	7			X				X	Bauxite, Sugar
Suriname	3.4	17				X			X	Bauxite, Gold
Oceania										
Solomon Islands	3.0	1.9	X						X	Fishing
Fiji	6.0	8		X			X		X	Sugar
Vanuatu	1.3	2.5		X					X	Fishing
Tonga	2.3	5		X			X		X	Fishing
Samoa	2.1	5			X				X	Fishing
Tuvalu	1.7	2.6		X					X	Fishing
Kiribati	0.8	1.7				X			X	Fishing
Nauru	10	5	X	X						Phosphate Exhaustion Prisons
Marshall Is	1.6	3.3			X		X		X	Compact of Free Assn w/ US, Copra
Micronesia	1.8	3.0		X			X		X	Compact of Free Assn w/ US
Palau	3.8	22				X	X		X	Compact of Free Assn w/ US, Textiles

Source: The World Almanac and Book of Facts 1996 and 2016. Arranged by author.

and other communication networks, viable ones for collection of taxes and managing revenue, maintaining order and rendering justice, and conducting relations with other countries. Deficiencies defeat development.

Basic to the lack of facilities and systems has been the lack of skilled manpower. Micro-countries have a shortage of effective community organizers and politicians, doctors and nurses, teachers, engineers, electricians and plumbers, accountants and lawyers, and managers and entrepreneurs. The lack of expertise and entrepreneurs adversely affects medical, educational, public works, and public safety. Many micro-country youth, like rural youth everywhere, migrate from their home communities; too few return. The shortage and the mediocrity of schools exacerbate the problem. The more promising youth are attracted by city life and opportunities, which worsens local recruitment and retention efforts. The inadequacy of governing institutions and public infrastructure has handicapped initiatives for the development. Misfeasance and malfeasance have, in many cases, waylaid the initiatives and crippled progress.

Most micro-countries maintain few diplomatic missions (see Table 4.3). All but five maintain a mission accredited to the EU; many of these also have one mission accredited to the EU, Belgium, and other European countries. Two countries (the Bahamas and Tonga) conduct their EU business from their embassy in London. Many micro-countries maintain a mission in Washington, D.C., or New York City (several share one suite) accredited to both the U.S. and the UN. The cost of maintaining offices and diplomats in overseas sites limits the extent and effectiveness of the representation of the country's security and trade interests.

INDUSTRY, INFRASTRUCTURE, AND INNOVATION

The revenue sources of the micro-countries vary widely. They include farming and fishing; food processing and manufacturing (mainly light); banking and expatriate remittances; foreign military base rentals; aid and assistance from foreign governments, international agencies, and NGOs; and, increasingly, tourism.

Table 4.3: Diplomatic Missions of 41 Micro-countries to US, UN, EU, and membership in (British) Commonwealth of Nations and Organization International de La Francophone (OIF)

Country	UN	US	EU	Commonwealth	OIF
EUROPE					
Vatican		X	X		
Republic of San Marino	X	X	X		
Monaco	X	X	X		
Andorra	X	NYC	X		
Liechtenstein	X	X	X		
Luxembourg	X	X	X		
Iceland	X	X	X		
Cyprus	X	X	X		
Malta	X	X	X		
Montenegro	X	X	X		
AFRICA					
Cape Verde	X	X	X		X
Equatorial Guinea	X	X	X		X
São Tomé and Príncipe	X	NYC	X		X
Comoros	X	NYC	X		X
Seychelles	X	NYC	X	X	
Djibouti	Dc	X	X		X
ASIA					
Maldives	X	NYC	X	X	
Bhutan	X	X	X		
Brunei	X	X	X	X	
AMERICAS					
Barbados	X	X	X	X	
Grenada	X	X		X	
Saint Vincent and the Grenadines	X	X	X	X	
Saint Lucia	X	X	X	X	X
Dominica	X	X	X	X	X
Antigua and Barbuda	X	X	X	X	
St. Kitts and Nevis	X	X	X	X	
The Bahamas	X	X	London		
Belize	X	X	X	X	
Guyana	X	X	X	X	
Suriname	X	X	X		
OCEANA					
Solomon Islands	X	NYC	X	X	
Republic of Fiji	X	X	X	X (Suspended)	
Vanuatu	X		X	X	X
Tonga	X	NYC	London	X	
Samoa	X	NYC	X	X	
Tuvalu	X			X	
Republic of Kiribati	X			X	
Republic of Nauru	X			X	
Republic of Marshall Islands	X	X	X		
Federal States of Micronesia	X	X			
Republic of Palau	X	X	X		

Source: The United Nations, Department of Public Information.
Notes: NYC = the mission to the United States is located in New York City
London=the mission to the European Union is located in London

Farming was long the major livelihood of most colonies.[10] Plantations and subsistence farming and fishing were long the mainstays of local economies. Private plantations, small plot farmers, and government or multinational corporations have produced cocoa, coffee, copra, cotton, cattle, bananas, and sugar. Fishing suffers from increased competition and is declining. Government corporations have not always been economically viable. Foreign firms send profits out of the country. Many micro-countries have lacked the economies of scale to compete successfully in the world markets, especially since trade barriers and internal subsidies still encumber trade in this only quasi-globalized world.

Extraction of oil, gas, coal phosphates, and other minerals provides the major revenue for a few micro-countries—most notably Equatorial Guinea and Brunei. A limitation especially evident in Nigeria is that the easily secured revenue from oil has diverted attention from other economic activities which whither. The second drawback is that extraction resources do not last forever. Few governments have prepared, as Norway has, for their exhaustion. Too often, extraction industries keep most of a county's inhabitants poor and prevent the country from developing new industries.

A few micro-countries have developed significant manufacturing industries. Liechtenstein has an electronics industry that supports the highest recorded Gaps in Europe. Malta has maintains a ship repair and construction industry. Samoa has developed a food processing industry. Others have developed handicraft industries, which supplement their GDP, but small geographically isolated countries are disadvantaged in exploiting a niche in a global market.

Foreign military base-site rent, and profits from the resulting commercial activity, generates a major income source for a few countries—Djibouti and Cyprus are examples. But foreign troops residing within the country may not be welcome neighbors.

Remittances by families that have migrated are a significant source of income to countries such as Suriname, Guyana, Barbados, Andorra, Cyprus, Cape Verde, Equatorial Guinea, the Marshall Islands, Palau, and the Federated States of Micronesia. The business of selling citizen and residence rights (CRBI) has become a lucrative source of revenue in several

---

10. The West Indies islands were the most coveted of European colonies in the Americas.

countries. They include Cyprus and Malta, which provide access to EU residency, five Caribbean islands, and Vanuatu and Tonga. Many consultants, lawyers, realtors, and accountants assist these wealthy clients who want an additional passport.

Banking havens, as they have developed, contribute significantly to the GDP of a few countries including Andorra and Luxembourg, Bahamas and Grenada, and Cape Verde. They face competition from established competitors such as Switzerland and Cayman Islands. Other countries yearn to follow suit, but there may be already more than enough to absorb the surging market already being tapped in North America and Western Europe, especially since the use offshore tax haven accounts is under scrutiny.

Tourism has become the major industry in many micro-countries. The significant increase of income of those in the upper income brackets not only in North America and Europe but also in the other continents will continue to fuel tourism. The development of bigger and faster planes and ships will speed up the process. More wealthy Asians and better transportation will especially help the Pacific micro-countries. However, tourism requires three essentials: attractions, accommodating travel and lodging, and, most importantly, people who have the time, money, and desire to visit. Attractions range from surf and sand, snow and slopes, scenery, museums, cathedrals, and other historic sites. Accommodating facilities may include hotels, modern air and seaports, convenient airline and ship schedules, and local transport options.

International assistance, another major source of income for most micro-countries, is highly dependent on the world economy, the flexibility of national and international organizations, and the state of global "cold wars" and trade wars. The Cold War between the Soviet Union and the United States stimulated such aid. China's more recent "Belt and Road Initiative"—reported by S. Horn, C. Reinhart, and C. Trebesch—has invested the equivalent of about $5 trillion (U.S.) in a total of 152 countries since the year 2000. The loans, mostly lent by Chinese state-owned banks, are especially concentrated in 50 smaller and less developed recipients whose debt to China grew from 1 percent of GDP in 2005 to 15 percent of GDP in 2016, which gives China significant leverage. Unlike loans emanating from other countries, about one-half of China's overseas loans to the developing world are "hidden." That is, they are not reported to the IMF or the World Bank. The extent of indebtedness provides China significant leverage. On a per capita basis, micro-countries have done well receiving bilateral and multilateral. But such aid only partially compensates for the revenue shortfall—a concern for most micro-countries.

Despite micro-country disadvantages of size, distance, and physical and social infrastructure, many micro-countries have been energetic in developing sources of revenue. They have been industrious in soliciting aid. Some have been perseverant in developing banking and tourism. A few have been innovative in developing new industries, from handicraft factories to the sale of passports and citizenship. Dr. Worrel points out that they could do better in managing their budgets and developing infrastructure supporting the development of the economy. But, however much they accomplish, threat of environmental disasters looms.

5. CLIMATE CHANGE DISASTERS

"In the last year alone my country has suffered through unprecedented drought in the north, and the biggest ever king tides in the south; and we have watched the most present devastating typhoons in history leave a trail of death and destruction across the region."

—Christopher Loeak, President of the
Republic of the Marshall Islands

RISING SEAS AND MORE FREQUENT FEROCIOUS STORMS

A few decades ago, the stunning increase in environmental disasters caused by climate change was not widely anticipated. Since then, the continuing contamination of the atmosphere has imposed more frequent and ferocious disasters than heretofore. Scientists predict that, over the next decade, the earth will warm by 1.5°C, and perhaps as much as 2°C if we fail to take drastic and sustained action against climate change. Even if dramatic action is taken now, the damage is already incurred. Even in the best-case scenarios, we will endure hurricanes, wildfires, droughts, rising seas, and other environmental disasters more often and more significantly than before. The consequences will be dire: loss of homes and businesses, lives and livelihoods, hunger and disease, the destruction of land and economies, and the continued existence of some countries.

The present climate change is the last of multiple ones that have imposed disasters, but it is the first that man was the principle cause. William Rosen, in *The Third Horseman: Climate Change and the Great Famine of the 14th Century*, calls attention to climate change and concomitant disasters that "put millions . . . In the path of apocalyptic destruction" during the early 14th century. The momentum of these disasters fundamentally affected many basic institutions including feudalism that became extinct, the Roman Catholic Church, which lost its primacy power, and many "proto-nations . . . [which had been] subsumed into larger [or luckier] opponents." Surviving countries developed their structure, their services, while omnipotent countries acquired sovereign stature.

The present threat has received so much critical attention that those who are not already concerned are immune to facts inimical to their interests. Among the recent books describing the crisis are *Carbon Ideologies, No Immediate Danger* and *Carbon Ideologies, No Good Alternatives* by William T. Vollman; *The Water Will Come* by Jeff Goodell; *The Water Problem* edited by Pat Mulroy; and *Losing Earth: A Recent History* by Nathaniel Rich. The increasing reality of climate change—the in-your-face impact of the release of fossil fuel contaminants causing global warming—cannot be ignored. Bill McKibben, in a *New Republic* article has pleaded that the only way the world can meet the challenge of climate change is to mount a mobilization similar to the one we did in Word War II.

The devastating impact of the warming air and oceans and drier land causes steadily rising sea levels, more frequent and ferocious storms, and fires. The National Aeronautics and Space Agency (NASA) has estimated rise of the annual average sea level to be 3.4 mm (1.3 in). Global warming has increased more rapidly than earlier anticipated. By 2050, the World Bank has reported that at least another 140 million people will be forced to evacuate and relocate because of the effects of catastrophic climate change.

The Paris Agreement on climate change was signed in 2015 by 195 signatories pledging to reduce their rates of contaminating the air with the CO_2 produced by burned fossil fuels. The Paris Agreement recommends "holding warming" from rising more than a dangerous 2°C (3.6°F) while attempting to limit it to 1.5 °C (2.7°F). The results, at least initially, have been disappointing. As of 2018, no country has achieved the set goal or has gotten close to doing so. U.S. President Donald Trump even withdrew the United States from the pact. In 2018, the United Nations' scientific advisory board, the Intergovernmental Panel on Climate Change, warned that extreme weather events and worsening scientific forecasts indicate that the reality of climate change are clearly mounting. Even 1.5 degrees of warming would be calamitous, it also pointed out that such warming has already become inevitable.

Waterfront land is not only being submerged and eroded, it is being rendered unusable because of salinization. Inundation threatens the economies of many island and coastal countries. Water comes in not only overland, in many cases it is also coming in underground via porous limestone. Even when the flood is only temporary, salinization of the land destroys its fertility. This submersion of whole islands is expected to escalate. Many islets have already disappeared, and others are preparing for submersion. When all, or most, of the built-up part of a country lies less than six feet above sea level, the country's very existence is highly vulnerable to rising sea levels. Many islands will be totally submerged. Repeated flooding and storms will cut off power, contaminate drinking water, and destroy commercial buildings, farms, and homes threatening the livelihood and the existence of more micro-countries.

Typhoons have ravaged the Pacific as the Marshall Islands' president attests. Hurricanes have devastated Dominica and other Caribbean islands. Among the micro-countries threatened with complete inundation are the Maldives in the Indian Ocean and the Marshall Islands, Kiribati, Tuvalu, and Palau in the Pacific. Volcanic eruptions and tsunamis are a related threat. These disasters occur unexpectedly. While the destruction of the buildings and land is thorough, the number of deaths are staggering. Dried land and their vegetation increases the risks of fire and devastation of urban and rural areas, demolishing homes, stores, factories, farms, and forests.

The short-term costs of inadequate action to reduce fossil fuel contaminants into the atmosphere multiply the costs of long-term destruction. The absence of sufficient action increases the devastating environmental consequences, escalates restorative fiscal challenges, and threatens the economic viability and existence of several countries. Given the extent of damage already done to the atmosphere and its projected long-term effects, no matter what is done now, it will take more than a minimum amount of action to reduce continued contamination of the environment. It is mandatory that action be taken to minimize the impact of rising seas and wreckful storms. The inevitability of inundation, or being otherwise being rendered uninhabitable, also raises existential questions: should the micro-countries invest in preventive measures that may only postpone the inevitable, in efforts to rebuild after repeated flooding, and/or in land elsewhere to relocate their countries' populations? More attention needs to be paid to how to contain and restrict the damage that is expected from seas that will continue rise and storms that will continue to devastate.

THE MORE VULNERABLE, LESS DEVELOPED, AND INSUFFICIENTLY PREPARED COMMUNITIES

The complete obliteration of low-lying island countries such as the Maldives, Solomon Islands, and Kiribati by rising sea levels and storms is catastrophic. Even islands whose land is not completely inundated with climate change catastrophes are threatened with extinction. When the center of the island is a mountain, vehicles flow around the mountain around which are the low-lying towns and villages with farmers, fishermen, and tourist vendors who depend on the uncontaminated seas, uneroded land, unspoiled sand and surf, unblemished scenery, and preserved historic sites to attract visitors and develop their economy. The warming seas rise and tides inundate the coasts. Storms erode farmland further inland. The destruction of businesses and homes is widespread. Eroded and unfertile farmland, warmed-seas and overfishing, and the destruction of beaches and other tourism attractions force many to face increased under and unemployment rates.

As seas continue to rise and storms come more often and impose more damage, more people will migrate. The residents will tire of the painful efforts of rebuilding repeatedly with insurance or aid. Insurance companies, banks, and aid agencies will become more prudent in setting the terms for such insurance, loans, and aid. As lands erode, as economies wither, and as inhabitants emigrate, countries become less capable of supporting a government and maintaining their existence.

The Netherlands, long a prosperous trading country, could afford to build and maintain dikes to protect their low-lying country with its well-developed farms and cities (whose economy as well as population is many times larger than any micro-country). Centuries ago, the Dutch confronted the threat of rising seas and storms and gained experience building sea-walls to protect their shores. Threatened cities such as Venice, New York, Boston,

Miami, Osaka, Hong Kong, and Shanghai have a sufficiently strong economic base to support such an effort. Boston, for example, has already proposed a plan to elevate 47 miles of shoreline, pointing out that the damage from rising sea levels would far outweigh cost of investments over the next few decades (estimated at as much as $1 billion).

Seawalls alone, however, do not protect beaches from erosion. Nor do they prevent seawater from coming underground through porous substrata to flood communities.

Beaches may be replaced by massive distributions of sand excavated from other sites. Floods may be prevented with extensive drainage systems and pumping stations. But, as seas rise and storms become more frequent and ferocious, these courses of action must escalate along with the seas and the storms. Even already well-developed seacoasts may find such costs prohibitive. The coastal communities and beaches most affected by the rising costs of climate change are those located in the poorer regions of the world where many micro-countries are.

Economics, politics, and time militate against the possibility of micro-country urban hubs protecting their developed coasts. Unless extensive action is taken to reduce the amount of contaminants released into the air, it is inevitable that the economy of many countries will suffer. Even worse, sufficient land will be so inundated that the existence of a few countries may cease. Several micro-countries have already recognized that their very existence is at stake. Under consideration are ex-situ conservation methods, which would exist outside the territory where the country had previously existed. The climate change issue has led to the formation of the 39-member Alliance of Small Island States (AOSIS), which has been remarkably successful in changing the tone and influencing policy at the UN and elsewhere. Several island countries are considering the purchase of land elsewhere. Kiribati has already bought land upon which it has relocated some of its people.

The primary effort of the Alliance of Small Island States, and that of their countries' at the United Nations and elsewhere, is to encourage action to reduce the extent of coal and other carbon contaminates reaching the atmosphere. Their representation at the United Nations considers that convincing the world to take stronger action is a primary objective. The planet's atmosphere will continue changing. Mankind's avoidance of short-term economic sacrifices in using fossil fuels militates against reducing the long-term cost of prolonged recklessness. The tragic irony is that the countries which have been most reluctant to recognize and reduce fossil fuel atmosphere contamination are the not the ones that will suffer the most from the lack of timely action. The island and coastal micro-countries will.

Arresting the acceleration of climate change damage appears hopeless. A UN scientific panel reported in 2018 that, to prevent 2.7°F of warming by 2040 and the damage that the warming would cause, greenhouse pollution must be reduced by 45 percent from the reported 2010 levels by 2030. The panel recognized that such a rapid transformation of the global economy was unlikely. The probability that global warming and its effects will accelerate over the next few decades therefore appears inevitable. Every country, every year, for half a century would need reductions of 4 to 5 percent to slow down the continuing acceleration of global warming. The world would not only have to develop, improve, and adopt energy innovations such as solar panels, windmill farms, and electric vehicles much more widely and far more quickly than what now seems predictable—as promising as these advances and new industries appear to be. Even leaving remaining coal and oil deposits un-extracted underground would not sufficiently alleviate the threat.

Self-serving myopia and potent lobbying by the "Coalburgs," "Gastowns," and "Oilvilles" to protect jobs as well as the Exxon-Mobils, Shell Oils, and Chevrons to protect profits will continue to persuade political leaders to postpone and pontificate. A few affluent Miami-Dades will build seawalls (over and underground). The less prosperous ones cannot afford constructing seawalls any more than reconstructing communities. With the loss of their commercial base, they face economic extinction—a death as real as the inundation facing the Maldives, Kiribati, and others. A comparison of the recoveries (and lack thereof) of Caribbean islands affirms the assertion: several months after the hurricanes the better developed, new ones have recuperated, but few efforts have been taken to repair the damage inflicted on less developed Dominica and even less developed Barbuda (which lost its one and only gas station). The magnitude of the devastation caused by fires in Australia in late 2019 and early 2020 and other destruction caused by the increasing frequency and velocity of storms in other parts of the world may focus more attention on the perils of continuing to deny the immediacy of the costs of the losses imposed by the escalation of climate change.

Contending with clashing values and budgetary woes, countries tend to postpone action rather take dramatic costly steps now. Too few micro-countries have access to the financial resources and/or the political acumen/courage to prioritize tomorrow's survival over today's realities. Even the wealthier countries are not taking sufficient action to slow down contamination of the atmosphere sufficiently to mitigate rising seas and more frequent fierce storms.

Attention may more pragmatically be focused on preparing seawalls and drainage systems to protect and preserve waterfront communities. But only communities with well-developed waterfront real estate have a realistic opportunity to secure the necessary investment from their country and foreign sources to protect and preserve their waterfront. Communities lacking already well-developed waterfront real estate will have difficulty securing the investment to protect it. The planet will adjust and survive this era's climate change disasters as it has in the past. The more vulnerable, less developed, and insufficiently prepared communities and countries, however, will not. Survival of the fittest applies to communities and countries as well as species.

The COVID-19 pandemic enveloped the world in 2020. Hundreds of thousands died, the economy floundered, businesses failed, and unemployment soared. Nevertheless, most countries looked forward to a recovery. But many micro-countries may not expect to recover. The handicaps of being small, economically disadvantaged, environmentally disaster–prone, resource-poor, isolated, and less developed in infrastructure may not survive this relentless and cruel scourge. Economies have been devastated by the loss of tourism. Curtailment of their exports and imports have strangled the commerce upon which they depended. They lack the infrastructure and financial base to mount a recovery—unless a coordinated global efforts, supported by rich countries concentrating on their own recoveries, consider them worth resuscitating. Some micro-countries may cease to exist.

6. ETHNIC, CLASS, AND OTHER IDENTITY DIVIDES

"The smaller the society, the fewer probably will be the distinct parties and interests composing it; the fewer the distinct parties and interests, the more frequently will a majority be found of the same party; and the smaller the number of individuals composing a majority, and the smaller the compass within which they are placed, the more easily will they concert and execute their plans of oppression."

—James Madison

POWER ELITES AND CROSS-FRACTURED MICRO-COUNTRIES

Identities matter, whatever the size of the country or community, because they affect how people behave, how they see themselves, how they relate and react to others, how they perceive the public good, and how they discriminate. The major ties binding humans into a sense of "we" include ethnicity, tribe, creed, distinctive culture, extended family, community (local to national), and social class.[11] Class distinctions generated by factors such as money, heritage, ethnicity, and position produce elites enabled to enhance their privileges, power, prestige, and wealth. These factors divide communities among ethnic and other identities, forming "power elites" that dominate economic, social, and political interaction and increase inequality of participation, power, and prosperity. Ethnic, class, and other identity divides extend the opportunities for dissension, discrimination,

---

11. Inheritance may be the most important factor affecting multiple aspects of 'elite' status.

deficits, and corruption. Critically, they minimize the prospect of adopting programs to minimize the economic competitive disadvantages and protecting their coasts against storms and flooding.

Revolutions in transportation, communication, and trade, and government have affected identity political behavior (i.e., whom we consider "us" and whom we call "them"). The evolution of human ties has created bonds that overlay local ones, creating a multiple fracturing of local political cultures. Instant communication, rapid travel, and increasing trade have cross-fractured communities. The threat is especially striking in areas and countries insulated from neighbors by water or mountains. This has affected not only the cohesiveness and central and local governments but also may affect the ways in which class domination and corruption opportunities affect governance.

The cross-fracturing of local communities (see Figure 5.1) has increased as the rapid advances in inexpensive and instant communication and costly and less time-consuming transportation has increased familiarity over longer distances. In more isolated micro-countries, which have more recently been exposed to and immersed in global affairs, this syndrome may have more evident consequences. Micro-countries have made economic and political gains, but many residents continue being disappointed with the pace of economic progress. Social instability, moral disorder, and the weakening of traditions lead many to look back nostalgically to what are remembered as less frantic times and less fractured communities. Increasing centralization has generated a local-mindedness counter-reaction. The phenomena drove not only decolonization but also the resistance of many colonies to considering merging or federating with neighbors. Increasing awareness of the outer world also increases an appreciation of a polity's distinctiveness and thus a negative approach to a polity containing outsiders.

Connections of blood, friendship, and acquaintance tie together diverse factions of society into sharing at least some common unifying interests. A plurality faction or party, even if only temporary, cannot ignore the opportunity to control selected government decision-making, which may extend to a wide range of activities. Depending upon the nature of the regime, this control may extend to the media—who votes and who gets the votes, who receives patronage, who receives contracts, and where government amenities are placed. How governments are organized and how antagonistic the factions or parties are affect the political climate and the extent of the nefarious political practices. Such conditions empower the rise of autocracies. Their prolonged ignoring of any interests other than their own increases the propensity for instability and the incentive for coups.

Diagram 6.1 Cross Fractured Micro-country Communities

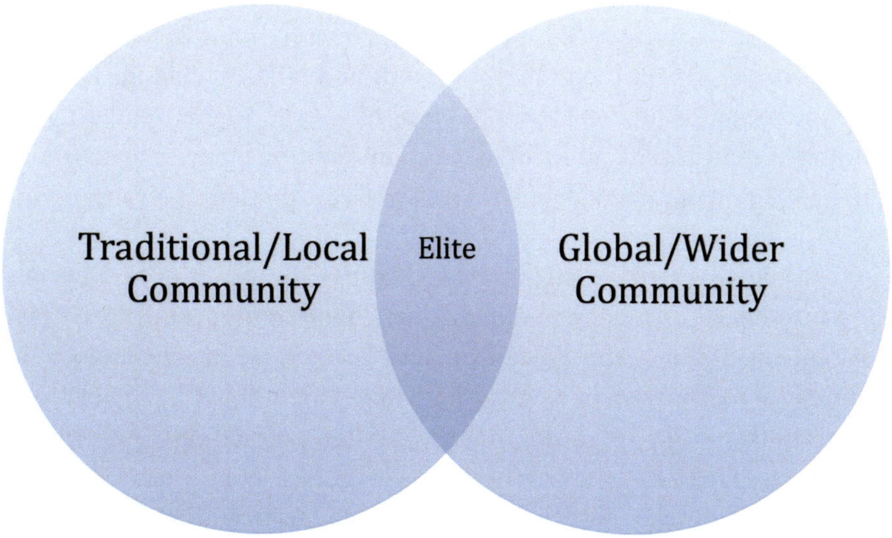

Traditional/Local Community   Elite   Global/Wider Community

Rapid travel and instant communication have sped the growth of cross-community ties, fracturing local communities with trans-local interests, perspectives, and connections. That professionals often communicate in a European language accelerates this fear of the unknown.

The influx of European settlers and labor imported or indentured from Africa, Asia, and Europe led to the emergence of mulattos (persons of European and African descent), mestizos (persons of European and Native American descent), and other mixes, which affected class and political distinctions. These manifold cleavages have complicated the political scene, undermining efforts not only to develop effective institutions of representative governance but also the local economy. Cross-fractured communities complicate the challenge of governing micro-countries. While tribal loyalties divided pre-colonization peoples, remaining tribal ties are only one set of the cleavages complicating the modern governance. The growth of commerce has introduced specialized trades and industries whose multiplication has driven the growth of towns and cities. Urban areas dominate the social, commercial, and political life of the region, rendering the countryside estranged and ignored. Faith in traditional power structures in rural areas remains as its people remain wary of modern institutions and politicians.

Traditional native faiths have been displaced and absorbed by religion introduced by migrants, military, merchants, and missionaries. These have included Hinduism, Buddhism, and Islam as well as the various denominations of Christianity including Roman Catholicism, mainstream Protestantism (e.g., Anglican, Baptist, Congregational, Methodist, and Presbyterian), and more evangelical sects (e. g. Mormon, Seventh-day Adventist, and Jehovah's Witness). Colonizers' languages and/or their dialects have become the *lingua franca* in many countries. Native languages and dialects, which may or may not be mutually understood, have, however, not been fully displaced. The impact of faith communities, especially the more evangelical ones, on the colonizing of micro-countries has deeply affected politics and governance.

Table 6.1 Languages and Ethnic Groups

MICRO-COUNTRY	Official Language	Other widely used Language	Major Ethnic Groups
Andorra	Catalan	French, Spanish	Andorran, Spanish, Portuguese, French
Antigua&Barb	English	Antiguan creole	black, mixed, Hispanic
Bahamas	English	Haitian creole	black, white, black and white
Barbados	English	Creole	black, mixed, white
Belize	English	Spanish, Creole	mestizo, , Maya, Garifuna, E. Indian, Mennonite
Bhutan	Sharchhopka, Dzongkha	Lhotshamkha	Ngalop or Bhote, ethnic Nepalese, indigenous or migrant tribes
Brunei	Malay	English, Chinese	Malay, Chinese, other indigenous
Cabo Verde	Portuguese	Creole	Creole, African
Comoros	Arabic, French, Shikomoro		Antalote, Cafre, Makoa, Oimatsaha, Sakalava
Cyprus	Greek, Turkish	.English, Romanian, Russian, Bulgarian	Greek, Turkish
Djibouti	French, Arabic		Somali, Afar, French, Arab, Ethiopian, Italian)
Dominica	English	French patois	black, mixed, indigenous
Eq. Guiana	Spanish, French	Fang, Bubi	Fang, Bubi, Mdowe
Fiji	English, Fijian	Hindustani	i-Taukei, Indian, European, Pacific Islanders
Micronesia	English	7 native	Chuukese, Mortlockese, Pohnpeia,
Grenada	English	French patois	African descent, mixed, E. Indian
Guyana	English	Guyanese Creole, Amerindian, East Indian, Chinese	East Indian, black, mixed, Amerindian
Iceland	Icelandic, English	Nordic languages, German	homogeneous mix of Norse-Celt descendants,
Kiribati	I-Kiribati, English		I-Kiribati, i-Kiribati/mixed
Liechtenstein	German		Liechtensteiner
Luxembourg	Luxembourgish, French, English	Portuguese	Luxembourger, Portuguese, French, Italian, Belgian, German
Maldives	Dhivehi	English	S. Indian, Sinhalese, Arab
Malta	Maltese, English		Maltese
Marshall Is	Marshallese, English		Marshallese, mixed Marshallese
Monaco	French	English, Italian, Monégasque	French, Monégasque, Italian
Montenegro	Montenegrin	Serbian, Bosnian, Albanian, Serbio-Croat	Montenegrin, Serbian, Bosniak, Albanian,
Nauru	Nauruan	English	Nauruan, other Pacific Islander, Chinese,
Palau	Palauan, English	Filipino	Palauan, Filipino, other Micronesian
Samoa	Samoan	English	Samoan, Europesian
San Marino	Italian		Sammarinese, Italian
São Tomé & P	Portuguese	Forro, Cabo Verdian, French, Angolar, English	mestizo, forros, servicais, tongas, European, Asian
Seychelles	Emglish, Fnech. Seychelles Creole		mixed French, African, Indian, Chinese, Arab
Solomon Is	Melanese, English	120 indigenous languages	Melanesian, Polynesian
St.Kitts&Nevis	English		mostly black; some British,Eortuguese, Lebanese
St. Lucia	English	French patois	black/African descent, mixed, E. Indian
Saint Vincent & Grenadines	English, French patios		African, mixed, E. Indian, European, Carib Amerindian
Suriname	Dutch	English, Sranang Tongo, Caribbean Hindustani, Javanese	Hindustani Creole, Javanese, Maroon, Amerindian, Chinese
Tonga	Tongan,English		Tongan
Tuvalu		Samoan	Polynesian, Micronesian
Vanuatu	local languages English, French		Ni-Vanuatu
Vatican	It., Latin, French		Italian, Swiss, other
		Official	**Widely Used**
English		20	5
French		7	2

SOURCE: World Atlas and Book of Facts (2016)

A pressing concern is the impact of towns and small cities. They serve as the magnets for the surrounding area, which has been marginalized as the land, sea, and underground resources withered or replaced by large-scale enterprises that do not depend on towns. Rural areas and small towns have neither the jobs nor the attractions to retain the younger generation. Larger cities have become the most productive places for business. They bring stimulating diversity: complementary businesses, specialized suppliers, and trade.

Insular isolated countries, though, do not generally have sufficient surrounding hinterland to support a city with the infrastructure that attracts sufficient business. Poor roads hinder getting produce to market, inferior schools and teachers retard educational achievement, and inadequate/mediocre workers and programs reduce productivity.

Chiefs are undercut, and are undercut by, the representative governing process, which slows change. Traditional power alignments may impede change in institutions and infrastructure. Ethnic, socio-economic, political stratification, and family alliances dominate economic, political, and social communities. Class, ethnicity, faith, and color stratification strengthen separatism. Insularity, traditional tribal governance, rural and urban uneasiness, the impending exhaustion of natural resources, and the difficulties of promoting new industries undermine prospects for improving living standards. Coping with these issues requires a degree of trust more likely to be present within extended families, tribes, and other in-groups, as contrasted with the distrust existent outside these circles. Voters distrust distant governments. The more distant the government, the more distrusted it is.

Micro-country publics, like all publics, demand more roads and utilities, schools and health, public safety, economic development, and emergencies met promptly. To meet these pressures, micro-countries need to overcome mindsets molded by isolation and nostalgia of the good old days, increasingly fragmented political landscapes, those committed to private-regarding interests, and decades-long cultivated aversion to over-centralization. Perhaps the most difficult is the temptation to corruption that not only disrupts pubic services and economic development but also undermines the public's confidence in government.

Micro-countries, with their smaller populations, may have may have fewer ethnic, faith, and other identity factions, but the intensity of the identity-divide schisms often complicate the countries' abilities to recognize problems and develop programs to alleviate them. Micro-country elites make the misdirection of public funds a norm. Representative governance provides a means of reconciling diverse interests through the development of policies and programs and the delivery of public services. Such skills require politicians with the courage and cunning to rally public support, to bargain, to strike compromises in no-holds-barred arenas, and to see through their implementation. Such politicians are gladiators whose combat not only determines the future of their polity but also that of their party and their own political mortality. Representative governance may not be a panacea, but, as Winston Churchill is reported to have said, it is better than all the alternatives.

Micro-countries continue to depend upon dedicated, dynamic, and demagogic leaders to arouse passions and secure the support required for survival. Micro-country governments must be sufficiently effective to manage the divisiveness of the local forces; they must be sufficiently self-confident to recognize that they cannot, in this increasingly interdependent world, manage without working alongside global and regional institutions.

CENTRAL AND LOCAL GOVERNANCE

How countries organize the three branches of central government (i.e., the executive, legislative, and judicial branches) and whether they organize elected and effective local governments varies significantly. With regard to the central government institutions, most new countries have been strikingly influenced by the example and the mentoring of the British, French, and American systems whose governments had the most colonies in the initial post-World War II years. The most striking difference is how the executive branch is organized in each country.

The British parliamentary system focuses gives executive power to the prime minister as the head of government as opposed to the British monarch, whose duties are delegated to a governor general in many former British dependencies or a president of sorts. The prime minister, as leader of the majority party or coalition, is selected by parliament and may be dismissed by it. Conversely, in the French system, the directly elected president appoints the prime minister. The president combines the roles of the head of state and head of government role whereas the prime minister is dependent upon the support of the prescient and parliament.

The directly elected American president also combines the roles of head of government and head of state but with no prime minister in effect. Where legislatures have some control of the executive there has generally been more collaboration between the executive and legislative branches of the government, avoiding a concentration of power. However, parliamentary systems often degenerate into executive strong-arming or political paralysis, observes Larry Diamond "when party control of their prime minister is weak and the form of cabinet government does not work in practice."

Whether legislators are elected by proportional representation (PR) or by single member districts (SMD) has depended principally upon their colonial inheritance. Countries influenced by Britain and America tend to use SMD for elections; French-influenced ones tend to use PR. PR has allowed a greater variety of interests and parties to gain representation and tends to produce coalition governments, but SMD has tended to limit the number of parties and has produced, consequently, more stable governments.

Judiciary arrangements vary widely. The executive ruler may dominate the judiciary branch of government—as it occurs in the Maldives, Equatorial Guinea, and Brunei. In others, such as San Marino and Monaco, the judiciary responsibilities are incorporated into that of the neighboring shelter country. The countries that formed the Organization of East Caribbean States (OECS) formed a common central bank and currency and a central court of appeal. Along with some other former British dependencies court cases may be appealed to the Privy Council in London.

The remoteness and insularity of many of the local communities, especially those remote from the capital, sometimes on distant islands or difficult terrain, stresses the case for local representative institutions. Advantages of local government include a unified sense of community identity, an opportunity for local involvement, and the means for areal initiative and coordination of areal governance. Paul Ylvisacker adds

additional opportunities and more readily available points of access, pressure, and control . . . serving to keep governmental power close to its origins . . . impedes the concentration of power . . . [and provides] additional assurance that demands will be heard and the needs will be served, potentially reducing frustration with capital-centric governance.

The values and features regarding local government accumulated over multiple centuries in Europe have been those used in developing or quasi-developing micro-country local government institutions. The new micro-countries' local institutions vary. In some cases, there are fully developed local governments with an elected council with a staff consisting of an executive leader who is council-selected, directly elected, or appointed by a central government. In others, elected councils may rely on the national or other layer of government to deliver the specific services. Countries have developed local government organizations similar to those in the colonial power.

The more rural a country—especially when all or major parts are isolated by water, mountains, and distance—the more likely it is be inbred, have value-entrenched traditions and institutions, and resist change. First-past-the-post (FPTP) and winner-take-all single-member electoral systems, as used in former colonies of the United Kingdom and the United States, enhance the legislative representation of the rural areas and the political power of rural voters. The reluctance of the more isolated communities to collaborate stymied the efforts made by the colonies to get their independence in the late 1900s by requiring a merge (or even a federation). In micro-countries where a major proportion of the population live in the leading city, the commercial and/or political capital, the concentration of economic, social, and political leadership the city fosters a sense of being ignored, "outsider-ism," and lower-class citizenship.

Table 6.2 Micro-country Capitals and Local Governments

CONTINENT Micro-Country	Population mid-2010s nearest 1,000	Capital City	Population of Capital city (2014) nearest 1,000	Local Governments
EUROPE				
Vatican	1	Vatican City	1	None
San Marino	32	San Marino	4	9 Communities
Monaco	38	Monaco	38	None
Andorra	81	Andorra La Vella	23	7 Parishes
Liechtenstein	37	Valduz	5	11 Communities
Luxembourg	543	Luxembourg	107	12 Cantons
Iceland	337	Reykjavik	184	8 Regions
Cyprus	847	Nicosia	251	6 Districts
Malta	431	Vailetta	197	68 Localities
Montenegro	622	Podgorica	165	23 Municipalities
AFRICA				
Cape Verde	508	Praia	145	22 Districts
Equatorial Guinea	799	Malabo	145	7 Provinces
São Tomé & Príncipe	203	São Tomé	71	2 Provinces
Comoros	770	Mohélii	56	3 Islands and 4 municipalities
Seychelles	94	Victoria	26	29 Administrative Divisions
Djibouti	900	Djibouti	529	6 Districts
ASIA				
Maldives	358	Male	156	7 Provinces and 1 Municipality
Bhutan	776	Thimphu	152	20 Districts
Brunei	429	Bandar Seri Begawan	14	4 Districts
AMERICAS				
Barbados	287	Bridgetown	90	11 Parishes
Grenada	106	St. George's	38	6 Parishes and 1 Dependency
St. Vincent & the Grenadines	109	Kingstown	27	6 Parishes
Saint Lucia	185	Castries	22	10 Districts
Dominica	72	Roseau	15	10 Parishes
Antigua and Barbuda	81	St. John's	22	5 Parishes and 2 Dependencies
St. Kitts and Nevis	55	Basssrlterre	14	14 Parishes
The Bahamas	388	Nassau	267	31 Districts
Belize	348	Belmopan	17	6 Districts
Guyana	808	Georgetown	124	10 Regions
Suriname	548	Paramaribo	234	10 Districts
OCEANA				
Solomon Islands	584	Honiara	73	9 Provinces and 1 City
Republic of Fiji	893	Suva	176	14 Provinces and 1 Dependency
Vanuatu	264	Porta Villa	53	6 Provinces
Tonga	106	Nuku'alofa	25	5 Island Districts
Samoa	193	Apia	37	17 Districts
Tuvalu	10	Funafuti	6	7 Islands and 1 Town Council
Republic of Kiribati	106	Tarawa	46	3 Geographical Units
Republic of Nauru	10	None Official	Na	14 Districts
Marshall Islands	53	Majuro	31	24 Municipalities
FS of Micronesia	104	Palikir	Na	4 States
Republic of Palau	21	Melekeok	Na	16 States

Source: For country population the United Nations, Department of Public Information

The greater the percent of the population residing in the major city (the country's political or commercial capital), the more likely it is to dominate a country's political as well as its socio-economic life. In the Vatican and Monaco, the capital embraces the whole country. In three other micro-countries, the capital city contains more than 50 percent of all the country's residents: Reykjavik (Iceland), Djibouti (Djibouti), and Majuro (Solomon Islands). In 15 other countries the capital comprises more the 25 percent of the population (see Table 6.2). Countries with between one-quarter and one-half of their population living in the capital are Cyprus, Montenegro, Malta, and Andorra in Europe; the Seychelles and Cape Verde in Africa; Maldives in Asia; and Saint Kitts and Nevis, Antigua and Barbuda, Grenada, Barbados, Bahamas, and Suriname in the Americas.[12] Among the Pacific island countries, other than the Solomon Islands, no city has more than one-quarter of its country's entire population.

The first-past-the-post electoral system discourages third parties and facilitates a two-party system. Under systems of proportional representation, as in former French and other former colonies, each party gets a proportionate share of the seats in the legislature, which encourages a multi-party system rather than a two-party one and, occasionally, fragile coalition governments. The representativeness effectiveness of governing institutions depends, to a major extent, on the degree at which private interests exert a disproportionate impact upon the development of public services. Private interests tend to drive the making of public policy. Without leadership, governance would be impotent. Private interests have always led and driven the making of public policy. Among the pressures and realities that confront government officials, especially elected ones, four stand out: personal convictions, constituencies (who votes), caucus (party leadership), and corruption. Having personal convictions or beliefs regarding policies and programs, placement of amenities, and political leadership is ideal for politicians and other government officials. The portion of their constituency to whom they listen most attentively, often an ethnic or elite group, exerts the most power. The direction of the majority party or coalition caucus leaders can be critical as well as the temptation of corruption. How these pressures converge and are reconciled determines the quality of governance.

---

12. Belize's capital Belmopan is an exception, as are the political capitals of the United States, Nigeria, Australia, and Brazil.

GRAFT AND GIFTS: JUSTIFIED AS GREASE AND MASKED BY GUILE

All countries endure corruption, which may be defined as "behavior, which deviates from the practices of a public person (elective or appointive) acquiring private (personal, close family, private clique, wealth, or status) gains." Every political system embraces a spectrum of practices enabling private interests to affect government policies and practices, threatening economic sustainability and political stability.

Graft and bribery pervade one end of the range of practices. The other end is the give-and-take of governance. Those who ingratiate decision-makers facilitate decisions made in their interest; it is as justified as greasing the machinery of governance-making. Guile employed in the give-and-take process partially masks the process. In-between are a variety of usages that attempt to avoid, confuse, and ignore issues such as tickets, dinners, gifts, investments, and kickbacks. Grease facilitates representative governance, graft corrupts governance, and guile shrouds the issues. Distinguishing between legitimate and illegitimate efforts to influence public policy and actions has long frustrated lobbyists as well as reformers.

It's a slippery slope. Opinions differ regarding even meal checks, tickets to cultural and sports events, and Christmas presents as outright bribery! Politics is too pressing and perilous a process for the public, the press, politicians, or even professors to determine what is or should be the border between legal and illegal practices. Corruption runs the gauntlet. Such practices include excessive spending and gift-giving (especially extravagant ones), transactions to expedite the passage of legislation affecting taxation, subsidies favorable to special interests, the selection and placement of projects, the influence of purchases and regulations, and the avoidance of the enforcement of industrial, criminal, and traffic codes.

Locating and funding new facilities such as stadiums, roads, sewers, schools, and clinics, determining loopholes in tax codes, and overlooking minor infractions are all examples of issues where perceptions of the public interest tend to vary. Where one sits affects where one stands on public issues. Such issues are the type that not only generate heated debate but also tempt proponents to the use of force, fraud, and favors while lobbying to secure the imposition of private interests, undermining government integrity.

The prevalence and potency of corruption stem from the reciprocal interaction regarding laws, their enforcement, the norms of public opinion, and tolerance for deviant practices and the extent and power of interest/pressure groups. To what extent are laws rendering practices illegal? To what extent are the laws enforced? To what extent does the public respect laws and expect them to be enforced? Private interests do tempt one to use gifts, graft, grease, and guile to influence others' decisions—and the political decisions of entire countries.

Diagram 6.2 Extent of Corruption: Four Dynamics

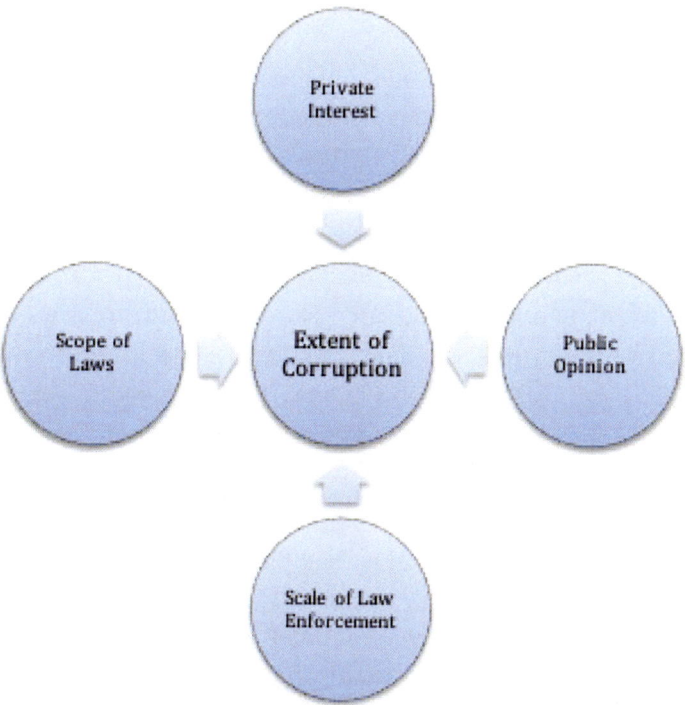

Corruption significantly reduces the integrity of governance. The opportunity for kickbacks on contracts, nepotism and other forms of favoritism in hiring; the prejudiced selection and placement of public projects; the seeking of favorable tax rates and loopholes; and the gratuities given to police and customs officials leads to waste and other inefficiencies, undermining fair and effective governance. No wonder public payrolls represent so large a fraction of many budgets. Corruption undermines a country's earning capacity. Theft redirects the amount of money available to invest in public facilities and services and, thus, defrauds taxpayers.

Micro-countries vary significantly in the extent of corruption. Transparency International ranked the extent of corruption in 175 countries, 17 of which are micro-countries (see Table 6.3). Their attempt to rank corruption, considering the extent to which practices that are deliberately not made easily visible, in environments that were, and still are, very different, is essential in comparing what is not easily comparable. Iceland, followed closely by other Nordic countries, ranked as the least corrupt country. Barbados, the Bahamas, Saint Vincent and the Grenadines, the United Kingdom, the United States, Singapore, Australia, New Zealand, and a few other European countries also rank low in terms of the amount of corruption. In contrast, Montenegro in Europe, Suriname and Guyana in South America, and Djibouti and the Comoros along with many of their African neighbors ranked among the most corrupt.

A Ugandan central bank executive has lamented that corruption and inefficiency in institutions is the biggest problem in African countries. If one could solve, or, at least, mitigate the extent of corruption, other problems would be easier to tackle. Adebayo Adedeji was Director of the Institute of Public Administration in Nigeria. Before he became the executive head of what is now the African Union (AU), Adedeji emphatically stated, "Everyone is corrupt!"[13]

Not only does corruption make projects more expensive and lacking in quality, it also discourages more responsible firms from bidding. Such deficiencies cripple administration and sabotage economic growth, as reflected in the GDP, GDPpc (average income), and worse, the median income (earned by the highest earner of the lowest half of the population). The greater and the more noticeable the corruption, the more undermined the government's creditability is—both within and outside the country. Not only is the respect of the country's residents impaired for their government's political institutions, administrative agencies, and its laws but the will of international institutions and donor countries to invest also suffers.

---

13. The statement was made during a conversation about corruption. At the time, he was my boss at the Institute. When I appeared not to fully appreciate the full significance of his assessment, he interrupted what I was saying and shouted, "Sam, listen to me! Everyone is corrupt!"

Table 6.3 Corruption Index
(Countries listed by extent of corruption by continent)

	Europe	Africa	Asia	Americas	Oceania
1-10	Denmark				
	Finland				
	Sweden				New Zealand
	Norway				
	Switzerland				
	Netherlands				
	Luxembourg		Singapore	Canada	
11-20	Germany				Australia
	Iceland				
	United Kingdom				
	Ireland		Japan	Barbados	
	Belgium		Hong Kong	United States	
	Austria			Uruguay	
	Estonia		United Arab Emirates	Chile	
	France		Qatar	Bahamas	
			Bhutan7	Saint Vincent & the Grenadines	
31-40	Cyprus	Botswana		Puerto Rico	
	Portugal				
	Poland		Taiwan		
	Spain		Israel		
	Lithuania				
	Slovenia			DOMINICA	
41-50	Latvia	Cape Verde	South Korea		
	Malta	Seychelles	Georgia	Costa Rica	
	Nicaragua	Mauritius	Malaysia		Samoa
	Czech Republic	Lesotho	Bahrain		
	Slovakia	Namibia	Jordan		
		Rwanda	Saudi Arabia		
61-70	Croatia	Ghana	Oman	Cuba	
	Macedonia	South Africa	Turkey	Brazil	
	Bulgaria	Senegal	Kuwait		
	Greece	Swaziland			
	Italy				
	Romania				
71-80	Montenegro	São Tomé 78 Príncipe			
	Serbia	Tunisia			
	Bosnia-Herzegovina	Benin	Mongolia	El Salvador	
		Morocco	India	Jamaica	
81-90		Burkina Faso	Philippines	Peru	
		Zambia	Sri Lanka	Trinidad and Tobago	
			Thailand	Colombia	
91-100		Egypt	Armenia		
		Gabon			
		Liberia		Panama	
101-110		Algeria	China	Suriname	
				Bolivia	
	Moldova	Niger		Mexico	
	Albania	Djibouti	Indonesia	Argentina	
	Kosovo	Ethiopia		Ecuador	
		Malawi			
111-120		Ivory Coast		Dominican Republic	
		Mali	Belarus	Guatemala	
		Mozambique			
		Sierra Leone	Vietnam		
		Tanzania			
121-130		Mauritania	Azerbaijan	GUYANA	
		Gambia	Kazakhstan	Honduras	
			Nepal		
		Togo	Pakistan		
131-140		Madagascar	Timor-Leste	Nicaragua	
		Cameroon		Iran	
		Nigeria		Kyrgyzstan	
141-148	Ukraine	Comoros		Lebanon	
		Uganda			
		Guinea	Bangladesh		
		Kenya	Laos		

Source": Transparency International for the data, arrangement by author.

The act of distinguishing (and overcoming) corruption has long challenged those concerned with public affairs. George Washington Plunkitt, the perspicacious New York politician and proletariat philosopher, once distinguished between the dishonest dollars pocketed by bungling politicians through graceless corruption and the goldmine to be made legitimately from what he called "honest graft." The humorist Peter Finley Dunne added, "the only thing to do is to keep politicians an' business men apart. They seem to have a bad influence on each other." Governing requires grease to develop the consensus and compromise among the diverse factions and interests to develop and deliver public services.

In terms, though, of their impact upon the feasance, misfeasance, or malfeasance of the operation of the system, grease and graft are at opposite ends of the spectrum. Whether a transaction is grease or graft, or shady, may depend on a variety of circumstances as well as differing from the point of view of those engaging in the practice. Lobbyists promote their special interest at the expense of the general interest. Whether a special interest contradicts a general one depends on one's particular interests.

The convergence of economic, environmental, and ethnic threats escalates the challenge that micro-countries confront! The more representative a political system, the more responsive, responsible, and respectful a government will be. Therefore, a government with a representative system may have a better chance of overcoming the challenge than with a non-representative system. Efforts to clearly differentiate illegitimate gifts, campaign contributions, and lobbying from legitimate activity will continue to be frustrated. Views vary too widely regarding what are legitimate public-serving interests, favor-seeking friendships, and public-interest lobbying; they vary significantly depending upon the interests and perspectives of those involved. The intermeshing of public and private interests and opportunities for greed in representative as well as autocratic governments does not facilitate efforts to define and reduce corruption.

7. THE COVID EPIDEMIC IMPACT ON ECONOMIES
by Hans Humes

IMPACT ON DEVELOPED EUROPEAN ECONOMIES

Small countries, like minorities and the poor, have disproportionately suffered from the effects of the Covid pandemic. In almost all cases, the actual incidence of infection and death due to the virus in these nations has been very low. The economic impact has been what has damaged and could possibly devastate countries that in many cases did not have a secure footing in the global community.

Countries with lower than one million population have suffered economically for different reasons. In many cases, these countries' economies are heavily reliant on tourism. In others, there has been damage due to lower commodity prices. And then there is the inevitable exogenous shock of the global economic slowdown. But to a nation, there has been a tremendous negative impact since the crisis began.

I will review the cases of the countries with which I am familiar: Iceland, Barbados, Surinam, Guyana and Belize. My company is and has been engaged in debt work out discussions with Barbados, Surinam and Belize so I can provide some insights as to how the Covid crisis has topped counties into situations where they have had to begin discussions with creditors so as not default on their external debt.

Let's start our cruise with a stop at the one "developed" country on my list: Iceland. My history with Iceland began with a visit in 2002 when I went to an amusing music festival in the Westman Islands. Since then, I' have been there more than twenty times. I've made a lot of friends in Iceland - once you know someone there, you seem to know everyone. I am now on the board of the Icelandic American Chamber of Commerce. I was due to take a trip in July from Iceland with friends in a submarine under the North Pole. That trip has been postponed.

Iceland has a population of 364,134 and a GDP of $24 billion. Fishing, aluminum smelting and tourism are mainstays of the economy. Iceland's GDP contracted in 2019 marginally, but the contraction will be far greater in 2020.

Iceland has controlled Covid very effectively. Prior to Europe restricting travel from the US on a broader basis, Iceland opened its airports and received travelers. Upon arrival, visitors could quarantine for two weeks or receive a free Covid test and isolate until the results came the next day.

The Covid statistics for Iceland are as 2,623 cases diagnosed and 10 deaths as of September 2020. The country reimposed some social measures recently as they have seen an upsurge in cases - particularly in the last month of September. While in the months of April through late July, daily diagnoses in Iceland over 10 new cases constituted an upsurge, 77 new cases were diagnosed on September 18. In general, however, according to my friends, life is relatively normal.

Iceland's economy is not faring as well. Iceland's credit outlook was revised to Negative on May 22 by Fitch while its rating was affirmed at A. The stated reason for the revision? "The coronavirus pandemic will severely affect Iceland's tourism industry, which accounts for around 9% of the economy's GDP (excluding indirect effects) and 35% of total exports. Passenger traffic had been in decline already in the first quarter of the year (-25% yoy), driven by a reduction in airline seat capacity following the bankruptcy of the budget airline Wow Air in 2019, and the grounding of Icelandair's Boeing Max jets. As a result, we forecast foreign tourist arrivals to decline by 80% this year. While we expect a recovery in 2021, we forecast arrivals to remain 40% below their 2019 levels."

While day to day life is relatively normal even with restrictions of gatherings of over 100 people, there is concern among the people. Unemployment (taking into account jobs with less than 100% participation) hit a peak of 16.7% % in April of this year from a norm of 5.1%.

Iceland remains a relatively wealthy country with a robust social safety net. It does not have the distortions it had prior to the economic crisis of 2007. Iceland's economy is anticipated to decline 9.1% on a year on year basis for the last 6 months of 2020. Iceland clearly is exposed to the ebbs and flows of the global economy. As a small economy, the shocks are magnified.

My friend Hlynur Gudjonsson, the Consul and Trade Commissioner in New York says this, "Icelanders and Icelandic companies have proved to be resilient, willing and resourceful dealing with hardship like the global economic downturn in 2008, the economy flexible, agile companies and people and supply chains are ready to deal with the pandemic. The authorities were as ready as could be to deal with an epidemic of this size and force and ready to execute a plan that matched the response model scenarios. Large-scale testing, protection of vulnerable groups, isolation and quarantine as needed, effective contact-tracing, remote care and timely intervention, and later economic assistance and recovery plan." After perusing the situation of Iceland, let's pay a visit to another island: Barbados. Barbados is a more traditional vacation destination than Iceland and is distinctly middle income. Barbados, like Iceland, has contained the Covid crisis well in health terms. But again, the impact in the economy has been severe, leaving policy makers scrambling to come up with solutions to underpin the economy. (The other European micro-countries, with well-developed economies well-integrated with their neighbors and the European economic system that took early action regard the Covid outbreak, may expect to have a recovery similar to Iceland. The less developed micro-countries, most of whom are heavily dependent on tourism, expect to have a more difficult recovery.)

IMPACT ON LESS DEVELOPED ECONOMIES

With a population of 286,400, Barbados had a 2019 GDP of US$5.209 billion and 2019 real GDP growth of -0.1%. The major economic sectors (% of GDP) are finance at 34.4%, tourism at 17.0%, transportation at 12.6%, distribution at 9.0%, manufacturing 6.1% and construction at 5.7%.

On the health side, the country has registered only 190 COVID-19 cases through September 2020 with 7 deaths. The infection rate has stayed lower than Iceland; there has been less travel into Barbados.

Returning to tourism's weighting in the economy, while the sector directly represents about 17% of the country's real GDP a much greater indirect percentage (~40%) via other economic sectors that are dependent on tourism activity.

Tourist arrivals effectively dropped to 0 by the end of March, with most hotels closing and extremely limited occupancy at those which remained open. Even in a scenario where tourism activity resumes later in 2020, which now appears unlikely, total arrivals for the year will be nearly 50% lower than 2019 levels.

Earnings of the tourism sector are expected to show an even sharper decline, given the various incentives and discounts that will need to be offered in order to promote activity later this year.

Barbados began to re-open its borders on July 12th, with scheduled international flights resuming over the following week.

As part of a unique program announced by the Prime Minister in early July aimed at jumpstarting tourism activity, Barbados will offer a 12-month visa designed to attract visitors who are able to work remotely. Similar initiatives have since been rolled out by a number of other tourism-dependent jurisdictions, including Bermuda, Estonia and Georgia

Unemployment jumped to 24% by the beginning of May 2020 primarily as a result of the halt in tourism, and has subsequently approached nearly 40% in July.

Part of the negative economic impact from tourism is being offset by lower global oil prices – as Barbados is a nearly 100% oil importer, this eases the strain on the country's current account.

Nevertheless, GDP is expected to contract by 11-12% in 2020, with the current account expected to register a 10% of GDP deficit.

The fiscal impact to the government associated with the pandemic is expected to be around 5% of GDP in 2020, driven by lower tax receipts from tourism and generally reduced economic activity, along with increased spending on state-owned medical facilities and social transfer payments.

Fortunately, Barbados benefited from strong multilateral support heading into the pandemic, having entered into an IMF program in 2018. The IMF upsized its lending program to Barbados by US$91 million in June 2020 to help the country accommodate pandemic response. Moreover, Barbados arrived at a restructuring agreement with its private sector creditors in December 2019, which afforded the country cashflow relief.

The gravity of the economic impact of Covid to Barbados is reflected in an Economic Letter written by DeLisle Worrell, a former governor of the Central Bank of Barbados.

Governor Worrell begins by saying: "Fear of the deadly Covid-19 virus has disrupted economic activity everywhere to an extent that calls for a change in economic strategy which is both radical and practical". He outlines how critical tourism is to the Barbadian economy and notes that tourism accounts for two thirds of foreign exchange. He continues by suggesting a modernization of the economy, public services, a change in energy policy and a possibility of replacing the Barbadian dollar with the US dollar. But as Governor Worrell points out, "Nothing can make up for that loss (of tourism dollars) in the near term." In short, like most countries with fewer than a million people, no amount of changing economic policy can insulate against an exogenous shock from the global economy.

Let's continue or trip with a short sail south to Suriname. With a population of 575,990, Surinam's GDP is US$3.84 billion, which makes it significantly less affluent than Iceland and a per capita GDP a third of Barbados. It's 2019 GDP growth was 2.3%, but projected 2020 GDP growth is -4.9%.

Surinam's economic sectors in % of GDP terms are Gold Mining and Processing and. Oil at 29.6%, Agriculture 9.4%, Government 9.4%, Wholesale / Retail Trade 9.1%, and Transportation at 7%.

Again, the health impact of Covid has been incidental: the country has registered 4,831 COVID-19 cases through September 2020 with 102 deaths. Like Iceland, the infection rate has accelerated in the last weeks.

Though tourism is an extremely small percentage of overall economic activity in Suriname, the negative impacts from the COVID pandemic have primarily been felt as a result of the lockdown-induced reduction in general economic activity, as well as the lower global oil price environment.

While Suriname is not a major net oil exporter, with the country averaging around 16,000 barrels per day of onshore oil production versus a domestic refining capacity of 17,000 barrels per day, the expansion of offshore oil exploration and production was expected to be a major driver of economic growth within the country over the next 5-10 years.

US firm Apache has made 2 significant offshore oil discoveries so far in 2020, and Exxon has announced plans to continue exploration activities later this year. However, the global oil market downturn that was partially induced by the COVID pandemic may mute or delay further investment.

Suriname is however a meaningful gold exporter, with gold production being the largest single economic activty – gold accounts for 75% of the country's total exports, and Suriname produces ~1.3 million troy ounces per year. In this sense, the nearly 25% year to date increase in gold prices has offset a portion of the COVID pandemic's impact on the country.

The pandemic hit Suriname at an interesting moment politically, as the country held scheduled national elections on May 25th 2020. Despite the various restrictions on movement and public gathering that were in place, voter turnout was 72%.

A coalition of opposition parties, led by the historically Indo-Surinamese Progressive Reform Party, was able to establish the required 2/3 majority in parliament, unseating incumbent president and former military ruler Desi Bouterse.

Surinam was able to negotiate a deferral of payment owed to private creditors in June of 2020. If economic indicators do not improve, it is impossible that the country will have to engage in a fell restructuring of its debt. The oil discoveries may not come on line early enough to absorb the shock of the global downturn.

It's a short hop next door to Guyana. Guyana is an outlier among the countries we are touring in that, while it is similar to other countries in terms of low impact on health by Covid - Guyana 2,725 total cases and 74 deaths - its economy has been able to avoid the downdraft caused by the pandemic storm. Let's take a look.

Guyana has a population of 782,000 and a 2019 GDP of US$4.11 billion. 2019 GDP growth was 4.4%. But projected 2020 GDP growth (projected) is 57%!

What makes Guyana such an outlier in the world? Let's look at its major economic sector. Agriculture accounts for 17.6%, Construction / Real Estate 16%, Mining (Gold) / Oil 15%, Wholesale / Retail Trade 7.5%, Administrative Services 6.6%, Government 6.3%, Manufacturing 5.1%. But let's look at the oil sector of the economy.

Despite the general economic slowdown associated with COVID, Guyana will experience by far the world's highest GDP growth rate in 2020, as mentioned above, is estimated by the Bank of Guyana at 57% (down from a pre-COVID level of 87%) and by the IMF at 52.8%. The next-highest projected growth rate is South Sudan, at 4.9%.

This is driven by the continued development of Guyana's offshore Stabroek oil block, operated by Exxon. Significant oil reserves were discovered beginning in 2015, with subsequent discoveries taking the total recoverable estimate to 8 billion barrels, and exploration activities continue.

The field began producing oil in December 2019, and is now pumping 120,000 barrels per day. This figure is expected to increase to 750,000-1 million barrels per day by the mid-2020s

Despite (or perhaps because of) this level of growth, the county's political environment has become extremely complicated. A no confidence vote by the country's National Assembly at the end of 2018 appeared to dissolve the existing government, though challenges to this were continued during 2019.

A general election was finally held in March 2020, with incumbent president David Granger claiming victory amid allegations of fraud during the tabulation of votes. A recount completed during June showed the opposition winning, though president Granger has yet to accept the results, and various legal challenges are still in progress.

In response, the US State Department imposed visa restrictions on certain members of the Granger government on July 15th.

In terms of economics, however Guyana has the luck of timing in a way Surinam has not. The oil discoveries were able to be developed in time enough to produce tremendous growth in the midst of the global slowdown. Given the political volatility, it remains to be seen whether a short term blessing becomes a long term curse.

The final stop in our trip is up through the Caribbean to Belize on the Atlantic coast of Central America.

Belize's population is 408,500, with a 2019 GDP of US$1.97 billion, a third lower than Surinam in terms of GDP per capita. 2019 GDP growth was 1.5% but 2020 GDP growth is projected to be down 12.0%. But this is an IMF estimate. The Belizean government's internal estimates have put the decline at anywhere between 18% and 30%.

Again, the health impact of Covid was initially incidental: the country had registered only 43 COVID-19 cases through July 2020 with 2 deaths. In the last 6 weeks, like in Surinam and Guyana, the infection rate has surged. In mid-2020 Belize now has 1,758 cases and 22 deaths.

Belize's economic sectors in terms of % of GDP are Wholesale / Retail Trade 24%, Tourism 20%, Industrial / Construction 16%, Government 13%, Agriculture 11%, Finance 10%.

As in Barbados, the tourism sector directly represents about 20% of the country's GDP, and an even greater indirect percentage (~37%) via other economic sectors that are dependent on tourism activity. Tourism also generates 60% of the country's foreign exchange earnings.

As a result, Belize is the most tourism-dependent economy in continental Latin America, with the total contribution of tourism to GDP more than double that of Uruguay, Mexico, Panama and Costa Rica.

Belize closed its primary international airport on March 23rd 2020, with a scheduled reopening on August 15th.

With the freeze in tourist arrivals and associated impact on the industry, more than 80,000 workers have applied for unemployment benefits, representing roughly 50% of the formal labor force.

The fiscal impact to the government associated with the pandemic is expected to be 7.6% of GDP in 2020 – driven by lower tax receipts from tourism and nearly all other economic sectors (revenue receipts to the government declined by 40% during the second quarter of 2020), and increased government disbursements for food assistance, unemployment payments, and medical equipment and supplies.

Belize was able to arrive at an agreement with its external creditors to defer payments in August 2020. With this deferral, it is hoped it is able to avoid a hard default. But Belize has suffered a severe economic shock and will unlikely be able to stabilize its economy without a return to normalcy in the rest of the world.

We have finished our virtual tour. We can see that the fact that we are unable to do this tour on a cruise boat has created more significate problems for most of our destinations than ourselves. The impact has not been on the health of the population (although more recent data is worrisome) but on their balance sheet.

For those of us in the United States with wanderlust, the Covid pandemic has left us with a sense of withdrawal and a passport that is unwelcome in most counties. The impact on those countries that rely on people like us coming and spending our dollars to underpin their economies is potentially catastrophic. We see the impact on Barbados and more profoundly in Belize. Iceland may not depend on American visitors to the same degree but the loss of tourism has hit as well. Surinam has felt the consequences of the dramatic global slowdown. Guyana stands apart as the exception that to some extent proves the rule. If not for the extraordinary windfall of oil wealth, it too would face an economy almost ground to a halt by the economic impact of a pandemic.

The Covid-19 pandemic has reinforced the threat to those economically disadvantaged micro-countries whose existence was already most endangered. They either lack the resources to cope with an outbreak of Covid cases or they have suffered from the loss of trade and travel upon which their fragile economy depends - or both. To add insult to injury the most economically advantaged countries have secured most of the vaccination supplies, leaving very little for others. Furthermore, the wealthier countries have become so concerned with their challenges at home to be generous in granting funds directly, or through international agencies, to less fortunate countries. Those countries whose continued existence was in jeopardy are even more in danger of extinction.

8. REPRESENTATIVE GOVERNANCE: INSTITUTIONS AND INTERDEPENDENCE

"The basis for a democratic state is liberty."

—Aristotle

"The possibilities for democracy are shaped by many grand historical and social forces: the failure of empires, the diffusion of models,
the movement of peoples, the change of generations, the transformation of values and class structures that comes with economic
development."

—Larry Diamond in *Developing Democracy.*

REPRESENTATIVE GOVERNANCE: THE THREE Rs

The rise of nationalism and anti-colonialism drove the challenging and laborious process of securing representative institutions, self-governance, and
independence. Robert Dahl says, "The greater the opportunities for expressing, organizing, and representing political preferences, the greater the number and variety of preferences and interests that are likely to be represented
in policy-making . . . [supporting] freedom as no feasible alternative can."

Pure democracy, or government of the people (from *demos* in Greek),
however, has limitations:

Dating back to Aristotle . . . the key shapers of democratic political thought have held that the best realizable form of government

is mixed, or constitutional, government, in which freedom is restricted by the rule of law and popular sovereignty is tempered by
state institutions that produce order and stability.

Aristotle saw that where the mass has supreme it might ignore the institutions and laws safeguarding the genuine public, allowing their government to degenerate to a form of despotism. Restraining the power of the
government can protect individual freedom.

This fundamental insight gave birth to a tradition of political thought,
what may be called republican governance (i.e., "in the respect to" the public's
governance). Others use the term "liberal democracy." The concept of representative governance requires criteria for evaluating the quality of its system of
governance. Three Rs—responsiveness, responsibleness, and respectfulness—
provide a basis for grading representative government (see Diagram 8.1).

Diagram 8.1 Republican Governance: Three Rs.

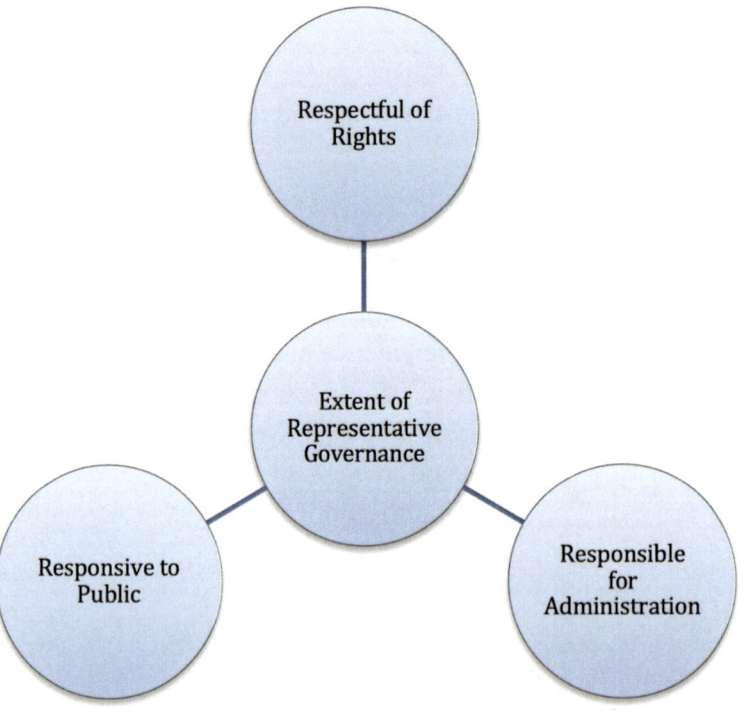

Table 8.1: Democracy Index

Rank	EUROPE (8)'	AFRICA (6)	ASIA (3)	AMERICAS (11)	OCEANIA (11)
'Liberal' Democracy(21)	Iceland			Bahamas	
	Cyprus	Cape Verde		Barbados	Kiribati
	Liechtenstein			Dominica	Micronesia
	Luxembourg			St. Kitts and Nevis	Marshall Is
	Malta			Saint Lucia	Palau
	San Marino			Saint Vincent and the Grenadines	Tuvalu
				Belize	Nauru
				Grenada	
'Flawed' Democracy(10)	Monaco	São Tomé and Príncipe		Antigua and Barbuda	Samoa
	Montenegro			Suriname	Tonga
				Guyana	Vanuatu
'Hybrid' Democracy (6)		Seychelles			Solomon Is
		Comoros	Bhutan		Fiji
			Maldives		
'Authoritarian' Regime (3)		Djibouti	Brunei		
		Equatorial Guinea			

Source: Freedom House, "Democracy Index (2014)"produced by the Economist Intelligence Unit. This Index divided countries into seven classifications; for simplification they have been collapsed into four classifications, which Democratic Index (2017), compiled by the Economist Intelligence Unit.

The two micro-countries not included in this Democratic Index were Andorra and the Vatican.

Among the other 'Liberal Democracies' are Australia, Canada, Germany, Mauritius, Netherlands, New Zealand, and the United States.

Among the other 'Flawed Democracies' are France, India, Indonesia, Italy, Papua New Guinea, Philippines, Portugal, and Trinidad and Tobago.

Among the other 'Hybrid Democracies' are Kenya, Madagascar, Malaysia, Nepal.

Among the other 'Authoratarian Regimes' are China, Guinea-Bissau, Cameroons, and Russia.

Responsiveness, in this context, refers to the extent that the legislators and chief executive are elected through free, fair, and frequent elections by universal suffrage. A responsive electoral system provides the voters the power to decide who represents them. Leaders use demagoguery, movements, parties, and other means to win. Electoral pressures increase the likelihood that, in the short-run at least, a party or movement will respond to incentives that motivate support.

Responsibleness, in this context, refers to the extent that government is held accountable by the public for the conduct of public affairs. Whether the system is parliamentary or presidential, facilitates majoritarian or consensual government, or concentrates power, a government's responsibility should be brought into question.

Respectfulness, in the context of political rights and civil liberties of the public, is the third characteristic an effective representative governance must possess. Ideally, a representative electoral system is not just a system in which representatives acquire the power to rule through a competitive struggle for the people's vote. The political system must provide for a rule of law and protect the right of individuals and groups to speak, write, assemble, demonstrate, lobby, and organize. Essential is a system in which legal rules are applied fairly regardless of class, status, or power by not only the judiciary, but also by prosecutors, auditors, and commissions overseeing regulations and laws.

The muscularity and interaction of these factors affect the politics of developing public support, passing legislation, adopting regulation, and implementation. Among the factors affecting the system in the newly independent countries are the presence of multi-party systems, the extent of the continuing vigor of feudal-like, tradition-bound chieftaincies or comparable contemporary oligarchies, and the rigidity of the ethnic separatism and racial stratification. They permeate politics and political and economic competitiveness, as well as cronyism and corruption. Such conditions are not limited to the developing world.

Measuring the quality of representative government is an inexact and perilous process. It may be viewed as a series of principles and practices that provide the institutions that ensure a government responsive to the public expressed in fair elections, an accountable administration, and protective of the rights of all. Using available information, Freedom House developed a Democracy Index with a four-level scale to label countries' governments—liberal democratic, flawed, hybrid, and authoritarian.

The "liberal democracies" label combines electoral systems with civil rights-infused pluralism in a constitutional system in which a universal suffrage electorate elects the legislative and, directly or indirectly, the chief executive. Twenty-one countries are considered liberal democracies based on this scale: six in Europe, Cape Verde in Africa, eight in the Americas, and six in the Pacific.

Flawed governments include three micro-countries—Monaco, São Tomé and Príncipe, and Nauru. Monaco's prince retains hereditary powers, São Tomé and Príncipe has not yet succeeded in discarding its colonial-cum-slave governance heritage, and Nauru faces isolated smallness, phosphate mining exhaustion, and dependence on Australian prisons.

Hybrid polities are governed by a single dominant one-party system, military domination, or continuing traditional leadership. Examples of hybrid polities are the Seychelles and the Comoros in Africa, the Maldives and Bhutan in Asia, and Fiji and the Solomon Islands in Oceania.

Authoritarian regimes are those with dictators (sometimes militaristic) who control whatever elections there are, dominate decision-making, share power and wealth with a limited few, and deny opportunities for opposing views and persons. Brunei, Djibouti, and Equatorial Guinea are examples of authoritarian countries.

How identity divides interact affects the extent or lack of representative governance. The more bitter the friction, the more divisive the politics, the less stable the governance, and the scarcer the development. More critically, a privileged identity often aligns with an oligarchy and/or an autocracy that monopolizes power, discriminating against other identities by repressing the vote, skimping on amenities and services, withholding employment and justice, and tolerating corruption.

That 21 micro-countries are considered *liberal* democracies is a tribute to the pressure and patience with which successive colonial and newly independent governments exerted. Nominated, or appointed, assemblies became elected ones; a limited voting franchise changed to universal suffrage; colonial power-appointed chief executive became popularly elected or selected by legislature. The rating of all of these countries is especially notable because of the poorer ratings of such nearby West Indies dependencies as

Jamaica, and Trinidad and Tobago, and African and Asian ones including Cameroon, Papua New Guinea, Indonesia, and Malaysia.

The nationalism that fueled the multiplication of independent countries after World War II similarly facilitated increasing pressures for representation governance. In the last three decades of the 20th century, the number of democracies increased, by one measure, from 31 to 81. The momentum of change may continue to drive more responsive, responsible, and respectful representative institutions.

The drive for national liberation and self-governance drove the evolution of representative institutions—creating elected bodies, granting them decision-making power, and expanding the suffrage. Factors hampering the maturation of institutions of representative governance and the political culture include the insularity of the country, the extent of ruralness, and the degree of tribal separatism, color stratification, chieftaincy strength, and corruption.

SHELTER COUNTRIES AND GLOBAL INTER-GOVERNMENTAL ORGANIZATIONS

The securing of independence by so many development-stressed countries after World War II has been accompanied by a fundamental change in the criteria for international recognition of sovereignty. As stressed earlier, countries no longer need to exert a monopoly of power over their affairs to be recognized as independent and admitted to the UN. This concession to the realities of late 20th-century global politics has introduced sovereign-lite independence as an acceptable standard of international recognition and UN membership.

The tensions of major power politics, the tempo of technological change, and rising aspirations dramatically increase the expectations of government far beyond what its people once expected. Many countries, not just micro-countries, cannot adequately support a military and foreign affairs establishment, nor can they develop and support an increasing demand for infrastructure and public services including such local services as police and fire protection, courts and community halls, schools and clinics, and local roads and bridges. Providing hospitals and central banks, universities and technical schools, airports, and seaports is even more difficult.

Governance is a tiered endeavor. While governance functions may be handled at multiple levels—including municipal, regional, and national—some require a larger base. Necessarily, the public structures and services of local (municipal and regional) and national overlap and intermingle depending upon the reality of how central and local services impinge upon and intermingle with each other. Globalization has forced larger as well as smaller countries to share resources, collaborate in common programs, and pool some of their sovereignty to meet global issues such as defense, trade, and economic development. No wonder sovereignty-lite independence has gained credibility for international recognition of independence and United Nations membership.

For many countries, their viability—indeed, their survival—has depended upon the continued support not only of a shelter country but also of global, more international, intergovernmental institutions; the number and scope of such institutions and their activities has increased exponentially since World War II. International support entities may be distinguished by 1) whether they are classified as general-purpose or limited-purpose, 2) whether government membership is globally inclusive, regionally, or otherwise selective, or 3) whether they are inter-governmental or non-governmental (see Figure 6.1).

The principal global organization, the United Nations, is comprised of 193 members representing almost a universality of the world's people[14], and have developed a spectrum of services and a potency far beyond that of its predecessor, the League of Nations. Its purpose is the maintenance of international peace, security, and friendly relations among countries. It provides an international forum for airing grievances and promotes collaborative programs assisting disadvantaged countries. Among the principal organs are the General Assembly, Security Council, Economic and Social Council, Court of Justice, Trusteeship Council, and Secretariat.

A host of specialized agencies work with the UN including the International Atomic Energy Agency (IAEA), the International Labor Organization (ILO), the World Bank, the World Health Organization (WHO), the World Trade Organization (WTO), the International Monetary Fund

---

14. However, the Vatican, Taiwan, which calls itself the Republic of China, Kosovo, and a few other breakaway polities do not belong to the UN.

(IMF), the International Fund for Agricultural Development (IFAD), the International Marine Organization (IMO), the United Nations Children's Fund (UNICEF), and the United Nations Educational, Scientific, and Cultural Organization (UNESCO).

Each of these specialized UN agencies and related organizations has a functional relationship or working agreement with the UN. Their financing comes from voluntary and assessed contributions. Their membership varies. Each plays a critical role with programs improving conditions throughout the world, assistance especially evident in the countries that more recently gained their independence. Non-government organizations such as the International Red Cross, Planned Parenthood, Doctors Without Borders, and Greenpeace play critical supportive roles.

Shelter countries, global institutions, and regional organizations of countries (ROOCs) played critical supportive roles sustaining the economic viability and political stability of younger independent countries, especially smaller ones. Among the shelter countries are Australia, Britain, France, Italy, Russia, Switzerland, and the United States. Such assistance continues in many of the pre-independence arrangements between the colonial power and the colony. In many cases, the assistance has been part of a mutually beneficial arrangement whereby the shelter country as well as the sheltered one derives some benefit.[15]

Assistance provided by the shelter country to a sheltered country may take many forms including the following: grants, loans, and rents, favorable trade arrangements, and essential public services, and defense and foreign affairs support. In return, the shelter country secures favorable terms for trade, military, or other support in international negotiations, etc. Such arrangements may infringe upon the sovereignty of the recipient country. The former colonial power may continue to control the defense and foreign affairs of the newly independent country. Shelter countries also provide essential local services. Italy has long provided San Marino with judges, and France assigns Monaco's judges and two of the four members of its executive party; similarly, Liechtenstein's postal service is provided by Switzerland.

---

15. Once, such arrangements were called "protectorates."

Diagram 8.2 INTERNATIONAL INSTITUTIONS ASSISTING MICRO-COUNTRIES (examples)

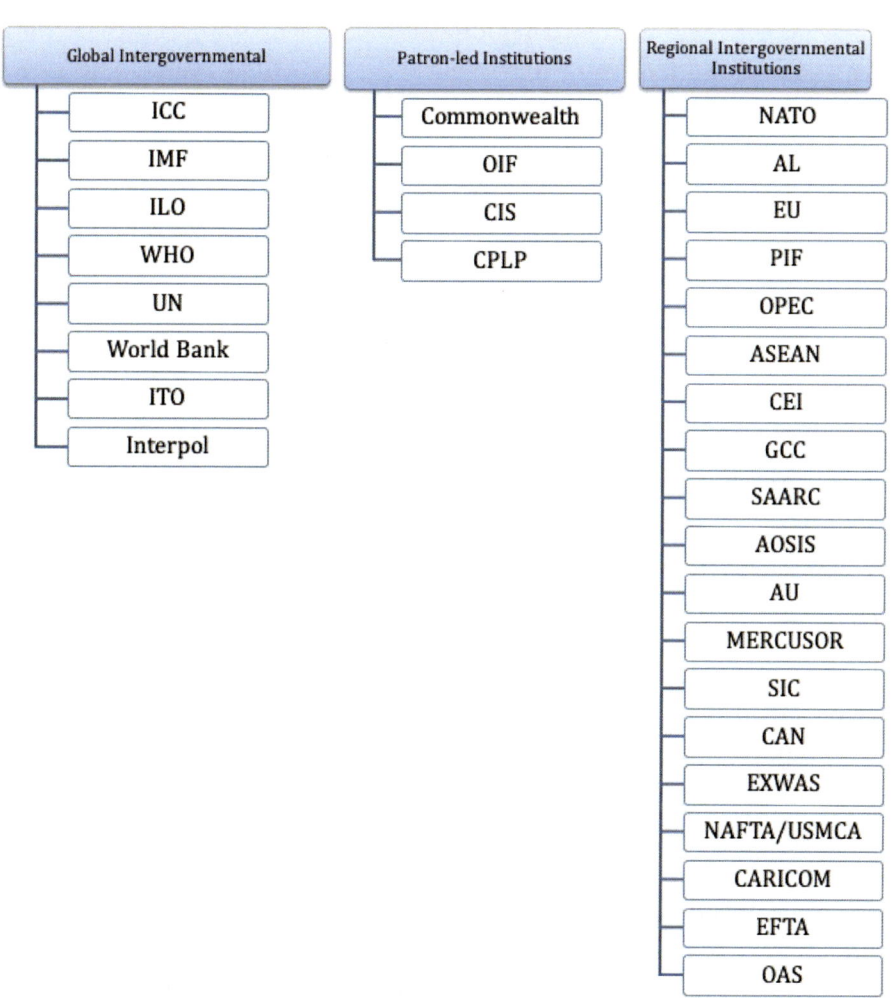

Patron-led associations of countries promote continuing relations between a former colonial power and its ex-colonies. Not only do they provide assistance and technical support but they also foster cultural and language ties. The United Kingdom promotes the Commonwealth (i.e., the British Commonwealth of Nations). The French government leads the Organisation internationale de la Francophonie (OIF). Russia is the enabler of the Commonwealth of Independent States (CIS), and Portugal leads the Community of Portuguese Language Countries.

More than 40 multi-purpose regional organizations of countries (also called "unions" or "associations") promote trade, collaborate on projects, mediate disputes, and see support for common interests. They vary dramatically in number of members, scale and scope of operations, and population. Some, however, more closely resemble a club whose members meet periodically for dinners and discussions. Each continent has a maze of regional organizations.

REGIONAL ORGANIZATIONS OF COUNTRIES

Multi-purpose regional organizations vary dramatically in number of members, scope of their activities, and their organization (see Table 4.1). A few have developed a common market, visa-less travel, and a common currency within the member countries; engaged in a range of common projects; lobbied for causes on behalf of their members; and developed a legislature, an appeal court, and a central bank. At the other end of the spectrum are ones that meet, greet, eat, take photos, and perhaps discuss doing more. Most are moderately developed; others are special-purpose organizations, such as those for defense (e.g., the North Atlantic Treaty Organization), oil (e.g., the Organization of Petroleum Exporting Countries), and lobbying (e.g., the Alliance of Small Island States). To illustrate the extent of growth of the overlapping maze of over 40 regional associations, the multi-purpose ones (and a few special-purpose ones) are briefly identified here:

The European Union (EU), the most developed general-purpose regional organization of countries, is the product of a series of treaties, each the result "of a myriad of compromises [and] an amalgam of generations' worth of visions . . . [which is] a form of government without precedent or parallel: the EU is a strange beast. Its uniqueness gives it a certain mystique." It began in 1951 with the establishment of the European Coal and Steel Community (ECSC) by France, Germany, Italy, Belgium, Luxembourg, and the Netherlands. In 1957, these counters morphed the ECSC into the European Economic Community (EEC). In 1992, the EEC was transformed into the European Union. The EU has many features of a federal government: a Parliament, a Commission directing EU staff, a Court of Justice, a Central Bank, an External Action Service (a foreign service), and the Council of the 28 heads of government which is "at the apex of collective decision-making among the governments." The EU oversees a single market, a customs union, social policy, agricultural subsidies, fisheries control, environmental policy, security, and, for 17 members, a common currency and the abolition of internal frontier controls according to the Schengen Agreement.

The EU has grown from 6 to 28 members.[16] Three micro-countries— Luxembourg, Malta, and Cyprus—are members of the EU; the others are embraced by the EU via the European Free Trade Association (EFTA) or shelter countries. Iceland and Liechtenstein participate in the EU through the EFTA. The Vatican and San Marino participate through Italy, their neighboring shelter country, and Andorra and Monaco participate through France. Montenegro has applied for EU membership. The EFTA consists of Iceland, Norway, Liechtenstein, and Switzerland. Additionally, the EFTA has full access to the EU single and common market through the European Economic Area (EEA), a free trade area which encompasses the EU members and three of the four members of EFTA (Norway and micro-countries Iceland and Liechtenstein).

---

16. In 2016, the United Kingdom voted by referendum to exit the EU. The exit was finalized December 31, 2020.

Table 8.2 Selected Regional Organizations of Countries Membership by Continent

Table 7.2 Selected Regional Organizations of Countries, Membership By Continent

Continent	Name	Acronym	Number	Microcountries
Europe	European Union	EU	28 X	Luxembourg, Malta, Cyprus
	European Free Trade Association	EFTA	4	
	Central European Free Trade Agreement	CEFTA	7	
	Central European Initiative	CEI	18	
	Nordic Counsel		5	Iceland
	Commonwealth of Independent State	CIS	11	
Africa	African Union	AU	54	All 6 African micro-countries
	Common Market of Eastern and Southern Africa	COMESA	16	Comoros, Seychelles
	Southern African Development Commission	SADC	15	Seychelles
	East African Community	EAC	3	
	Tripartite Free Trade Area	TFTA	COMESA+EAC+SADC	Comoros, Seychelles
	Economic Community of West African States	ECOWAS	15	Brunei
	Indian Ocean Commission		5	
Asia	Association of South-East Asian Nations	ASEAN	10	Brunei
	Asian Cooperation Dialogue	ADC	36	Comoros, Djibouti
	Asian-Pacific Economic Cooperation	APEC	21	
	The Leage of Arab States	Arab League	8	
	Organization of Oil Exporting Countries	OPEC	7	Comoros, Djibouti, Maldives
	Shanghai Regional Cooperative Council		8	Brunei, Maldives
	Organization of Islamic Cooperation		57	
	South Asian Association for Regional Cooperation	SAARC	8	
	Gulf Cooperation Council	GCC	8	11 micro-countries
	Turkish Council		4	6 Leeward islands micro-countries
Americas	Organization of American States	OAS	35	11 Carribean micro-countries
	Organization of East Caribbean States	OECS	6	
	Association of Caribbean States	ACS	25	
	Carribean Community and Common Market	CARICOM	14	
	Southern Common Market	MERCUSOR	5	
	United States-Mexico-Canada Agreement	USMCA (NAFTA)	3	
	Bolivian Alliance for People of our America	ALBA	11	
	Andean Community of Nations	CAN	4	
	Central American Four	CA-4	7	Belize
	Community of Latin American and Carribean	LIMUN	33	9 American micro-countries
	Pacific Alliance		4	
Oceana	Pacific Islands Forum	PIF	18	11 Pacific micro-countries
	Melanesian Spearhead Group	MSG	4	Fiji, Soloman Islands, Vanuatu

Compiled by author   X: In 2019 the United Kingdom was considering withdrawal and Montenegro was considering joining.

Completing the European maze of regional organizations of governments are the Nordic Council that has a parliament and secretariat; the Council of the Baltic Sea States; the Central European Free Trade Agreement (CEFTA) that emerged from Yugoslavia plus Albania and Moldova; the Central European Initiative (CEI) formed to promote East-West European relations; Visegrád Group (Visegrád Four or V4) consisting of the Czech Republic, Slovakia, Poland, and Hungary, the Commonwealth of Independent States (CIS) consisting of Russia and former components of the Soviet Union, and the Eastern Partnership is an EU launched initiative with six former Soviet Republics. The 47-member Council of Europe (COE) promotes "economic and social progress by further realization of human rights."

Africa has not only a continent-wide organization but also several regional ones. The African Union (AU) was founded in 2002 as the Organization of African Unity (OAU). All 54 African countries except Morocco, which withdrew after South Sudan was admitted, belong to the OAU, including its six micro-countries.[17]

Among the maze of less than continent-wide regional organizations are the Economic Community of West African States (ECOWAS) including Cabo Verde, the Arab Maghreb Union composed of five North African countries, and the Indian Ocean Commission (COI) which consists of Madagascar and Mauritius, the Comoros and the Seychelles, both micro-countries, and Réunion (a French overseas region). The Common Market of Eastern and Southern Africa (COMESA), the Southern African Development Commission (SADC), which includes two micro-countries, the Comoros and the Seychelles, and the East African Community (EAC). In 2015 an agreement was made to form the Tripartite Free Trade Area (TFTA) which merges these three regional organizations into one which covers 26 African countries, creating the biggest free-trade area on the continent from Cairo to the Cape.

A prominent feature of Asian regional organizations is the extent to which they overlap. The Gulf Cooperation Council (GCC) consists of six of the seven Arabian Peninsula countries—all except Yemen. The Turkic Council, formally the Cooperation Council of Turkish-Speaking States, consists of Turkey and three Turkic-speaking countries which became independent when the Soviet Union broke up. The 10-member Economic Cooperation Organization (ECO) extends from South Asia to countries that emerged from the dismantled Soviet Union. The 19-member Organization of Islamic Cooperation is the collective voice of 57 Muslim countries. The 22-member League of Arab States (or Arab League), founded in 1945, which includes the Comoros and Djibouti. The 54-member Organization of Petroleum Exporting Countries (OPEC) was formed in 1960.

---

17. Nkosazana Diamine-Zuma, the outgoing AU head, and South African president's ex-wife, raised hopes for the development of the AU. Troops organized by the AU did a creditable job in Somalia, but repression in Burundi and South Sudan remain untamed. The AU condemned blatant coups but its monitors have approved elections that were far from free and fair.

Table 8.3 Micro-country Participation in Selected Regional Organizations of Countries

Regional Organizations of Countries	African Union	APEC	ASEAN	CARICOM	EFTA	EU	OECS	League of Arab States	OAS	Pacific Islands Forum	SAARC	COMESA
Number of Member Independent Countries	54	21	10	14	4	28	6	8	35	18	8	16
Founded	2002, 1963	1989	1967	1973	1960	1951	1981	1945	1908, 1945	1971	1985	1994
EUROPE												
Vatican												
San Marino												
Monaco												
Andorra												
Liechtenstein					X							
Luxembourg						X						
Iceland					X							
Cyprus						X						
Malta						X						
Montenegro												
AFRICA												
Cape Verde	X											
Equatorial Guinea	X											
Sao Tome and Principe	X											
Comoros	X											X
Seychelles	X							X				
Djibouti	X							X				
ASIA												
Maldives											X	
Bhutan											X	
Brunei		X	X									
AMERICAS												
Barbados				X					X			
Grenada				X			X		X			
St. Vincent and the Grenadines				X			X		X			
St. Lucia				X			X		X			
Dominica				X			X		X			
Antigua and Barbuda				X			X		X			
St. Kitts and Nevis				X			X		X			
Bahamas				X					X			
Belize				X					X			
Guyana				X					X			
Suriname				X					X			
OCEANA												
Solomon Islands										X		
Fiji										X		
Vanuatu										X		
Tonga										X		
Samoa										X		
Tuvalu										X		
Kiribati										X		
Nauru										X		
Marshall Islands										X		
Micronesia										X		
Palau										X		

Compiled by author

The eight-member South Asian Association for Regional Cooperation (SAARC) and the 36-member Asia Cooperation Dialogue (ACD) both include Bhutan and the Maldives. The 21-member Asian-Pacific Economic Cooperation (APEC) includes Brunei. The 1967-founded Association of South-East Asian Nations (ASEAN) is comprised of 10 members, one of which is micro-country Brunei, yet China is not a member. ASEAN has negotiated with six other countries (Australia, Japan, India, New Zealand, South Korea, and China) to form the 16-member Regional Comprehensive Economic Partnership (RCEP).

The Shanghai Cooperation Organization (SCO) was initially composed of China, Russia, Kazakhstan, Kyrgyzstan, Tajikistan, Uzbekistan—India and Pakistan joined in 2017. Asia-Pacific Economic Cooperation, an association of 21 members on both sides of the Pacific (including Australia, Brunei, Canada, Chile, China, Indonesia, Japan, (South) Korea, Malaysia, Mexico, New Zealand, Papua New Guinea, Peru, Philippines, Russia, Singapore, Thailand, and the United States) took office in 2018 and has been called "A perfect excuse to chat."

The proposed Trans-Pacific Partnership (TPP), initially a trade deal among 12 countries from 4 continents around the Pacific Rim including micro-country Brunei would have affected 40 percent of world trade. The pact, which the United States initiated and led the negotiations for, was abandoned by President Trump. Despite the loss of what was once considered the indispensable partner (i.e., the U.S.), the 11 remaining countries have finalized the pact with a new name, Comprehensive and Progressive Agreement for the Trans-Pacific Partnership (CPTPP). China and the CPTPP expect to work more closely than the TPP and China would have.

The Pacific Islands Forum (PIF) consists of the 11 Oceania micro-countries, Australia, New Zealand, Papua New Guinea, and several dependencies. The Pacific Community (SPC) includes the 11 Oceania micro-countries, Australia, New Zealand, the United States, France, Papua New Guinea, and 8 dependencies. The Melanesian Spearhead Group (MSG) consists of Papua New Guinea and micro-countries Fiji, Solomon Islands, and Vanuatu.

The two Americas embrace more regional organizations than Europe, Africa, and Asia do. The bi-continental Organization of American States (OAS), which traces its lineage to the International Union of American Republics in 1890 and reorganized as the Pan-American Union in 1908, was established in 1948. The Organization of American States (OAS) was organized in 1948 but traces its lineage to 1890 and has 35 members, 11 of which are micro-countries. Its 35 members include all 11 micro-countries in the Americas. The Central American Four (Ca4) includes Belize. The Central American Integration System (SICA) has eight countries, including Belize. The Community of Latin American and Caribbean (LIMUN) includes six micro-countries (i.e., Dominica, Grenada, Guyana, Saint Lucia, Saint Kitts and Nevis, Saint Vincent and the Grenadines, and Suriname).

The Organization of East Caribbean States (OECS), formed in 1981, consists of six micro-countries in the Lesser Antilles along with British dependencies Anguilla and Montserrat as affiliates; its East Caribbean Central Bank (ECCB), manages the East Caribbean Dollar. The Caribbean Community and Common Market (CARICOM), established in 1971, includes micro-countries Belize, Suriname, Guyana, and the Bahamas as well as the six OECS members. The Association of Caribbean States (ACS) is composed of 25 Central American and Caribbean members including all 11 American micro-countries.

The Andean Community of Nations (CAN), a free trade area formed in 1969, comprises Bolivia, Columbia, Ecuador, and Peru. The Southern Common Market (MERCOSUR), formed in 1991, is made up of Argentina, Brazil, Paraguay, and Uruguay; Venezuela joined later but was suspended in 2016. The Pacific Alliance was formed in 2011 by Chile, Columbia, and Peru. The Union of South American Nations (UNASUR), founded 1969 as a free trade area, once had 12 South American members, but most had withdrawn by 2019.

The Bolivarian Alliance for the Peoples of Our America (ALBA) was formed by 11 countries, including 5 OECS members, to head off U.S. efforts to create a free trade area in the Americas. The Lima Group was founded in 2017 by 12 South and Central American countries (later joined by Guyana and Saint Lucia) concerned with escalating issues in Venezuela.[18]

---

18. In 2019, over 50 countries terminated diplomatic relations, including the U.S. and Canada, but not Mexico and Guyana.

Table 8.4 Activities of Selected Regional Organization of Countries

Continent	Regional Organization Of Countries	Free Trade Area	Economic and monetary union			Free Travel		Political pact	Defense pact	Members
			Customs Union	Single Market	Currency Union	Visa-free	Border-less			
Europe	EU	in force	in force[7]	in force[2]	in force[1]	in force	in force (Schengen[1,7], NPU and CTA[1])	in force	in force (NATO[1,7] and CFSP/ESDP[1])	28
	EFTA	in force		in force[2,7]		in force	in force[1,7]		in force[1,7]	4
	EEU	in force	in force[1]	in force	Proposed	in force			in force[1]	5
	CEFTA	in force								7
Africa	ECOWAS	in force[1,3]	in force[1]	Proposed	in force[1] and proposed for 2012[1]	in force[1]	Proposed	Proposed	in force	15
	ECCAS	in force[1]	in force[1]	Proposed	in force[1]	in force			in force	11
	EAC	in force	in force	proposed for 2015	proposed for 2015	Proposed	?	proposed for 2015		6
	SADC	in force[1]	in force[1]	proposed for 2015	de facto in force[1, proposed all for 2016]	Proposed				15
	COMESA	in force[1]	proposed for 2010	?	proposed for 2018					19
	Common	proposed for 2019	proposed for 2019	proposed for 2023	proposed for 2028			proposed for 2028		54
Oceania	Pacific Island Forum	in force					in force			4 members 44 observers
	Pacific Island Development Forum	proposed for 2021[1]								18
Americas	MERCOSUR	in force	in force	proposed for 2015			in force	proposed for 2014		5 members 5 associated 3 observers
	CAN	in force	in force[1]	proposed[1]			in force			4
	OECS	proposed for 2014[4]	proposed for not after 2019	proposed for 2019	proposed for 2019	in force	proposed for 2019	proposed	in force	12 members 2 observers
	CARICOM	in force	in force	in force[1]	in force[1] and proposed common	in force[1]	Proposed	Proposed		15 Full 5 Associate
	NAFTA/ USMCS	in force							in force[1,7]	3
West Asia	GCC	in force	in force	Proposed	proposed[1]	in force			in force	6
	Common	in force[1]	proposed for 2015	proposed for 2020	proposed			Proposed		18
Asia	ASEAN	in force[5]		proposed for 2015	proposed[8]	in force		proposed for 2015	proposed for 2020	10 members 2 observers
	SAARC	in force[1,6]	Proposed	Proposed		in force[9]				8 members 9 observers

SOURCE: UN Department of Public Information[1]

[1] Only 17 members participate. [2] Involving goods, services, telecommunications, transport (full liberalization of railways from 2012), energy (full liberalization from 2007) [3] Telephone, transport and energy – proposed. [4] Sensitive goods to be covered from 2019. [5] Least members to join from 2012. [6] Least members to join from 2017. [7] Additionally some non-member states also participate (the European Union, EFTA and NATO have overlapping membership and various common initiatives regarding the European integration). [8] Additionally some non-member states also participate in ASEAN plus Three. [9] Limited to "entitled persons."

In North America the North American Free Trade Agreement (NAFTA) of Canada, Mexico, and the U.S. negotiated in 1994, was renegotiated in 2018 as the United States, Mexico, and Canada Agreement (USMCA).[19]

Associations that transcend the boundaries of one geographical region promote the interests of ex-colonies. The African, Caribbean, and Pacific Group of States (ACP), consists of 79 members.[20] The Alliance of Small Island States (AOSIS) has 39 members and 5 observers (27 are micro-countries) who share similar concerns about the destructive effects of climate change. BRICS is the acronym of an association formed by five major emerging countries: Brazil, Russia, India, China, and, later, South Africa.

The European Union and a few other regional organizations including the Organization of East Caribbean States have not only developed an impressive range of activities but also features of a federal government such as an administration (secretariat), court of appeal, a legislature, and a central bank issuing a common currency—nurturing the framework of an emergent form of governing polity. The path to developing this type of polity has been difficult; further development will be as challenging.

The presence, proximity, and potency of a regional organization and population of a country affect whether micro-countries use a currency issued by a regional organization's central bank, the currency of a shelter, or their own currency. Seven European micro-countries use the European Union's euro: three as members (Luxembourg, Malta, and Cyprus) and four others that are non-members (the Vatican and San Marino sheltered by Italy, Andorra and Monaco sheltered by France, and Montenegro without a shelter country).[21] Liechtenstein uses the Swiss franc, and Iceland uses its own currency.

The six OECS members use the East Caribbean dollar. Equatorial Guinea uses the France-sponsored Central African franc. The Marshall Islands and Palau, in "free association" with the U.S., use its dollar; Micronesia, Tuvalu, and Nauru have a similar arrangement with Australia and use its dollar. The Seychelles (92,430), Tonga (106,501), and Kiribati (106,

---

19. In 2019, USMCA awaited ratification.
20. All members of ACP were former colonies of EU members who, according to the terms of the Lomé Convention agreement signed in 1975, received aid from the EU.
21. Montenegro uses the euro in anticipation of joining the EU although none of its neighbors do.

711) are the only countries with less than 190,000 people that use their own currency.

The maze of over forty regional organizations of governments, to a varying extent energized by the advantages to be gained by cooperative action, are proliferating, merging, and increasing trans-country efforts. The imagination, energy, and salesmanship of the regional organizations has increased and will continue to increase the extent and effectiveness of inter-country collaborative efforts, including more effectively participating in the global arena on requiring global action. Climate change, coupled with the so far inadequate countermeasures, threatens the existence of some micro-countries.

K. C. Wheare, in his respected book *Federal Government*, asserts that regional organization in which independence (sovereignty) remains with the member countries, but "where 'league' or 'alliance' is not sufficient to describe an association, 'confederation' is the only suitable term left." In contrast, a system of government that embodies a division of powers between members and union, in which the union exercises the predominant power and has sovereign status, the term "federation" is used. The European Union's organization, scope, and potency is clearly more than that of an association, league, or alliance, as the complexity of the Brexit negotiations demonstrates. The OECS and EU may already have achieved this status. "Confederation," as a term to describe an emergent form of governing polity, provides means by which countries may collectively address a range of issues while preserving their sovereignty.

9. RETROSPECT AND PROSPECTS

"[Not] only the map of the world but the nature of states has changed dramatically in the last few decades."

—Alberto Alesina and Enrico Spolaore

THE RISE AND NEAR-FALL OF COLONIALISM

Characteristics associated with "modern" countries—including bureaucracies, professional military establishments, and revenue systems—emerged in Europe in the late 15th and early 16th centuries. This transformation fueled the beginning of European intercontinental explorations, European multi-continental imperial expansion, and the advent of such energetic monarchs who unified and centralized royal power and launched the growth in the scope and scale of governance that spanned continents. The late 20th and early 21st centuries are witnessing a shift in global governance.

Spices, ships, and sovereigns expedited the ages of exploration, colonization, and European multi-continental colonial empires. Ships, trains, cars, and planes, combined with the telegraph, telephone, television, and the internet, have made extra-continental prospective neighbors. The growth of global commerce, multi-national companies, and international cities fostered middle and professional classes, capitalism, and nationalism. European empires, aided by aggressive military, adventurous merchants, and zealous missionaries, expanded overseas, exploring, exploiting, and experimenting with the cultivation and governance of colonies. The world dominated by European capital-centric empires, which ruled for half a millennium, is being replaced with a global governance system in which more countries participate.

World War II propelled a wave of anti-colonialism, fueling national-isms and driving demands for self-government. More than 120 new in-dependent countries were established—breaking up the British, French, Dutch, Portuguese, Spanish, Soviet, and Japanese ones. Micro-countries are adjusting to ever-changing socio-technical-economic innovations and adapting to governance challenges. The era of imperial-domination of world governance, introduced five centuries earlier, is undergoing a funda-mental change, bringing an end to empires as we have known them. Em-pires have mutated, and economic hegemony has replaced political rule.

Micro-countries are caught between relentless globalism and unrelent-ing localism. Separation from neighboring communities and suspicious of those who do not share their ethnicity, language, faith, and mythology did not contribute to the multi-community harmony that would favor com-mon governance. No wonder aspiring islands and enclaves fought for keep-ing governance as close to home as possible despite the widely accepted pundits' wisdom regarding the hazards of smallness. Nationalism drove many of the colonies to reject not only rule from a colonial capital but also from one or more neighboring communities. Instead, these communities established independent micro-countries.

Accompanying the nationalist drive for self-determination and inde-pendence was the push for representative institutions. An expanding suf-frage drove the desire for amenities, escalating public safety, public works, and human services. The decline of the pre-war empires has prompted the rise of international institutions, a principal purpose of which was lowering trade barriers. Progress in alleviating trade barriers has promoted quasi-glo-balization. Freer trade reduces the challenge facing the micro-countries competing in the competitive world markets. Shelter countries, global in-stitutions, and regional organizations have facilitated this transition.

The rush of countries securing recognition as independent forced the issue: whether, given the number of social, political, and economic disad-vantages small, underdeveloped countries faced, what chance did they have of achieving viability? Given the relatively few independent countries be-fore the 20th-century World Wars, how would the international commu-nity cope with a near tripling of independent countries? The near tripling of the number of independent governments was concomitant with the international community effectively redefining independence. The newly independent countries, rejuvenated by the enthusiasm that generated their securing of independence brought renewed spirit to developing their representative institutions, infrastructure, and industrialization.

The rise of international institutions has assisted countries in securing money, manpower, technology, and cooperation in common projects. But, while independent countries have shared their sovereignty, they remain responsible for services within their borders. These phenomena, plus the catalyst of two 20th-century, globally encompassing World Wars, precipitated the decline of colonial empires and multiplied the number of independent countries—albeit with a changing criteria of independence. European trans-continental empires no longer exist as we once knew them.

The economic bonds and political ties of the expanding empires have generally survived. The old empires remain, but dismembered and with diminished means of exerting their will. Empires no longer exert political suzerainty, but they have retained economic hegemony. Not only has the global map changed, the nature of the world's polities has transitioned, and international politics are conducted differently. No longer is so much of the world colored red for the British Empire. Today's world maps display a more checkered assortment of colors. The concomitant multiple increase in independent governments and international institutions is shifting world governance from hegemony by a few empires to more pluralism, albeit one with managing partners. Though, even an authoritarian country's powers are constrained by threats of sanctions, loss of aid, use of force, etc. Even within their own borders treaties, covenants, and other phenomena restrict the exercise of absolute power within every country's borders.

Despite concerns regarding their smallness in population, expertise, land, and resources, the micro-countries have maintained their viability sufficiently to support public services and their global presence. This viability, though, has generally depended on participating in patron-like shelter arrangements and the creation of regional organizations of governments. Somewhat ironically, the political accommodation that has enabled this viability has compromised the essential sovereign attribute of an independent country, as defined by Jean Bodine and reiterated by Max Weber. Shelters, regional organizations of governments, and global institutions with whom sovereignty has in fact been shared, have fatally crimped the traditional concept of sovereignty, long the *sine qua non* of independence.

Many newly independent countries, with a "nudge and a fudge" have thus been able to gain their long longed-for independence. Minor modifications of the constitutions of land custom song autonomous and loosening the criteria by which countries are considered sufficiently independent to be admitted to the UN paved the way for Andorra, Liechtenstein, Monaco, and San Marino to receive international recognition of independence after World War II. The fact that San Marino's judges are Italian and two of the four-man executive team in Monaco are appointed by and from the French administrative corps would seem to preclude international recognition of their independence.

Shelters and regional organizations of governments, albeit subtly and imperfectly, provided assistance that maintained these newly independent countries. In fact, though, as many of the country profiles have demonstrated, the extent of dependence belies the traditional definition of independence. Countries are now recognized as independent if they have relationships with a shelter country that resemble those of what were once called protectorates. The very success of these micro-countries may encourage more dependencies to follow.

The number and impact of global and regional institutions has grown substantially. Their inter-country scope, scale, and impact will continue developing. The fact that principal executives of Monaco are appointed from the French administrative corps compromises their sovereignty—at least by traditional criteria. The fact that the international affairs of Micronesia, Palau, and the Marshall Islands are handled by the United States, with which they are in free association, mocks the traditional concept of independence.

Adjustments in non-European countries have sometimes been subtler but just as significant. These phenomena herald the end of the age of empires as we have known them. In a world of more independent governments, continuing dependence, and more globally inclusive system world governance may have come to resemble a general partnership. Granted, partnerships have junior, senior, and managing partners.

A country's central government embraces the "pyramid" of a country's maze of ministries and their countrywide bureaucracies directing its internal and external affairs. Its chief executive is the vital linchpin who connects a country's vertical mosaic with the array of international institutions.

The aftermath of World War II changed not only the map of the world but also the nature of governments and inter-governmental relations.

The increased number of independent countries and the growth of global and regional organizations have transformed global governance. The growth of global institutions and regional organizations provides the means for small countries to participate more effectively in an increasingly interdependent world.

To what extent will favorable political and economic conditions, including the continuing quasi-globalization of world trade, supported by the growth in number and potency of supportive international institutions, enhance micro-country viability? The growth of trade among nations is among the most consequential and controversial current development issue. Is supporting protectionism becoming entrenched? Support for global causes such as containing disease, minimizing conflict, tempering environmental damage, and improving the infrastructure and the economy. Not only the vitality but also the very existence of many micro-countries is at risk!

The increasing myopic nationalism driving the populist-driven protectionism exhibited by many prosperous and powerful countries may negatively affect the extent of foreign aid. Those countries relying upon vulnerable tax havens, dwindling mineral resources, and climate change threatened tourism have done little to find alternate sources of income. Aid programs, reduction of trade barriers, addressing climate change, and other efforts that have facilitated micro-countries' viability and survival. Without their continuance, what are the third-world micro-countries' prospects? The new micro-countries, with help, have survived! Given the growing economic, environmental, identity threats, however, will they continue to be viable?

INDEPENDENCE-LITE AND REMAINING DEPENDENCIES

The number of independent countries has almost tripled since the World War II. The initiative and ingenuity of micro-country leaders, working with the colonial powers, facilitated political and economic steps that enabled their survival and jump-started global governance. Building on the strength of the post-World War II drive for independence, they not only expedited independence but also secured political and economic assistance. Reliance upon the former colonial power, the UN, and other global organizations, and sharing of power with a neighboring country or former colonial master may curtail the exercise of the sovereign monopoly of power in managing its internal and external affairs. In such cases, a country's government may realistically be described as independence-lite.

Who comprises "we"? Reassessing how we identify ourselves—noting what is exceptional about our heritage, culture, and ethnicity—energizes our quests for understanding with whom we consider "we." The more we know about others, the more conscious we become of the ever-enlarging concentric circles of who we are and who they are. This affects with whom a "we" wants to share a government. We identify more easily and completely with government closer to home. The more comfortable many are with their "we" distinctiveness, the more uncomfortable they are with the "them."

Separatism is the principal factor leading to the creation of new independent countries.

The drive for countries to separate is more likely where parts of countries have distinctive linguistic, historical, and/or cultural identities. Countries and regions with long-term internal ethnic tensions include Scotland, Wallonia, Quebec, and Catalonia, and smaller ones such as Mayotte. In several countries, separatists have continued to attract global media coverage—despite efforts to discourage such separatism—for precedent appears to encourage countries' own separatist movements.

A second and larger source of possible new independent countries are dependencies that have remained dependent by their own choice or by the choice an administering power—or both (see Table 8.5). Three dependencies have over one million inhabitants—Puerto Rico, Palestine, and Hong Kong—each of which have fractious relations with their colonial powers. Thirty-three dependencies have populations of between 1,000 and one million, 21 of which have less than 100,000; among these are Greenland and the Cayman Islands off the coast of North America, Mayotte off the coast of Africa, and the Cook Islands in the Pacific.

Table 9.1 DEPENDENCIES WITH POPULATIONS OF OVER 1,000

Dependency	Pop n nearest (1000s)	Disputed	BRITAIN	CHINA	DENMARK	FRANCE	MOROCCO	NETHERLANDS	NEW ZEALAND	USA	Status
		A Dependency of									
Tokelau	1								NZ		Self-administering territory
Niue	1								NZ		Self-governing in 'free association'
Falkland Is	3		BR								Overseas territory (also claimed by Argentina)
St. Helena	4		BR								Overseas territory
Montserrat	5		BR								Overseas territory
St. Pierre and Miquelon	6					FR					Self-governing territorial collectivity
Wallis and Futuna Is	13					FR					Overseas territory
Anguilla	15		BR								Overseas territory
Caribbean Netherlands	20							NL			Overseas territory
Cook Islands	21								NZ		Self-governing in free association
British Virgin Is	29		BR								Internal self-governing overseas territory
Gibraltar	29		BR								Overseas territory (also claimed by Spain)
Turks & Caicos Is	34		BR								Overseas territory
Sint Maarten	45							NL			A constituent country
Faroe Is	49				DM						Self-governing overseas administrative div
Nor. Marianas	55									US	Commonwealth in political union
American Samoa	56									US	Unincorporated and unorganized territory
Greenland	57				DM						Self-governing overseas administrative division
Cayman Is	60		BR								Overseas territory
Bermuda	66		BR								Overseas territory
Isle of Man	87		BR								Crown Dependency
Aruba	104							NL			A constituent country
US Virgin Is	107									US	Organized unincorporated territory
Channel Is	164		BR								Crown dependency
Curaçao	164							NL			A constituent country
Guam	170									US	Organized unincorporated territories
Mayotte	234					FR					Territorial collectivity
French Guiana	262					FR					Overseas 'department'
New Caledonia	263					FR					Overseas territory
French Polynesia	283					FR					Overseas 'department'
Martinique	406					FR					Overseas 'department'
Guadeloupe	470					FR					Overseas 'department'
Macao	584			CH							Special Administrative Region (SAR)
Western Sahara	604						MO				Claimed and administered by (in dispute)
Réunion	895					FR					Overseas 'department'
Puerto Rico	3683									US	Commonwealth associated with the US
State of Palestine	4436	IS									Israel-occupied Palestinian Authority
Hong Kong	7260			CH							Special Administrative Region
Totals	38	1	11	2	2	9	1	4	3	5	

Source: United Nations "List of Countries by Population (2015)"

While the UN lists the French Guadeloupe, Martinique, French Guiana, and Réunion as dependencies, they are actually French local governments. Similarly, the Caribbean, Netherlands, Curaçao, and Aruba have been granted a status that is not considered by the Dutch as dependent.

The British continue to possess the most dependencies, including the Isle of Man, the Channel Islands, Gibraltar, the British Virgin Islands, Montserrat, Anguilla, the Cayman Islands, the Turks and Caicos, Bermuda, and the Falklands. The United States' dependencies include the American Virgin Islands in the West Indies, and American Samoa, Guam, and the Northern Marianas in the Pacific. New Zealand has the Cook Islands, Niue, and Tokelau; Australia governs Norfolk Island and Christmas Island in the Pacific. Two Danish dependencies are located in the North Atlantic: Faroe Islands and Greenland.[22]

The French govern Saint Pierre and Miquelon Islands dependency (at the mouth of the Canadian Saint Lawrence River), Saint Barthélemy, and Saint Martin in the Americas, and French Polynesia, Wallis and Futuna, and New Caledonia in the Pacific.[23] French Guinea, Guadeloupe, Martinique, Réunion, and Mayotte have become French overseas local governments with the representation in the French parliament instead of becoming independent. The Netherlands made Curaçao, Sint Maarten, and Aruba special municipalities and autonomous countries within the Kingdom of the Netherlands, but the United Nations (most of whose members are former colonies) classifies them as dependencies. France and the Netherlands do not consider these governments to meet the criteria of dependencies.

Micro-countries have secured independence and viability despite the early, and not unrealistic assessment of the potential disadvantages of "going it alone." Several colonial powers and colonies exercised dexterity in negotiating the terms leading to independence. The survival of the young micro-countries may encourage a few more dependencies to pursue independence. Regional organizations have developed to assist in this sharing. Reduction of global and regional trade barriers, along with the expediting of assistance and aid, has significantly eased the trials of the newly independent countries' initial years.

---

22. Mineral-rich and strategically located Greenland is adjacent to Canada between the Atlantic and Arctic Oceans. The United States built the Thule Air Base there in 1943 and continues to use it to this day. A 1946 U.S. offer to buy Greenland was firmly rejected by Denmark. President Trump's public consideration of purchasing Greenland was soundly and immediately rejected by officials in both Denmark and Greenland.
23. France's threat to cut off aid to New Caledonia may have affected its 2018 vote to reject independence.

The *sine qua non* of independence is to be accepted as members of the United Nations. Some countries quite evidently do not possess full control of their affairs. Prime examples are a few European micro-countries (e.g., San Marino, Monaco, and Andorra) for which a neighboring "shelter" country provides critical services. Others are the few Oceania micro countries (e.g., Palau, Marshall Islands, and the Federated States of Micronesia) that have compacts with the United States to be responsible for their defense and international affairs. For the Marshall Islands, Palau, and Micronesia, free association treaties provided by the United States allow the U.S. to manage their defense and foreign affairs, yet they are UN members. But the Cook Islands and Niue, which have similar arrangements with New Zealand, remain dependencies.

Within Europe, Monaco and Andorra have long worked closely with France; San Marino with Italy; and Liechtenstein with Austria and then Switzerland, enabling the larger country to assume the role of the shelter country. The substantive redefinition of independence, and its obligations, may tempt countries that have not opted for independence to reconsider their earlier choice. The *de facto* weakening of standards for gaining international recognition and UN membership has sped up the granting of independence and related of admission to the UN and will continue to do so. Now, admittance generally comes almost immediately after the granting of independence, even when there are many strings attached to the country's independence.

How many dependencies will opt for independence? Not many! Several of these micro-dependencies have already considered and rejected the independence option. However, the way is now open to consider one of the variants of independence that some of the 41 have successfully espoused – which have facilitated funding, expedited trade agreements, assisted in foreign and defense affairs, and collaborated on programs—and avoided some of the risks and the responsibilities of independence once associated with it. But the pace of dependencies becoming independent has slowed. External and internal pressures may facilitate Palestine, Gibraltar, and Greenland in gaining recognition of their independence, but the exigencies of their circumstances may prescribe a version of independent-lite.

IS THE PRESENT PROLOGUE?

Localism, nationalism, globalism, and "hegemony-ism" drive conflicting local, national, and global values with which we live. Since our ancestors engaged as social animals, many have expanded their horizons. The more far-sighted led the migrations that populated the continents. In the last few centuries, improvements in transport and communication have exponentially increased the percent of humans that reconcile an expanded view of our community with the local one. How we define our communities, the ones in which we live and work, depends upon our circumstances, opportunities, and means of transport and communication. *We* defend "us" and oppose "them." Maturation tends to extend the boundaries of whom we include among our "we."

Nation extending and globalization has enlarged the boundaries of whom we include among "we" and "they." For centuries, extending trade and conquering, colonizing, converting, and moving people have advanced a limited globalization. In recent decades, the development of multiple associations and countless pacts and agreements have facilitated an increasing movement of products, capital, jobs, culture, people, and ideas around the world has cultivated a more inclusive quasi-globalization. The century's long efforts to cultivate a sense of empire-centric nationalism generated a colony-localized nationalism among communities once politically absorbed into empires.

The convergence of a yearning for spices, the advent of navigation technology, the invention of the printing press, the expansion of commerce, the cunning of monarchs, the courage of navigators, and the covetousness of merchants drove off European empires expanding into Africa, Asia, the Americas, and the Pacific. Nationalism, fostered by aggressive rulers to expand nationalization, in time, became the driving force used to dismantle empires, ending imperial rule as the world had known it. The number of independent countries has multiplied to almost 200. The United Nations and related global organizations such as the World Bank, IMF, WTO, and WHO play a significant role in world governance. A few former colonial powers continue to shelter their former dependencies. Many regional organizations of countries have emerged, some ambitious, and a few, notably the European Union and the Organization of East Caribbean States, with significant achievements.

Concerns regarding the viability of the new micro-countries facilitated the expansion of the number and potency of global institutions, the role of a few former colonial powers to continue substantial aid, and the emergence of a maze of regional organizations. How much more will change in the next several decades?

Will future scholars and statesmen write favorably of a more potent United Nations, even more active global organization, and the presence of more energetic regional organizations promoting trade, environmental issues, and other common concerns for international affairs? How will 21st-century economic empirical hegemonies behave in a world in which many more countries participate more effectively in global governance? How will global governance deal with protectionist-driven trade and climate change challenges, whose devastating effects threaten their economies and very existence?

At this juncture of history, it is tempting to herald the development of regional organizations of governments into confederations comparable to the earlier development of federations. The dream and drive of European leaders to develop "an ever-closer union," binding together a continent of disparate histories and languages has succeeded in forming a successful confederation. Regional organizations in Africa, Asia, the Americas, and Oceania are taking steps toward evolving into confederations, acting cohesively in confronting global issuers, undertaking collective endeavors, and mitigating conflicts among its members. Ahead lies the challenge of rationalizing the overlapping maze of regional organizations and successfully increasing the scope and scale of their efforts facilitating the free movement of goods, capital, people, and ideas.

Is the present prologue? Will any of today's dependencies seek independence? Will the members of the UN continue to recognize them? Will the UN and related global organizations continue supporting developing countries and enabling cooperative programs? More speculatively, to what extent will regional organizations develop carrying out operations at the behest of its member countries that can more effectively manage collectively? To the extent that they do develop ways to cooperate more effectively, micro-countries will more effectively advance their interests in such global concerns as freer trade and climate change.

Strengthening their participation in the world governance has enhanced the viability and prospects of the micro-countries. The aftermath of World

War II introduced several decades of global efforts including providing aid, reducing trade barriers, addressing climate change issues, and implementing other embracive global programs. These global efforts were facilitated by the support of the more prosperous, powerful countries. Will such support, exerted through shelter countries, the UN and its affiliated agencies, and regional organizations continue?

The rise of nationalist-driven populist, protectionist demagoguery threatens the support of globally minded efforts. Aid programs may decline. Efforts to promote freer trade may not only deteriorate, they may go into reverse; witness the unfavorable prospects for completing the Trans-Pacific Partnership that would have reduced tariffs among twelve Pacific Rim countries, but not China. Efforts to address climate change are no longer appearing to gain the support necessary to prevent the inundation of several island countries.

Efforts facilitating micro-countries' viability encourage optimism regarding their continuing progress. But an increase in trade wars and decreased efforts to address tensions threaten that progress. Lack of adequate support for protecting the economics and environment do not elicit optimism for the continued growth of threatened micro-countries. Without continuing international efforts by the more prosperous and powerful countries, the challenges confronting micro-countries appear threatened. The rise of nationalism brings threats to trade, aid, and climate control efforts upon which many micro-countries depend.

The micro-countries' profiles demonstrate that—while their demographics, economies, and politics are distinctive—they face similar challenges. It is striking that, as their economies and politics have become more globally interdependent and their people have become more globally aware over time, they have become more conscious and protective of what they feel makes them distinctive. Micro-countries with longer histories of independence, more resources, and/or more prosperous adjacent neighbors face fewer challenges. In contrast, the more isolated/island African, Asian, American, and Oceanic micro-countries (i.e., those without developable resources and industries), confront economic and environmental threats that challenge not only their people's prosperity and political potency but also their very existence.

PART TWO

THE 41 MICRO-COUNTRIES—
BY CONTINENT AND MICRO-COUNTRY

"The task of the historian, therefore, goes beyond the duty of tending the generalized memory. When a few events in the past are remembered pervasively, to the exclusion of equally deserving subjects, there is a need for determined explorers to stray away from the beaten track and to recover some of the less fashionable memory sites."

—Norman Davies

The individual descriptions of each micro-country enables a comparison of their often tumultuous histories, the extent of their natural resources, the variation in their governance, and the severity of the threats to their viability and their existence.

10. DEPENDENCE, INDEPENDENCE, AND INTERDEPENDENCE

"The Westernization of the world within the last 500 years has been the work of a number of separate rival local Western states. Their competition with each other has been one of the major driving forces behind the West's expansion."

—Arnold Toynbee

THE CONTINENTS

Micro-countries have transitioned from various forms of dependency (with labels such as territory, colony, crown colony, and protectorate) to independence. Many secured their independence from Britain, others from other colonial powers including Portugal, Spain, France, and Denmark. Most gained limited autonomy prior to independence—at different paces, under different conditions, and with different stratagems.

The extent of geographical isolation, degree of ethnic/tribal separatism, and the influence of colonial power settlers, elite natives, and foreign corporations were among the factors that influenced the extent of success in the stages of securing limited autonomy, gaining independence, cultivating credible representative institutions, developing the infrastructure and economy, and control of international and internal affairs. Circumstances affected the vigor with which dependencies were ruled.[24] These same factors affected when, how, and on what terms independence was secured.

---

24. Many British dependencies used indirect rule in which a colonial officer (e.g. Resident) 'advised' the nominal native ruler (e.g. sultan, oba). The ruler accepted the advice.

Mainland European micro-countries descended from feudal fiefs. They shared ethnic backgrounds with the countries with which they had once shared sovereigns. Their gradual securing of self-governance before World War II was accompanied in several cases by arrangements with the neighboring country to provide selected basic services such as defense, international representation, judiciary, and post.

The six African micro-countries occupy sites on the continent's west and east coasts. They were among the first explored and colonized by Portugal and Spain. The brutality of the Atlantic and Indian Ocean slave trade and the colonial occupation has left a cruel legacy. All the African micro-countries belong to the African Union. Four of them remain among the least prosperous economies in the world.

The three South Asian micro-countries, stretching from the Indian Ocean (Maldives) to the Himalayas (Bhutan) to the China Sea (Brunei), are as ethnically different as they are geographically disparate. All three have long had autocratic governments.

Eight of the 11 American micro-countries are islands with a British colonial heritage. Six Caribbean island countries belong to the Organization of East Caribbean States.

Three are coastal enclaves (English-speaking Belize and Guyana, and Dutch-speaking Suriname) isolated by language and culture from their continental Latin neighbors, which have motivated them to cooperate more with the island micro-countries than the countries with which they share borders.

The Pacific Ocean remained a source of colonial rivalry into the 20th century. Long distances, multiple ethnicities, countless languages, and innumerable religious groupings divide oceanic micro-countries.

Table 10.1 41 Micro-countries: Urban and Migrant Populations, Literacy, and Growth Rates by Age

Continent	Country	Per capita GDP (2016) (rounded)	Urban %	Migrant %	Literacy %	Growth from previous year %	Unemployed %	Under 16 %	Over 65 %
Europe	Vatican		100	100	100	N/A	N/A	N/A	N/A
	San Marino	60,700	10	15	94	0.6	N/A	16	19
	Monaco	78,700	100	64	99	0.1	N/A	11	20
	Andorra	37,200 (2011)	85	56	100	0.1	N/A	15	15
	Liechtenstein	89,400 (2009)	14	33	100	0.8	N/A	16	17
	Luxembourg	92,000	90	43	100	2.1	69	17	15
	Iceland	43,000	94	10	99	1.2	5.6	20	14
	Cyprus	30,800	67	18	99	1.4	17	16	11
	Malta	33,200	96	8	99	0.3	6.5	15	14
	Montenegro	16,000	69	8.2	99	0.4	19.8	15	19
Africa	Cabo Verde	6,300	60	3	86	1.4	7.0	30	3
	Equatorial Guinea	32,300	10	95	95	2.6	8.0	41	4
	São Tomé and Príncipe	3,200	65	53.3	75	1.8	N/A	43	3
	Comoros	3,000	28	1.7	89	1.8	4	41	4
	Seychelles	25,000	54	13	92	0.3	N/A	31	7
	Djibouti	3,100	77	14	N/A	2.2	N/A	33	4
Asia	Maldives	14,100	46	24	46	-0.1	11.6	21	4
	Bhutan	7,600	39	7.7	65	1.1	2.1	27	6
	Brunei	73,200	77	48	46	4.3	3.8	24	4
Americas	Barbados	16,200	32	11	N/A	0.3	12.2	23	7
	Grenada	12,000	36	11	N/A	0.5	N/A	24	10
	Saint Vincent and the Grenadines	10,000	51	9	N/A	-0.3	N/A	21	11
	Saint Lucia	21,000	19	7	N/A	0.3	N/A	21	11
	Dominica	10,600	70	8.9	N/A	0.2	N/A	22	11
	Antigua and Barbuda	22,600	24	32	99	1.2	N/A	24	8
	St.Kitts and Nevis	21,000	32	11	N/A	0.8	N/A	21	8
	Bahamas	25,000	83	16	N/A	0.9	14	23	7
	Belize	8,200	44	15	70	1.9	15	35	4
	Guyana	6,900	29	2	89	0	11	28	7
	Suriname	16,000	54	13	92	1.1	7.6	6	6
Oceania	Solomon Is	1,900	22	1.4	84	2	3.9	36	4
	Fiji	8,200	54	2.6	N/A	0.7	8.8	38	6
	Vanuatu	2,600	26	1.2	85	2	N/A	37	7
	Tonga	4,000	24	5.2	99	0	N/A	35	6
	Samoa	5,200	19	3	N/A	0.6	N/A	33	6
	Tuvalu	3,500	60	1.5	N/A	0.8	N/A	29	6
	Kiribati	1,700	44	2.5	N/A	1.2	N/A	31	4
	Nauru	5,000 (2005)	100	21	N/A	0.6	N/A	33	2
	Marshall Is	3,300	22	3	N/A	N/A	N/A	36	4
	Micronesia	3,000 (Subsidy)	22	2.5	N/A	0.5	N/A	31	4
	Palau	16,300 (Subsidy)	60	1.5	100	0.4	N/A	33	6

Source: Statistics from 2016 World Almanac

LOCATION, PROXIMITY, AND RESOURCES

Micro-countries are significantly affected by their location, proximity to wealthy neighbors, and access to resources, which also remarkably affect average incomes of citizens within the countries (as approximated by per capita GDPs) and, consequently, the range of average incomes of micro-countries on each continent (see Table 9.1). Most of the Oceania and African micro-countries have average incomes below $6,500.[25]

Urban-rural splits affect polity cohesiveness and policy making (or the lack of it). Micro-countries outside Europe with urban populations of 60 percent of their total populations are Cabo Verde, São Tomé and Príncipe, Djibouti, Brunei, the Bahamas, Dominica, Tuvalu, Nauru, and Palau. Migrants affect polity cohesiveness; micro-countries with 30 percent or more migrants are the Vatican, Monaco, Andorra, Liechtenstein, and Luxembourg in Europe; Cabo Verde and São Tomé and Príncipe in Africa; Brunei in Asia; Antigua and Barbuda in the Caribbean; and Nauru in Oceania.

Every European micro-country has a smaller ratio of citizens younger than 16 years of age and a larger percent of citizens over 65 years of age than every non-European country—a factor that will affect future fiscal planning.

Micro-countries differ significantly in the extent of the economic disadvantages, environmental disasters, and ethnic/class divides they will confront in the coming decades. The countries will be described in continental groups generally in the order in which European countries explored and colonized them. Within the continental groupings, the micro-countries are generally presented with others with similar characteristics.

---

25. The exceptions are Equatorial Guinea, whose main resource is oil, the Seychelles, who relies on tourism, and Palau, which has an exceptionally large United States subsidy.

11. EUROPE: THE FIRST MODERN MICRO-COUNTRIES

"The Treaty of Westphalia (1648) is often heralded as . . . signaling the acceptance by major European powers of the principle of territorial sovereignty . . . At Westphalia, European powers . . . attempted to end a prolonged period of religious and dynastic conflict—known as the Thirty Years' War . . . The treaty recognized the sovereignty of some three hundred princes on territories under the Holy Roman Empire, but the empire remained an overarching political entity, for another 158 years . . . They pursued and were subjected to imperial ambitions for the next three centuries."

—Jane Burbank and Frederick Cooper

Map 11.1 European Micro-Countries

Source: World Map by M. Ruskin Co. LLC; the author inserted the names of countries.

FEUDAL SURVIVORS

The 10 European micro-countries are feudal survivors. The six Western European mainland micro-countries survived the amalgamation of feudal fiefs that gave rise to the jurisdictional geography of modern Europe.[26] A combination of factors allowed the Vatican, Andorra, Liechtenstein, Luxembourg, Monaco, and San Marino not to be merged or absorbed into a larger entity.

---

26. Their origin as feudal polities remains evident in the names of four micro-countries: the Principality of Monaco, the Principality of Andorra, the Principality of Liechtenstein, and the Grand Duchy of Luxembourg.

The adroitness of local lords and leaders, political accommodations with an adjacent power, and political "nudges and fudges," hereditary restrictions, imperial rivalries, and sea and mountain imposed isolation. The mountainous terrain of San Marino, Andorra, and Liechtenstein, the sea frontiers of Iceland, Malta, and Cyprus, and the rivalry of neighboring powers of all the European micro-countries facilitated their efforts to gain and preserve their autonomy and ease their way to independence. The Dutch king was the Luxembourg grand duke from 1815 to 1831; after Belgium won its independence from the Netherlands, the Belgian king held the position from 1831 to 1890. The last official link with the Belgian throne was severed because laws affecting Luxembourgian ducal succession forbid a female inheriting the duchy (when Queen Wilhelmina succeeded to the throne of the Netherlands). Since a daughter inherited the Dutch throne and a female was not eligible to become the Luxembourg Grand Duke, the posts were separated. Imperial rivalries facilitated Monaco's autonomy. The stature of the Roman Catholic Church facilitated the Vatican's survival. Each of these European micro-countries navigated through uncharted political terrain from self-government to *de facto* independence to *de jure* independence, affirmed by general recognition and admittance to the UN, which has, for practical purposes, become the collective arbiter of statehood through the process of admission to the UN whose members use a flexible criteria for admission.

Four European microstates gained independence during or after World War II. They include a North Atlantic island, a Balkan coastal enclave, and two Mediterranean islands. Iceland, colonized by Viking expeditions and ruled by Nordic monarchs since 1262, severed its umbilical cord with Denmark in 1944. Malta and Cyprus, caught in the conflict for control of the Mediterranean, succumbed to the British naval strength. Malta, which had been ruled by Phoenicians, Romans, Arabs, Normans, the Knights of Malta, France, and Britain (since 1814) became independent in 1964. Cyprus, a part of the Ottoman Empire from 1571 to 1878 until it yielded control to Britain, became independent in 1959. Montenegro, for centuries a quasi-autonomous part of the medieval Kingdom of Serbia, was independent before World War I, absorbed into Yugoslavia after World War II, and recovered its independence in 2006 when Yugoslavia disunited.

Map 11.2. EUROPE 1815

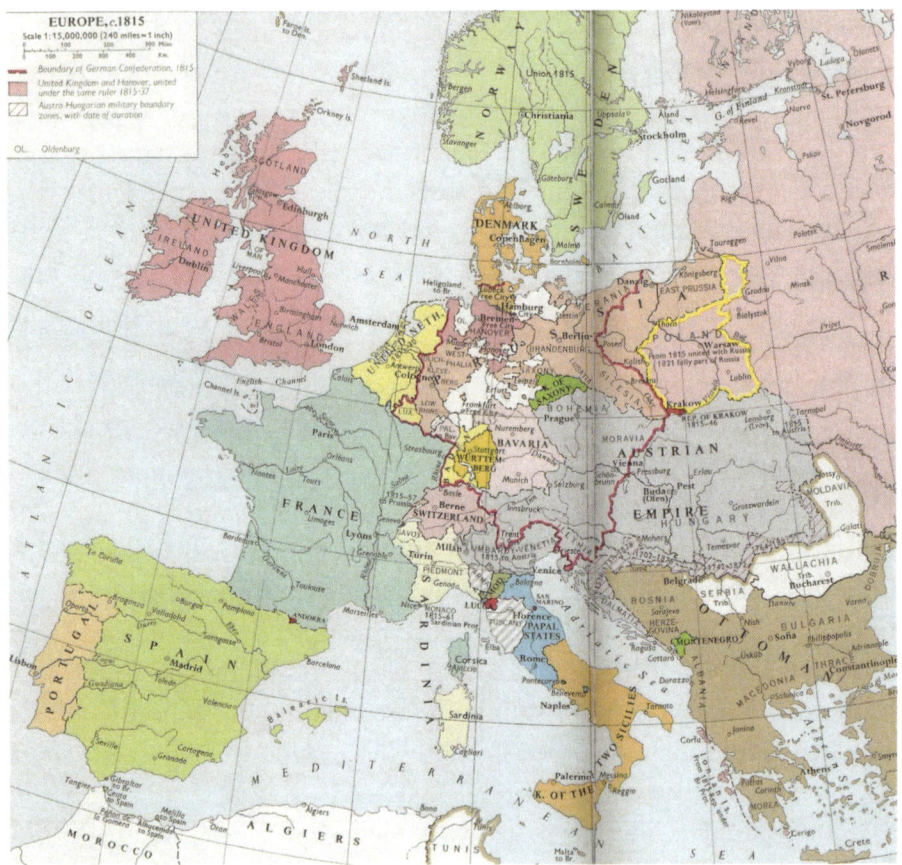

Source: Darby, H.C. and Harold Fullard (editors), *The New Cambridge Modern History Atlas*, Cambridge University Press, Cambridge, 1970, pg. 78-79.

European microstates vary widely in population: the Vatican has fewer than 1,000 citizens; San Marino, Monaco, and Liechtenstein have less than 40,000; and Andorra has less than 90,000.[27] In the mid-2010s, the GDPs per person ranged from $33,000 in San Marino and $38,500 in Iceland, to $63,000 in Monaco, $81,000 in Luxembourg, and $89,000 in Liechten-

---

27. Compare to Cyprus, Montenegro, and Luxembourg with more than 400,000 citizens in total.

stein. Clearly, the European micro-countries did not confront economic disadvantages as seriously as those in Africa, Asia, America, and Oceania— neither did the European micro-countries face as serious environmental threats and ethnic/identity divides.[28]

The distinctive cultures, eventful histories, and independence of these enclaves stimulated the national spirit of these countries. The theocratic focus of the Vatican differs significantly from Luxembourg's finance-focus, the casino-centeredness of Monaco, the charm of the historic mainland enclaves of Nordic Iceland, and the mixed Malta and Cyprus heritage of the Mediterranean islands, both of which recovered their independence within two decades after World War II—and Montenegro only recovered its independence from Serbia within the past two decades.

The 10 European micro-countries participate in the EU common market directly or indirectly. Four of them are members of the EU itself: Luxembourg was a founding member, but Malta, Cyprus, and Montenegro have since been admitted. Iceland and Liechtenstein[29] customs unions vary in regard to the specific goods exempt from duties; in 2015, they entered into negotiations with the EU to create an "Associate" status for countries by 2020. The Vatican may qualify, for example, because it is so small and its borders with Italy so porous that no duties are collected.

The decision of the United Kingdom to leave the European Union has highlighted the variety and complications of arrangements of European countries not members of the EU but working with it. Once the Brexit matter is settled, there may be an effort to rationalize further the relationships binding the member countries of the European Union.

---

28. While the least populous European micro-countries—Andorra, Liechtenstein, Monaco, San Marino, and the Vatican—did not have an airport, they were served by one adjacent to their land borders. All island micro-countries, by virtue of the lack of land borders, have an airport.
29. Iceland and Liechtenstein, along with Norway and Switzerland, belong to EFTA, which is associated with the EU in the EEA. The EEA is a free trade area but not a customs union. Andorra, Monaco, and San Marino are also members of the European Economic Area.

THE STATE OF VATICAN CITY (THE HOLY SEE)

"The [spiritual and administrative] center of the Catholic Church is a 109-acre complex, built largely during the Renaissance, the spiritual and administrative headquarters of a global institution with 1.2 billion followers."

—Alexander Stille

The State of Vatican City (also known as the Holy See), a less than 0.2 sq. mi. enclave of holy edifices, encircled by the city of Rome, is the world's geographically smallest and least populous country. It is the site of the Holy See, the monumental campus headquarters of the Roman Catholic Church. Its 800-odd residents include the Pope and other functionaries of the congregations and prefectures that comprise the headquarters bureaucracy, which oversees the operations of hundreds of dioceses spread throughout the world. The Pope delegates the management of the public affairs of the State of Vatican City to a pontifical commission, which administers the courts, police, public works, and other temporal affairs of the corporate campus. The Holy See itself manages the church's international affairs.

Even before the recognition of Christianity as a corporate body by the Roman Emperor Constantine in 395 CE, the Apostolic See had gradually become a territorial entity, mostly by donations of land. In the late 700s, the Popes became dependent upon Charlemagne for protection of the Papal States from the Lombards, a relationship that led to Pope Leo III's presence at Charlemagne's crowning as Holy Roman Emperor in 800 CE, thus legitimizing his assumption of the Merovingian crown from the *rois fainéants* (i.e., lazy kings); Charlemagne was descended from "mayors of the palace" who long were the effective leaders of the empire.

Jorri C. Duursma writes, "After 962 and until the 11th century the Ottonian emperors, [Otto I, II, and III of the Holy Roman Empire], had taken practical control over the Papal States though recognizing the theoretical supremacy of the Pope." While the Papal States suffered continuing attacks, especially from the French, it was not until 1807–09 that Napoleon successfully invaded and annexed them. The Congress of Vienna restored their independence in 1815. Following the 1848 unification of Italy, the Pope fled. Only with the assistance of Napoleon III did the Pope return in 1850 and, with the help of French troops, regain control of the Papal territories. But, when the French troops were withdrawn in 1870 for the Franco-Prussian War, Italy annexed the Papal States. The Italian Law of Guarantees of 1871 provided that the Pope could freely use the Vatican, the Lateran palaces, and the Villa of Castel Gandolfo, granting him in addition the freedom of communication and the right to receive diplomatic missions accredited to the Holy See, but temporal sovereignty of the Vatican City was abrogated. The pope never accepted these terms.

After many negotiating efforts, the Kingdom of Italy and the Holy See signed the Lateran Agreements in 1929 by which the Holy See acquired its present limited walled enclave of land in which Italy recognized full and exclusive sovereignty. The Holy See possesses a number of buildings outside the Vatican City in Italian territory, which are exempt from Italian expropriation and fiscal legislation. Some of these properties enjoy the same immunity as that accorded to seats of accredited diplomats of foreign countries. The Vatican depends upon Italy for postal, telegraph and telephone communications, customs, and energy.

The uniqueness of the Vatican as a micro-country extends significantly beyond its minute area and population. Its uniqueness stems singularly from its status as the corporate headquarters campus of one of the largest and influential organizations in the world. All of its inhabitants work for the Holy See. Unlike other countries, citizenship is granted by *jus officii* (prerogative of the state), not by *jus sanguini* (your birthright) or *jus soli* (your parent's birthright).[30]

Of those residents who work for the church, few, if any, are native to the Vatican; the residents have immigrated from elsewhere. Its independence did not stem from self-determination or nationalism, but, rather, the will of the Holy See with its support and recognition throughout the world. While many of its higher-ranking clergy and Swiss guards are not natives, the population is largely Italian. Italian, French, and English are

---

30. Most of those who work in the Vatican, many of whom are clergy, live outside the Vatican walls.

the spoken languages in the Vatican—and even Latin, spoken by those who work in the Vatican.

The Vatican is a theocracy; it does not possess even the pretense of a representative government. The Pope possesses sole executive, legislative, and judicial powers. He is an elected autocrat, whose electorate is limited to the College of Cardinals whose members were selected by his papal predecessors. He selects the Pontifical Commission, composed of five Cardinals, which is charged with the temporal legislative and executive tasks of governing the Vatican City and appoints those responsible for the executive, legislative, and judicial functions and retains the authority to override their decisions. The Pope also appoints the judges of the temporal as well as the ecclesiastical courts.

The budget of the Vatican City was $326 million in 2011. The income from worldwide dioceses that support the Holy See as headquarters of the Roman Catholic Church is supplemented by worldwide banking and financial services as well as printing and manufacturing coins, mosaics, and postage stamps. Continuing revelations in a series of revelations about the misuse of church funds. *The Economist* has reported the following:

> the latest financial scandal to rock the Vatican . . . the Vatican's gendarmes, on orders from its prosecutors, raided the offices of the Financial Information Authority (AIF), the banking regulator, and the Secretariat of State, which combines the roles of prime minister's office and foreign ministry in the Vatican administration . . . The Vatican's secretive culture and sovereign privileges make it ideal for dubious transactions.

The Vatican's focus on the sacred appears to have distracted their attention to the secular.

The Vatican City does not belong to the European Union or the United Nations, but it does belong to a number of UN specialized organizations. The Holy See's Secretariat of State—not of the State of the Vatican City—manages the country's international relations, maintains diplomatic offices abroad, concludes treaties, and becomes the member of an international body. Italy is responsible for the Vatican's defense. Swiss guards, according to a long arrangement, provide local security.

The Vatican City thus differs from other micro-countries in critical respects: 1) its demography is not one with a well-established genealogical affinity; its affinity is a common faith-based employer, 2) it was created solely for the convenience of and to guarantee the independence of the Holy See, and 3) it is, nevertheless, an entity that because of the Holy See's role is recognized by 183 countries, most of whom receive representatives from the Vatican and maintain representation there.

REPUBLIC OF SAN MARINO

"The survival of San Marino as a separate entity throughout history is probably due to its political insignificance and its will to preserve its autonomous institutions."

—Jorri C. Duursma

The Republic of San Marino (short for the Most Serene Republic of San Marino), a mountainous enclave surrounded by Italy, traces its existence as an autonomous polity to the founding of the Papal States. It has survived for more than a millennium through the traumas and termination of the Papal States, the advent of Napoleon's Roman Republic, and the unification of Italy. The principal Marinese governing institutions trace their existence through these transitions.

This minute, 24 sq. mi. northern Italian enclave, 14 mi. from the Italian city of Rimini, is the geographically smallest independent country in Europe after the Vatican and Monaco. With 32,000-plus inhabitants, the Republic of San Marino rivals the smallest of the small micro-countries— only the Vatican and three Oceania micro-countries have fewer inhabitants. The enclave embraces nine communes; the largest is the quaint mountainside capital town, San Marino, with 4,000 inhabitants. The people are Italian or Sammarinese, as those who live in San Marino are called. The national language is Italian, and the common religion is Roman Catholic.[31]

Local legend credits a pious stonecutter, Saint Martinus, with the 301 CE founding of the settlement of San Marino on the slope of the Titano Mountain, which had allegedly been given to him by a noblewoman for saving the life of her son. In mid-16th century, feudal Italy San Marino was a quasi-autonomous fief of the Papal States.

By the early 1700s, the extent to which San Marino had become a refuge for criminals had become a concern of many Sammarinese, as well as the Cardinal papal legate in the neighboring Romagna province. The cardinal convinced Pope Clement XII to send troops to occupy and di-

---

31. The presence of picturesque San Marino soldiers and other tourist attractions distinguishes the community from other Italian hill towns.

rectly govern the town. The international community, though, pressured the Holy See to hold a local plebiscite. An overwhelming pro-autonomy vote in 1740 enabled the Republic of San Marino to regain its previous status under the protection of the Holy See. In 1798, Napoleon's Roman Republic concluded a treaty with the Republic of San Marino and offered to include it within an extended France. San Marino graciously refused the offer.

When, in 1815, the Congress of Vienna took no action regarding San Marino's status, its government claimed that the lack of action implied recognition of its *de facto* independent status acquired between 1798 and 1815. But the Holy See never accepted this interpretation and continued its hold on San Marino's restricted autonomous administration. The strained relations between the Holy See and San Marino continued until the Risorgimento, the unification of the Italian State in 1861. A treaty concluded between San Marino and the Italian kingdom in 1862, accepted the Italian king's protective friendship, and recognized San Marino's independence. During both 20th-century World Wars, San Marino asserted its neutrality. San Marino has survived its transition from medieval feudalism to become an internationally recognized independent state; it not only preserved its traditional autonomous institutions but also suffered satisfied international criteria for it to be recognized as independent. Its relative political insignificance also facilitated its encircling neighbor's willingness to accept its independent status.

The governing institutional framework of San Marino essentially stems from the Statutes of 1600 and retains features from the Roman Empire era. San Marino's parliament, its Great and General Council composed of 60 members directly elected every five years, possesses the legislative power and selects the two Captains Regent (the heads of state) for six-month terms each. Their powers trace back to the two consuls of the Roman period referred to in the Statutes of 1660; their powers include the right to approve or disapprove legislation.

The executive leaders, the State Congress, consisting of the 10 ministers who constitute the plural head of government, usually based on a coalition of the two political parties, are elected by and responsible to the Great and General Council. Heading the State Congress as the head of government is the Secretary of State for Foreign and Political Affairs. Duursma points out that "one of the remnant institutions of the Statutes of 1600 is the 'Arengo,' an assembly of the heads of all San Marinese families. Historically, the 'Arengo' had the right to present claims or propositions to the Captains General who could submit the requests to the 'Great and General Council.'"

Since 1974, Article 15 of the San Marino fundamental law stipulates that the judges in San Marino cannot be San Marino nationals, unless by virtue of statutory exceptions. This provision acknowledges the fact that the Italian judicial system extends into San Marino. While the Great and General Council has the nominal appointment power, the judges are from the Italian judicial system. The small size of San Marino has led it to delegate postal functions to Italy. Its defense is also guaranteed by Italy.

Since World War II, San Marino has prospered economically, especially by promoting tourism and banking, which its independence facilitates. Other enterprises include that of textiles, electronics, ceramics, cement, and wine. The GDP in the mid-2010s was $1.3 billion. The extent to which the country is economically integrated with Italy is demonstrated by the fact that over 80 percent of its exports and imports were to Italy. The labor force includes 0.2 percent in agriculture, 33.5 percent in industry, and 66 percent in services.

At the beginning of the 20th century, San Marino did not have the financial resources to support extensive diplomatic relations. The post-World War II improved economy has enabled the micro-country to enter into diplomatic and consular relations with more than 70 countries and become a member of several international organizations, thus accentuating the international recognition of its independence. It maintains an honorary consulate in Washington, D.C., and a mission to the European Union. Not only is it a member of the United Nations, but it also belongs to several related organizations including the OSCE, FAO, UNESCO, WHO, and ILO.

San Marino appears to have its cake as well as eat it, too. It has the attraction, distinction, and economic/tourist advantages of independence as well as the services of Italy!

THE PRINCIPALITY OF MONACO

"Due to lacking means of international judicial control, Monaco's actual independence will not be fully guaranteed in practice, even if it has formal independence and over 60 countries have recognized it and it has been admitted into the United Nations. It cannot impose a restrictive and correct interpretation of the treaty of 1918."

—Jorri C. Duursma

The Principality of Monaco lies on the Mediterranean coast surrounded on three sides by the French Alps-Marines department. The principality has belonged to and been ruled by the Grimaldi family since 1297 and claims independence since 1641, except during the French Revolution. It covers a land area of 0.77 sq. mi. and a sea area of 71 sq. mi. Its 30,000-plus population consists of only 16 percent Monégasque, 47 percent French, and 16 percent Italian. French is the official language, and Roman Catholicism (accounting for 90 percent of all the practicing religions in the country) is the official faith.

Phoenicians from what is now Marseilles founded a colony in what is now Monaco in the 6th century BCE. Monaco came under Roman control in the 1st century BCE and remained under their control until the fall of the empire in the west in 476 BCE. For the next seven centuries, Monaco was successively controlled by Phoenicians, Lombards, Franks, and Saracens. In 1171 CE, the Roman Emperor Henry VI granted suzerainty to Genoa.

In 1215, a detachment of Ghibellines—a pro-emperor faction opposed to the pro-papal Guelphs—began construction of a citadel atop the Rock of Monaco. The Guelphs were expelled from Genoa when they lost their protracted struggles with the Ghibellines. Among the Ghibellines were members of the Grimaldi family. Duursma tells of the beginning of the Grimaldi rule of Monaco: "On January 8, 1297, Francois Grimaldi, disguised as a Franciscan monk, managed to penetrate the fortress of Monaco held by the Ghibellines. The fortress was seized, and the Grimaldi family established its authority over Monaco." In order to protect his territory from outright annexation, the Lord of Monaco concluded a treaty with the Spanish king, making Monaco a Spanish vassal.

In 1633, the Spanish Court recognized the self-declared change of title of the Lord of Monaco to prince of Monaco. In 1641, the prince, perceiving the Spanish suzerainty becoming overly dominant, negotiated with King Louis XIII of France the Treaty of Péronne, which protected Monaco against aggression and granted the prince of Monaco sovereignty over his territory.

In 1793, the French Revolutionary Government annexed Monaco. In 1814, after the first exile of Napoleon, the Principality of Monaco regained sovereignty under the same conditions as granted in the Treaty of Péronne. But, in 1815 the Congress of Vienna, ignoring the prince of Monaco, agreed with the Kingdom of Sardinia that Monaco should not continue under French oversight but be placed under Sardinian protection, again under the Treaty of Péronne conditions. In 1861 the king of Sardinia renounced his protectorate over Monaco.

In 1911, Prince Albert granted the Principality a constitution providing for an elected national council. In 1918, the prince entered into a further treaty of friendship with France, and, in 1930, another Convention regarding the nationality and selection of some executive and judicial posts. In 1951, Monaco and France updated earlier agreements by signing a convention of administrative assistance concerning customs, taxes, postal affairs, telecommunications, and monopolies. In 1962, France protested Monaco's refusal to impose a direct income tax, claiming that the tax exemption gave an unduly privileged position to firms located in Monaco; consequently, France threatened to abrogate the 1951 convention and set up customs controls on the borders.

After lengthy consultations, a new convention was signed in 1963 providing for income taxes to be paid by Monaco residents of five years or more and Monaco firms to derive more than 25 percent of their revenue from outside the principality. In 1949, Prince Rainier III succeeded his grandfather and later married actress Grace Kelly. In 1959, he suspended part of the constitution and dissolved the national council. Pressure from France led him to restore the council and install a more liberal constitution in 1962. Prince Albert II succeeded his father in 2005.

The executive, the government, is comprised of the Minister of State and three government counselors, which are appointed by the prince who may dismiss them; the government, therefore, is not responsible to the Monégasque National Council. A Franco-Monégasque Convention of 1930 provides that the Minister of State and the Government Counsellor for the interior have to be French nationals; the French Government proposes several candidates from its own administration, from whom the prince of Monaco makes his choice. The other two government counselors, one for finance and economy and the other for public works and social affairs are Monégasque nationals.

The National Council comprises 18 members directly elected for five-year terms. The National Council approves the budget, consents to any direct or indirect contributions to the budget, except those stemming from international conventions, and considers legislation proposed by the executive. However, the Council may not initiate legislation. The prince has the right to dissolve the National Council. The prince and the government exercise executive power under the prince's authority. Succession to the throne passes in direct line to legitimate descendants of the prince, with priority to male descendants: "Judicial power is held by the prince who delegates it to the courts and tribunals, which administer it in his name. By virtue of the . . . Convention of 1930, the majority of judges in the Monégasque courts and tribunals have to be French nationals seconded from the French judicial administration, except for the Criminal Tribunal." Convicted persons serve their sentence in a French prison.

In accordance with the Treaty of 1918, Monaco's right to establish diplomatic and consular relations with third-world countries as well as the choice of the head of the mission is subject to prior agreement with France. France can place its diplomats at Monaco's disposal, who become seconded from the French diplomatic corps. In addition, certain public offices concerning the security, public order, and external relations of Monaco are reserved to seconded French civil servants. France, too, manages Monaco's postal and telegraph services, but Monaco does maintain its own telephone services.

Since the late 1800s, Monaco entrepreneurs began developing its economy, opening a casino, which the government has since taken over, and promoting luxury tourism. Since World War II, the inflow of foreign capital and residents and the development of tourism have contributed to a significant surge in prosperity. The GDP is $6 million; the GDP per person exceeds $85,000. In the mid-2010s, the budget was $1.1 billion.

Monaco belongs to the United Nations and several other global intergovernmental organizations including the FAO and WHO. Among its missions are ones to the EU and UN. Like San Marino has with Italy, Monaco appears to have worked out a satisfactory mutually beneficial *modus vivendi* with France. In the future, will the interaction between Grimaldi and French interests require further adjustments?

THE PRINCIPALITY OF ANDORRA

"The objective characteristics which distinguish the Andorran nationals from Catalan people are tradition, culture, and history. Catalan being the official language, there is no specific Andorran language."

—Jorri C. Duursma

The Principality of Andorra is a 182-sq. mi. micro-country located in the Pyrenees Mountains, adjacent to France on the north and Spain on the south. The principality's population grew from less than 10,000 in the late 1950s to about 46,000 in 1989, and to over 85,000 in 2016. Its growth has been driven primarily by tourism and an export-driven economy (over 60 percent France-directed) in which banking, tobacco, cattle, timber, and furniture manufacturing predominate. The 2011 per capita GDP was $37,200.

About 50 percent of the population is Andorran; 25 percent are other Spanish, 14 percent are Portuguese, and about 4 percent are French. The official language is not Spanish or French, but Catalan, the language spoken in the separatist-prone Spanish province of Catalonia.[32] The national song, which eulogizes Charlemagne, is sung in both French and Catalan.

According to disputed legend, Charlemagne conquered Andorra when he recovered Barcelona from the Moors in 803 CE. His son, Charles II, in 819 CE, granted Andorra to the Count of Urgel. In 988 CE, his descendants exchanged Andorra with the Bishop of Uriel for other possessions. In the early 1000s, the Bishop of Urgel (located in Catalonia) granted Andorra to the Caboet family of France. Subsequently, in 1208, the feudal rights of the Caboet family passed by marriage to the Count of Foix. Disputes between the Count of Foix and the Bishop of Urgel regarding the extent of each other's feudal rights led to an arbitrated pariatge in 1278, amended in 1288, in which the Andorra became a principality with an elected council and the Bishop of Urgel and the Count of Foix as co-sovereigns.

---

32. Catalan is a Romance language akin to Provençal and Spanish, but quite different from both, especially from Spanish.

A descendant of the Count of Foix, who inherited the county of Béarn and the Kingdom of Navarre, married Antoine de Bourbon from one of the French noble families of France. In 1589, when the young Navarre prince began the Bourbon dynasty of France as Henry IV (and may have said "Paris is worth a mass"), the Andorran co-sovereignty passed to the French throne. When the monarchy was abolished, sovereignty was vested in the French presidency, whatever oversight is exercised by the prefect of the department of Ariège.

With the transition of France from a kingdom to a republic, the French co-ruler of Andorra passed from the monarchy to the presidency of France. Self-determination, driven by the desire to have their country's independence internationally recognized, led to the drafting of a new constitution, adopted in 1993, which clearly vests the sovereignty with the Andorran and limits the role of the former co-rulers. With its independence so clarified, Andorra promptly received its admittance to the UN in 1993.

"The 'pariatges' are the founding documents of the present institutions of Andorra . . . Due to the lack of a clear written document the rights of both co-princes in Andorra, . . . an unresolved and disputed question" remained, which would finally be resolved by the reform constitution of 1993. In 1931, tensions arose when the civil war erupted in Spain. That same year, the co-princes cancelled the vote of the Andorran General Council (Andorra's parliament) to establish a gambling house—a controversy that led to the adoption of universal male suffrage. Efforts to initiate institutional reform began in 1975 and took 18 years to complete. Since 1990, the reforms were taken to strengthen Andorra's international status so as to permit a greater international representation and activity—a goal, that, before 1990, had always been opposed by the French co-prince.

The 1993 reform constitution introduced major changes: removing traces of feudal tradition, clarifying the international status of Andorra, transferring powers from the co-princes to the General Council and its government, and specifying that sovereignty resides in the Andorran people. The power of the co-princes is limited to that of constitutional heads of state, not heads of government. The French co-prince no longer controls the international representation of Andorra, an issue that had long stirred controversy.

Legislative power is vested in the General Council composed of 28 members, elected by Andorrans for four-year terms. Two members are elected from each of the 7 parishes; the other 14 are elected at-large. Responsible to the General Council is the executive of Andorra, composed of six ministers appointed by the Head of Government, after having been elected by the General Council who is appointed by the co-princes. The General Council may dismiss the Head of Government, who may ask the co-princes to dissolve the General Council. With regard to judicial affairs, one Supreme tribunal has been established in Andorra to replace those formerly located in Urgell, Spain, and Perpignan, France.

Since 1993, with France no longer controlling Andorra's foreign affairs, Andorra maintains its own diplomatic missions, including ones to the European Union and the United States. Where Andorra does not maintain a mission, it may charge either France or Spain to act on its behalf. The Principality belongs to the United Nations and several affiliates including FAO and WHO. The Andorran postal service is maintained by both France and Spain. The Principality is embraced by the European Union customs union. Will future Spanish-Catalonia friction facilitate Andorra changing its relations with its co-sovereigns? Will it become a European Union member?

THE PRINCIPALITY OF LIECHTENSTEIN

"The place which a micro-state like Liechtenstein reserves for foreigners in its legal system is of particular interest, as usually a relatively high percentage of micro-states' inhabitants are aliens."

—Jorri C. Duursma

The Principality of Liechtenstein escaped the vicissitudes of feudalism to become a sovereign state in 1806. Like other early emancipated mainland European micro-countries, the fact that the principality was not absorbed into a larger polity stems from its mountainous location and the initiative and ingenuity of generations of its rulers taking advantage of rivalries between European powers. The Liechtenstein family has long governed Liechtenstein. In the past few decades, the balance of power has shifted significantly from the prince to representative institutions.

This landlocked, mountainous enclave, squeezed between Switzerland on the west, Austria on the east, Germany on the north, and Italy on the south, covers 62 sq. mi (160 sq. km) and embraces 11 communes. Two-thirds of the 37,000 residents are Liechtensteiners. More than three-fourths of the population is Roman Catholic. Its economy relies principally upon electronics, metal manufacturing, dental products, optical instruments, ceramics, pharmaceuticals, and tourism. Its mid-2010s GDP of over $3 billion provided a GDP per capita of almost $90,000, the highest in Europe. Before World War II, this micro-country identified primarily with—and worked closely with—Austria. Since then, it has developed close working relations with Switzerland.

The County of Vaduz and the Seigniory of Schellenberg were united in 1434 when the Holy Roman Emperor placed these two imperial feudal fiefs under his direct authority and dismissed the then deep-in-debt ruling count for mismanagement. Prince Johann von Liechtenstein, whose family had held high offices in the Empire and had been granted the title of prince in 1608 even though he did not possess a principality, bought the property (the Seigniory in 1699 and the County in 1712) and gave the combined entity his family name, Liechtenstein.

Because Prince Johann, as ruler of the combined territory, was subordinate directly to the Emperor, he sought for a seat in the Council of Electors and became an Elector, which gave him the right to vote on the election of an emperor.[33]

In 1799, Napoleon's troops occupied Liechtenstein. In 1806, Napoleon included the sovereign Principality of Liechtenstein within the Rhine Confederation with Napoleon as its protector. In 1815, the Congress of Vienna placed Liechtenstein in the new German Confederation, comprising 39 sovereign German states. When the German Federation was dissolved in 1866, Liechtenstein was free of any federal affiliation.

During World War I, the Principality of Liechtenstein remained neutral; but, after the war, it severed its customs and postal links with Austria, which had allied with Germany. In its place, the principality secured its international interests through the diplomatic channels of Switzerland. It entered into a postal union with Switzerland in 1920, gradually introduced the Swiss franc as legal currency, and signed a customs union treaty with Switzerland in 1923.

As a result of popular demands for representation in the affairs of the Principality, the first direct male elections to the Landtag (the Liechtenstein parliament) were held in 1918; a new constitution was drafted in 1921. Throughout World War II, Liechtenstein again, along with Switzerland, remained neutral. Since the war, the economy has prospered dramatically due to the development of high-tech industry, including electronics, precision and optical instruments, pharmaceuticals, and tourism.

The Principality of Liechtenstein belongs to the European Free Trade Area, which works closely with the European Union. By the EU's agreement with Liechtenstein (as with Norway and Iceland), it is included in the European Economic Area, which is a single market and free trade area, abiding by the European Union rulings. Liechtenstein maintains in Brussels a mission (accredited to the EU as well as several other European countries) and in Washington, D.C. (accredited to the UN as well as the United States).

---

33. An emperor's heir did not automatically receive the throne. Candidates would bargain, bully, and bribe the Electors to secure the emperorship. A sitting emperor would similarly scheme to have his heir selected as his successor.

The Principality of Liechtenstein may be described as a hereditary constitutional monarchy with a democratic parliament. The Government of Liechtenstein consists of the prince (Hans-Adam II) whose critical role in governance is substantially that which his predecessors exercised. The prime minister (Adrian Hasler) and four Government Councilors are appointed by the prince as proposed and approved by the parliament (the Diet).

The prime minister generally handles internal, external, and financial affairs; his deputy (one of the four Government Councilors) handles economy and justice. The other government councilors are part-time. The Government Councilors, who must have been born with Liechtenstein nationality, are responsible to both the prince and the Diet and may be dismissed by the prince at his own request or at the request of the Diet. The Diet consists of 25 members elected by universal and direct suffrage for four-year terms. In conformity with the provisions of the constitution and of other laws, every law, decree, or ordinance is issued by the prince issued jointly with the prime minister.

How sustainable/vulnerable is this tradition-based joint arrangement? Will it continue indefinitely?

THE GRAND DUCHY OF LUXEMBOURG

"The small size of The Grand Duchy of Luxembourg belies its influence and success in international matters, especially in the European theater."

—Jeanne A. K. Hey

The Grand Duchy of Luxembourg borders Belgium to the northwest, Germany to the east, and France to the south. The enclave covers 998 sq. mi (2,586 sq. km) and has a population of over 570,000 as of 2016. Citizens of other countries comprise 57 percent of the population; others are Portuguese (16 percent), French (6 percent), Italian (4 percent), Belgian (3 percent), and German (2 percent). The fact that the Secretariat of the European Parliament, the European Court of Justice, the European Court of Auditors, and the administration of eight Directorates-General are located within the Grand Duchy has facilitated its cosmopolitanism sand prosperity. Luxembourgish is the national language, but French and German are also official languages; all three are widely spoken.

Luxembourg was part of the Frankish Kingdom of Austrasia—and, later, the Charlemagne Empire. It traces its own jurisdictional history to the 963 acquisition of the Luxembourg Castle by Sigfried, Count of the Ardennes. His descendant Conrad took the title Count of Luxembourg about 1060. In 1308, the count of Luxembourg was elected the Holy Roman Emperor as Henry VII. He strengthened his Luxembourg dynasty by marrying his son John to the heiress of King Wenceslaus of Bohemia.

The Luxembourg dynasty continued on the imperial throne with Henry VII's grandson, Charles IV (r. 1346–78), and two of Charles' sons, Wenceslaus (r. 1378–1408) and Sigismund (r. 1410–37).[34] In 1354 Emperor Charles IV made the county a duchy and gave it to his younger half-brother, Wenceslaus, who pawned it to his niece, Elizabeth of Gorlitz. When Wenceslaus did not repay the loan, the duchy became hers.

---

34. The four Luxembourgian emperors were an interruption in the long chain of Habsburg emperors that began in 1273 with Rudolph I and ended with Francis II in 1806.

In 1443, Elizabeth, lacking an heir, ceded the duchy to her nephew Philip the Good, Duke of Burgundy. Having extended his Burgundian duchy throughout the Low Countries, he headed one of the most prosperous and prestigious courts of Europe. In 1516, along with the rest of the Low Countries, Luxembourg passed by inheritance to Spanish rule when Charles V, inheritor of the Burgundy duchy, became the Spanish king in 1516 upon the death of his grandfather.

When the 1568–1648 Dutch Revolt against Spain (initially led by William "the Silent") split the Low Countries, the northern half became the Netherlands; the southern half, including Luxembourg, remained a possession of the king of Spain. The French captured Luxembourg in 1684, but it was returned 13 years later following the War of the Grand Alliance and its Treaty of Rijswijk. The War of Spanish Succession, which concluded in 1713 with the Treaty of Utrecht, made Luxembourg part of the Austrian-ruled Low Countries.

In 1795, France made Napoleon-occupied Luxembourg, along with the rest of the Low Countries, a French "departement" (i.e., a local government with communes within it). Following Napoleon's defeat in 1815, the Congress of Vienna awarded Luxembourg as a grand duchy to William I, king of a newly-created independent Kingdom of the Netherlands as well as grand duke of Luxembourg. The eastern part of Luxembourg became Prussia. When the southern provinces of the Netherlands revolted in 1830, they became Belgium.

In 1838, William I of the Netherlands finally agreed to recognize Belgium's independence; the larger part became the Belgian province of Luxembourg. But the part in which the city of Luxembourg is located remained a grand duchy with the Dutch king as grand duke. William III died in 1890 without a son; Salic law precluded his daughter succeeding him. A male relative, Adolphe of Nassau, succeeded; his descendant is the present Grand Duke.

Despite Luxembourg's declared neutrality, Germany occupied it in both World Wars. It was a founding member of the European Iron and Steel Community in 1951, which became the European Economic Community in 1957 and the EU in 1993. Luxembourg's parliamentary government has a prime minister, named by the grand duke upon recommendation of the Chamber of Deputies, comprised of 60 members directly elected for five-year terms. The present prime minister is Xavier Bettel. A Council of State, whose 21 members are appointed for life by the grand duke, assists the Chamber in drafting legislation. The Grand Duchy has 3 districts and 106 communes.

Luxembourg has a stable, growing, and prosperous economy, supporting one of the highest GDPpcs in the world ($92,000) and a national budget of more than 25.5 billion in the mid-2010s. Until the 1960s, the grand duchy depended primarily upon iron and steel. The economy has diversified into banking and financial services, including investments, insurance, and reinsurance. These have grown to make the grand duchy one of the world's major fiscal centers—and a tax haven. The extent of Luxembourg's participation in international organizations illustrates its diplomatic clout. Luxembourg belongs to the European Union and the United Nations and its global affiliates such as FAO, IBRO, ILO, IMF, WHO, and WTO.

European Union institutions and a major banking industry, and a major tax haven, concentrate a well-paid workforce and wealth within the highly urbanized Grand Duchy. It is basically a medium sized city that draws commuters from three adjoining countries. The GDP of over $51 billion sustains a $25 billion national budget, supporting international aid and foreign affairs missions.

The significant role of Luxembourg, its dynastic leaders during the Empire, and the fact that three of Luxembourg's prime ministers later served as the president of the European Commission (Gaston Thorn, Jacques Santer, and Jean-Claude Juncker) clearly indicates that, as in medieval history, in modern history Luxembourg has punched well above its weight.

REPUBLIC OF ICELAND

"Iceland with its past in Europe and its future in North America bridges two continents. That is our challenge and our opportunity."

—Sigmundur Davíð Gunnlaugsson,
former prime minister of Iceland

Iceland, a North Atlantic micro-country, sits 155 nautical miles from Greenland, 430 nautical miles from Scotland, and about 600 nautical miles from Denmark. This long-time Danish colony re-asserted its independence in 1944, shortly before the conclusion of World War II.

Irish monks briefly inhabited Iceland before the arrival of the Vikings, roughly from the late 7th to the early 8th centuries. According to Morrison (1971), "several hundred monks, outraged by . . . efforts to bring the Irish church into line with Rome . . . sailed to Iceland." They spread out, forming monastic cells on the, heretofore uninhabited, island.

Viking voyages and visits began in the 9th century. In 874, a west Norwegian chief, Ingólfur Árnason, moved there with his extended family. Later, colonists came from western Norway and the Orkney, Shetland, and Hebrides Islands. From Iceland as a base, the long Viking warships, driven by oar and sail, planted settlements on the Greenland and North American coasts (the Vinland of Icelandic sagas)—more than five centuries before the voyage of Columbus.

Unified government began its development when local chiefs convened an Althing in 930 CE.[35] Before 1000 CE, Christian missionaries arrived in Iceland. When the clash between Christianity and paganism threatened civil war, the Althing reached a settlement resulting in the whole Icelandic settlement accepting Christianity. Since the 1100s, Iceland and Greenland were under the nominal suzerainty of the Norwegian crown and economically dependent on Norway. In an age when church and secular governance were intricately intertwined, the Norwegian kings sought to control Iceland indirectly, and directly, through the Norwegian archbishopric.

---

35. Iceland claims this Althing is the oldest parliament in the world.

In 1262, taking advantage of the internal conflict that arose among the Icelandic local chiefs, the Norwegian king, Haakon IV ("Haakon the Old"), succeeded in persuading Iceland to accede to his hegemony. Nordic scholar Knut Gjerset has noted "the reconciliation between the King and the Icelandic chieftains seems to have been effected toward the end of the period of colonization." The formation of the Kalmar Union of Norway, Denmark, and Sweden in 1415 led to it taking possession of Iceland. In 1465, Iceland was included in the Union of Denmark, Norway, and Sweden. After Sweden broke from the union in 1523, Iceland remained a Danish-Norwegian dependency. As it was a personal fief of the king, the religious and secular rulers of Iceland were monarch-appointed foreigners. Infertile soil, volcanic eruptions, deforestation, an unforgiving climate, and the Black Death (that infected the country twice)—along with the imposition of foreign rule—led to Iceland enduring the next four centuries as one of the poorest in Europe. During this period of continuing economic decline, the Nordic king tightened his authority. By 1800, the Althing, the last vestige of independent Iceland, was abolished.

In the early 1800s, the Napoleonic wars and trade blockades brought continued hardship to Iceland. In 1814, the breakup of the Denmark-Norway union made Iceland a Danish dependency. Not until the 1830s, stimulated by nationalistic ideas emanating from mainland Europe, did agitation for the revival of the Althing slowly gain momentum. Iceland's long struggle for autonomy, then independence, from Denmark focused on relations between Icelandic institutions and Danish ministerial power. The 1874 Icelandic constitution increased the competence of the Althing but provided that a Danish member of the Danish cabinet would occupy the post of minister for Iceland. This issue dominated the next round in the struggle for independence.

The struggle over the following decades focused on "Icelandization" of the governmental power by having the minister for Icelandic affairs be an Icelander and reside in Iceland. In 1884, the constitutional issue gained momentum with efforts to develop political parties. Gunnar Karlsson points out "all these attempts failed, but the fruit was reaped by the informal nationalist movement . . . In the decade to come it concentrated on the revision of the constitution." In 1918, decades long contention between Iceland and Denmark led to a treaty (which would remain valid for 25 years) recognizing Iceland as a self-governing polity under the Danish king with Denmark continuing to handle Iceland's foreign affairs and its embassies throughout the world displaying flags of both countries.

The Icelandic Althing in 1937 declared that the treaty would not be renewed upon its expiration in 1943. The advent of World War II and the threat of Germany occupation led to Iceland appointing a regent and severing its connection with Denmark.[36] Following the expiration of the 25-year-old Danish-Icelandic Act of Union Treaty in 1943, Icelanders voted in a plebiscite to terminate the union, abolish the monarchy, and establish a republic. In 1944, Iceland declared its independence.

At the start of its independence, fishing was the primary vocation; shepherding soon developed as the secondary occupation. The Icelandic economy has long been based principally on fishing and fish processing; from the 1960s to the present, export earnings from fish products has declined from 90 percent to 40 percent. Its economy has diversified into aluminum smelting, ferrosilicon production, tourism, and finance. Iceland recovered remarkably quickly from the financial disruption of 2007–2010. Its mid-2010s GDP was $13 billion, supporting a GDP per capita of $40,700. Its exports of over $4 billion are mainly to the Netherlands and Germany; its imports of over $4 billion are mainly from Norway and the United States.

Glaciers cover 11 percent of Iceland. They have grown and contracted over history, perhaps reaching their maximum extent in the late 1800s— the result of global warming, whose consequences have been traumatic throughout the past two decades. The continuing disappearance of glaciers is on its way to transforming Iceland into an ice-less Iceland. Its tourist industry is already threatened: the airline (WOW), bringing one-quarter of its tourists, has ceased flying, tour companies and hotels lament lost bookings, and the central bank warns that the economy is likely to suffer.

Iceland is a parliamentary democracy, which has been governed by multi-party coalitions. They consider the Althing the re-establishment of assembly founded in 930 CE and later suspended. Since World War II, Iceland has moved from its long-term status as a Danish shelter country to a more independent role. It is a member of European Free Trade Association

---

36. The advent of war also led the United States to station troops in Iceland.

(EFTA). Iceland also belongs to the UN, NATO, EFTA, and Organization for Economic Cooperation and Development (OECD). Its ties with the Nordic countries, Germany, the United States, and Canada are especially close due to cultural, economic, and linguistic similarities. Iceland has long belonged to the Nordic Council. Since 1960, along with Liechtenstein, Norway, and Switzerland, Iceland has been a member of the European Free Trade Association.[37] A major goal has been developing its tourist industry.

Iceland has a president (head of state) and prime minister (head of government). Sigmundur Davíð Gunnlaugsson became prime minister in 2009 when his center-right Independence Party secured the most votes and formed a coalition, and again in 2013. He subsequently resigned after the release of the Panama papers, which raised questions regarding his investments and tax evasion.

The 2007–2008 financial crisis hit Iceland and its banks hard. When short-term financing for the banks' over-extended loans dried up in 2008 and capital rushed out of the country, the currency tanked and inflation soared, leading to de-evaluating its currency. By 2014, the economy had recovered—unemployment was down to 5.6 percent, and the budget was in surplus.

In the 2016 elections, the Independence Party came first in the general election; after two months of negotiations, it formed a coalition with two others but with a fragile a one-seat majority (and with Bjarni Benediktsson, the Independence Party's new leader, as prime minister). The coalition folded when the Bright Futures party left the coalition government. In 2017, the Independence Party withstood a challenge from the Left-Green Movement. A new Center Party won the most votes, but a diminished number of seats.

Iceland's future will be affected by how its tourist industry adjusts to the changing dynamics of global warming. Its unique straddle of Europe and North America and record of sound fiscal management gives reason for optimism.

---

37. Since 1994, EFTA has been incorporated into European Economic Area (EEA), which provides the country access to the EU single market, protecting its fishing rights from competition and international restrictions, and its livestock from foreign infections.

REPUBLIC OF CYPRUS

"Cyprus is a scene of the continued conflict between East and West, which is always to be reinterpreted, but never to be ignored."

—H. D. Purcell

Cyprus, an island at the eastern end of the Mediterranean between western Asia and northeast Africa (40 miles south of Turkey, 50 miles east of Syria, 200 miles north of Egypt, and more than 400 miles southeast of the Greek mainland) is considered a European country. Its proximity to three continents has led to influxes of different peoples, initial waves coming from Greece more than 3,000 years ago. Cyprus' strategic location, close to the Suez gateway to India and the East Indies, encouraged multiple Mediterranean powers to clash over its control.

Although the UN recognizes the Republic of Cyprus as sovereign over the whole island, control of the island has been divided by Turkish intervention. While the Republic of Cyprus possesses the main part of the island, the Turkish Republic of Northern Cyprus (TRNC), Turkey's ward, controls the northeastern portion of the island. The United Nations considers the northeastern part of the island governed by the TRNC as illegally occupied by Turkish forces. Negotiations regarding the separation of these two entities, separated since 1974, sporadically continue.

The internationally recognized Republic of Cyprus belongs to the United Nations and the European Union. Over 902,000 inhabitants occupy the Republic of Cyprus-controlled part of the island; more than 280,000 inhabit the Turkish-controlled Turkish Republic of Northern Cyprus (TRNC) part of the island.[38] The diverse population includes 71 percent Greek and 18 percent Turkish inhabitants. Eighty-nine percent are Greek Orthodox. Greek and Turkish are official languages; English, as well as Romanian, Russian, and Bulgarian are also spoken. Nearly all of those in the TRNC are ethnically Turkish.

---

38. If Cyprus were not divided, its 1,189,197 population in 2016 would exclude it from this book.

Many powers and peoples have occupied the strategic island of Cyprus over its long history. They include Phoenicians, Mycenaeans, Hittites, Persians, Greeks, Egyptians, Romans, Byzantines, Crusaders, Genoese, Venetians, Ottomans, and the British. This succession of rulers introduced an assortment of customs, a medley of European, Asian, and African ethnic backgrounds, and a diversity of Orthodox Christian, Roman Catholic, Protestant, and Islamic faiths. Its settlement and association with neighboring lands, successive conflicts and rulers, and its struggle for independence from British rule parallels that of Malta. The Greek pressures for unifying Cyrus with Greece, by the movement known as enosis, and Turkish partition pressures have escalated tensions long endured by Cyprus. Conflicts, consecutive occupations, and the coming together of disparate elements produced a fascinating, unusual history.

Cyprus gained importance in the Bronze Age not only because of its strategic location but also because it was a major source of copper. A Mycenaean invasion and migration with the longest-lasting impact began around 1400 BCE. By the end of the century, it had colonized Cyprus, introducing the Greek culture and language, which continues to dominate most of the island. Along with Assyria, Cyprus had become one of "the two last major players of the late Bronze Age in the ancient Near East."

Mycenae power collapsed between 1200 and 1150 BCE. The Assyrian King Sargon II asserted dominance over multiple Cypriot kingdoms around 800 BCE. Following the breakup of the Assyrian Empire, Egypt seized the island in 560 BCE. When, in the wars between the Greeks and the Persians, Alexander the Great won a victory in 333 BCE, the Cypriot kings welcomed him as their overlord. After Alexander's death, his successors struggled for decades for control of Cyprus. In 323 BCE, Cyprus passed, with the support of the Cyprus kings, to one of Alexander's generals who, as king of Egypt, took the name Ptolemy I. In 58 BCE, the Roman Empire seized Cyprus. Under Roman rule, Greek remained the language of Cypriot literature as well as the *lingua franca* of trade in the eastern Mediterranean. After the Roman Empire split in 395 CE, Cyprus became subject to the Byzantine emperors for 700 years, with only brief interruptions. In the 7th century, Muslims invaded Cyprus twice and made it pay tribute.

When the Byzantine governor of Cyprus resisted the Crusaders, the Crusader King Richard I led English crusaders in the conquest of Cyprus, and then sold it to the Knights Templar. When the Knights did not pay, he gave it to Guy de Lusignan, the French noble who had been dispossessed as king of Jerusalem. As trade grew and the Genoese and Venetian merchants exerted more influence, the power of the Lusignan dynasty faded and control passed to Venice in 1473. The Ottoman Turks conquered Cyprus in 1571. Under Turkish administration, there were uprisings in 1764, 1804, and 1821 followed by attempts at self-governing reforms.

In 1878, the United Kingdom asserted control over Cyprus—a step that provided it a base securing the seaway to the Suez and beyond. While Cyprus remained under Turkish sovereignty, this change in administration marked an increase in pressure by Greek residents (80 percent of the Cypriots at the time) to press for enosis (a union) with Greece. In the first year of their rule, the British created a Legislative Council of four to eight appointed members, half of whom were officials. In 1882, when responding to a petition from the Greek Archbishop and other Greek representatives, the British set up a Legislative Council with 12 elected members (9 Greek Cypriots and 3 Turkish Cypriots) and 6 official ones.

When World War I broke out, Turkey allied with the Central Powers, and Britain annexed Cyprus. Under the Treaty of Lausanne, ratified in 1924, Turkey formally recognized the annexation. In 1925, Cyprus became a British crown colony. In 1931, demands by Greek Cypriots for enosis incited riots, which led to the British abolishing the Legislative Council.

Following World War II, Archbishop Makarios III, patriarch of the Orthodox Church in Cyprus, emerged as leader of the enosis movement. Supported by various groups in Greece as well as Cyprus, the campaign for enosis gathered strength. In the escalating atmosphere of distrust and terror, the EOKA (National Organization of Cypriot Struggle), headed by Colonel Grivas, a former Greek army officer, engaged in many attacks, including ones on British bases. In 1955, a new governor, Field Marshall Allan John Harding, imposed emergency regulations and opened negotiations with Makarios III regarding self-government.

After failure to reach an agreement, Makarios III was exiled in 1955. In 1956, Britain supported the adoption of a new (more liberal) constitution. By 1958, the Turkish community (about 18 percent of the island's population) led by Fazıl Küçük, aroused by the escalated demand for enosis, demanded partition. Distrust escalated, terrorism renewed, and the British imposed fresh restrictions. Responding to Makarios' proposal for self-government, the Greek and Turkish governments (without Britain) negotiated an agreement, which Britain approved, calling for an independent republic that would not engage in political or economic union with any other state nor be subject to partition—with Britain retaining sovereignty over its two bases.

Executive power under the new constitution was vested in a Greek Cypriot president and a Turkish Cypriot vice president, to which Makarios and Küçük had already been elected.[39] They were assisted by a 10-member council of ministers—7 Greek Cypriots and 3 Turkish Cypriots. The civil service was also to be 70 percent Cypriot Greek and 30 percent Cyprio-teTurkish. Legislative power was vested in a 50-member House of Representatives, with 70 percent elected for five-year terms from Greek Cypriot communities, and 30 percent from Turkish Cypriot ones. Independence was achieved in August 1959. In the general elections held in July 1960, 30 of the 35 seats were allotted to the Greek Cypriot Community; 30 seats were won by the party led by Makarios, and 5 others were allotted to the communist AKEL; all 15 of the Turkish Cypriot seats were won by the party led by Küçük.

Cyprus became a member of the United Nations and the British Commonwealth in 1960. Cyprus belongs to the European Union, the Commonwealth, and the UN and its affiliates. Its per capita GDP of $27,500, and its $10.9 million national budget is supported by tourism, which it has tried to expand, offshore banking, ship repair, food and beverage processing, textiles, light chemicals, cement and gypsum products, and metal products. Its principal trading partner is Greece, who both imports and exports more than 20 percent of its trade. In part because Cyprus held large amounts of Greek bonds, Cyprus suffered a banking crisis in 2013, critically affecting three of the largest banks. The crisis forced the intervention of the International Monetary Fund and the European Central Bank to help to restructure the banking system. The GDP shrunk in 2012, 2013, and 2014, and the 2014 unemployment rate was 16 percent.

Since 1963, Greek and Turkish communities have escalated to rioting. British troops tried to keep order. In 1964, the UN Security Council sent

---

39. Both posts were granted veto powers relating to security defense and foreign affairs.

more troops. The situation was further enflamed by Greek Cypriot raids and Turkish air raids repelling them—during which many Turkish Cypriots fled to Turkish-held Cyprus. Efforts of the UN, Greek, and Turkish governments to mediate the crisis failed. The Turkish Cypriots withdrew from the Cypriot government; the Cypriot Greeks extended the presidential term of Makarios, and the Turkish Cypriots did the same for Küçük.

In 1968, Makarios and Küçük were reelected—with Küçük heading the "transitional administration" the Turks had established to govern the affairs of the Turkish Cypriot community. Tensions continued, negotiations were unsuccessful, and the UN mandate was repeatedly renewed. In 1974, Greek military deposed Makarios and attempted a coup to unite Cyprus to Greece. Five days later, Turkey invaded and took control of the northeast part of the island. International pressure restored Makarios to power in 1975; when he died in 1977, the posts of archbishop and president were separated. Subsequent enosis efforts led to Turkey in 1983 to declare the Turkish-occupied part of the island the Turkish Republic of Northern Cyprus.

The United Nations' plan, proposed by then Secretary-General Kofi Annan for uniting the island in 2004 was approved by the Turkish Cypriots but rejected by the Greek Cypriots. In 2008, Demetris Christofias, the candidate of the Progressive Party of the Working People, (the Cypriot Communist party known in Greek as AKEL) was elected as president. Despite his orthodox Communist ideology, he won over Cypriots with promises to reunify the Greek and Turkish sectors of the island. But talks stalled in 2011 when containers of gunpowder and other munitions, which had been confiscated on their way from Iran to Syria, exploded. In 2013, the conservative candidate, the leader of the Democratic Rally Party, Nicos Anastasiades, was elected president. Negotiations recommenced in 2014, but the stalemate lasted. Neither the Turkish president, nor the Greek Cypriot Communist party, nor the Cypriot Orthodox Church, which strongly supported the "no" vote in 2004, has relaxed its opposition. The tensions are so deep that few expect a Cyprus reunification.

REPUBLIC OF MALTA

"L'mir tal bahar u min andu Malta." (Who holds Malta rules the sea.)

—Maltese proverb

Malta's strategic site on trade routes in the Mediterranean Sea—halfway between Gibraltar and the Suez Canal, and midway between Italy and Tunisia—accounts for its contentious history. Over the past 3,000 years, it has endured many battles and successive masters. Since 800 BCE, major powers, appreciating Malta's location at the Mediterranean crossroads, recognized that control of their harbors was the key to the control of the routes tying together most the trade routes. As early as 800 BCE, many successive Mediterranean powers successfully invaded and occupied the Maltese islands.

The languages spoken reflect the multi-ethnic melting pot of over 400,000 people that live on this 122 sq. mi archipelago of islands. The Maltese language combines 10th century Arabic with substantial French, Italian, Spanish, and later English—with a Latin rather than an Arabic alphabet. The tumultuous history of successive invasions and administrations has bequeathed Malta a multi-ethnic heritage, a cosmopolitan outlook, and a determined desire for self-rule. Most Maltese are descended from Phoenician, Carthaginian, Italian, and other Mediterranean forebears; a few reflect the more recent British occupation. The heterogeneity of the micro-country is both a challenge and an asset as it faces the future.

Initially, these islands were successively ruled by the Phoenician, Carthaginian, Roman, and Byzantine empires. From the 9th through most of the 18th centuries, they were ruled by Sicily, Norman Knights, the Germanic House of Hohenstaufen, the House of Anjou, the kings of Aragon, and knights of the Sovereign Military Hospitaller, the Order of Saint John of Jerusalem. The Ottoman Empire, under its greatest ruler, Suleiman the Magnificent, raided Malta several times in the 1540s and '50s. The Ottomans then undertook what is known as the Great Siege of Malta, which engaged the heroic efforts of all the Maltese men, women, and children as well as the knights. Following the failed siege, the knights of Saint John completed impressive fortifications. But, as the Islamic threat diminished, so did the Order, which was ill prepared for French Revolutionary forces.

Napoleon captured Malta in 1790, thus securing sea lanes to Egypt before his Battle of the Nile. The French imposed new taxes to support the local French garrison and closed several churches, selling their decorations to enhance the treasury. These actions, and others, led to a Maltese revolt; with support from Britain and Naples, the French were evicted. Determined to hold onto the Mediterranean sea lanes, the British insisted on controlling Malta, which, together with Gibraltar, safeguarded their access to the Suez Canal (initially the isthmus crossing, and from 1869, the canal), the gateway to India and the East Indies.

British rule of Malta lasted more than 240 years, until the post-World War II rise of nationalism, the loss of Britain's empire, and the consequent diminished need to support a mid-Mediterranean base. The love-hate relationship between Britain and Malta was marked and sustained by repeated British-Maltese constitutional struggles, marked by repeated concessions to and retreats from Maltese demands for participation in their governance. In 1838, an appointed council of government, in which only two of its eight members were Maltese, was established to advise the governor. In 1849, continuing agitation led to the creation of an 18-member legislature—eight elected and 10 appointed by the governor—with the power to enact legislation.

But, in 1883, the British, uneasy with even this limited concession, reconstructed the advisory Council of Governments. To appease the Maltese protests in 1887, a new constitution provided Maltese control of the legislature and British control of the executive. But, in 1903, the British reneged, re-adopting the 1849 model of a largely appointed legislature. A post-World War I depression and accompanying unrest led to another new constitution in 1921, which created what in effect were two separate governments; one for domestic affairs led by the Maltese and the other for military and related strategic matters run by the governor. In addition to the issue of local representation in Maltese governance, other concerns touched on the subject of language.

When the British arrived, there were two languages spoken in Malta: Italian, used by the intellectuals and higher business classes, and Maltese, used by everyone else. Unlike Italian, however, Maltese was not a written language. The British succeeded in making English the official language in Malta, which would be taught at the schools and the university, alongside Italian. Proficiency in English was a major factor in civil service staffing. Since 1921, the two major political parties in the country were the strongly pro-British Constitutionalist Party that promoted the use of the English language and the Nationalist Party that promoted the use of Italian. Despite the proximity of Malta to Italian-speaking Sicily (*ca.* 50 miles) and the long-shared political ties with Sicily, the British colonial government successfully initiated the process of displacing Italian, the long and well-established language, as one of the two major languages of their colonial possession.

In June 1940, Mussolini joined World War II and began to bomb Malta. The surrender of France put Malta in jeopardy. Malta endured constant bombardment in 1941 and 1942, averaging more than one raid a day for many months. The arrival of a British supply convoy, through contested waters, saved Malta. After World War II, Britain was no longer a major sea and colonial power, nor did it need Malta to protect what had been its vital seaway to Asia and the Pacific any longer. It reduced its Mediterranean Fleet (by 1969 Royal Navy ships were not stationed at Malta) and closed its Malta dockyards—the major local employer.

Rising nationalism facilitated the Nationalist party, and its leader, the once-deported Enrico Mizzi, to come back into power in 1950. Initially, much discussion ensued over Malta's constitutional future: whether to pursue union with Britain or independence. In 1956, the Nationalist party opposed and successfully called for a boycott of a referendum, proposing a union with Britain. When the talk of a union fizzled, independence became the objective. Concurrently, the Constitutionalist Party faded and was replaced by the Malta Labour Party.

In 1960, a British government commission, chaired by Sir Hilary Blood, recommended first that Malta's religion, Roman Catholicism, must be respected and, second, that Malta should be granted independence. In October 1961, Malta's new system government provided for a Legislative Assembly of 50 members, with full powers in every area except defense and foreign affairs, which were to be shared with the British. In 1964 the Maltese flag was raised, demonstrating Malta's independence. In exchange for rent, the British were allowed to maintain troops at Malta for 10 years; the arrangement was subsequently renegotiated with the United Kingdom and NATO.

The Socialist Labour Party governed from 1971 to 1987, the Nationalist Party from 1987 to 1996, and the Socialist Labour Party again from 1996 to 1998. Though, the Nationalist Party has won elections in 1998, 2003, 2008, and 2013. Joseph Muscat became prime minister in 2013. In 2017, he called and won a snap to election—despite accusations by Daphne Caruana Galizia, an anti-corruption blogger, and others—of kickbacks received by Muscat and other senior politicians related to the Panama Papers, which he denied. Later, in 2017, she was assassinated by a car bomb. The seriousness of the situation has led Malta's government to ask the United States Federal Bureau of Investigation to aid in investigating her murder. Muscat resigned amid an outcry that he protected friends involved in the murder. Robert Abela replaced him as party leader and was elected as prime minister in 2020.

Malta has diversified its economy. Ship repair is no longer Malta's primary industry, tourism, is! Its long strategic tri-continental history and culture has made the island a major tourist attraction.[40] Despite "the search for economic assistance from the Arab World, China, North Korea, Eastern Europe and the USSR, Malta's ties still lie with its Western trading partners." In the mid-2010s, its GDP was $11.2 billion, its GDP per capita was $29,100, and its budget was $4.3 billion. Malta's economic renaissance has depended heavily on migration. Twelve percent of the labor force are foreign workers who have been drawn to Malta to work in its growth industries such as financial services, generic pharmaceuticals, and online gambling.

Malta has no defense ministry; its armed forces, consisting of three regiments (currently 1,500 troops), report directly to the prime minister. Malta does not belong to NATO, which it expelled in 1979. Malta achieved independence, was admitted to the UN and joined the British Commonwealth, the EU, and several global organizations. It maintains an embassy in Washington, D.C.

---

40. Its capital, Valetta, has been named by UNESCO as a World Heritage Site.

MONTENEGRO

"The Montenegrins are, despite provincial and historical differences, quintessential Serbs, and Montenegro [is] the cradle of Serbian myths and of aspirations for the unification of Serbs."

—Milovan Djilas

Montenegro (trans. "Black Mountain"), was an integral part of the Federal Republic of Yugoslavia from 1918 to 2006. The country lies along the eastern Adriatic coast between Croatia and Albania. Once, this mainly mountainous enclave marked the border between the eastern and western Roman Empires. Since then, this southern Balkan land has been repeatedly contested by powers including Venice, Serbia, the Byzantine Empire, the Holy Roman Empire, the Ottoman Empire, Austria-Hungary, and the Soviet Union. The 20th-century World Wars and the subsequent Cold War aggravated tensions that have enabled conflict.

The heterogeneity of Slavic and other ethnicities, the diversity of Slavic and other languages spoken, and the variety of versions of Christian and Islamic faiths reflect the plurality of its multi-cultural heritage. The population consists of 45 percent who consider themselves Montenegrin, 29 percent Serbian, 9 percent Bosnian, 5 percent Albanian, and 12 percent other/unaffiliated. Montenegrin is the official language but Serbian, Bosnian, Albanian, and Serbo-Croatian are spoken. Orthodox Christianity claim 72 percent of the population; 19 percent are Muslims, 3 percent are Catholic, and 6 percent are other/unaffiliated. From the 11th to the 16th centuries, successive dynasties ruled what is now Montenegro, alternating between periods of local Montenegrin rule and as part of Serbia. Its mountainous terrain helped preserve its autonomy; its ethnic-linguistic-religious heritage contributed to its fractious history.

The Principality of Zeta (or Duklja), near Kotor, was merged into the Serbian empire in the late 1100s. After the disruption of Serbia in 1355, the Balsich family ruled in Zeta until 1421. In 1455, the Crnojević family became rulers of an enlarged area, making it the last free monarchy in the Balkans until the Ottoman Empire invaded the peninsula in 1496. In the

1500s, under the Ottoman Empire, Montenegro developed a degree of autonomy; by the 1600s, Montenegrins had begun rebelling repeatedly, culminating with a victory over the Ottomans in the Great Turkish War at the end of the century. From 1697 the Petrović-Njegoš dynasty of celibate prince-bishops (called *vladikas*), who descended from uncle to nephew, governed Montenegro. The first *vladika*, Danilo I (r. 1697–1735), created the office of *vladika*, making it hereditary, and centralized authority in its office. While he saw himself as the leader of a revived Serbian state, his reign is often seen as the beginning of a specifically Montenegrin history.

In the long effort to clear the Balkans of the Ottoman Empire, Montenegrin forces won the Battle of Carev Laz (1712) against overwhelming odds; the victory is still celebrated as the great national triumph. In the late 1800s, Prince Nicholas I enlarged the principality, fighting the Ottomans in the Montenegro-Turkish Wars, whose battles included the immortalized one at Grahovac in 1858. He won international recognition of his country's independence in the Treaty of Berlin in 1878. In 1882 its status changed from a principality to a kingdom. In 1910, Nicholas I assumed the title of king.

Montenegro sided with Serbia against the Central Powers in World War I. In early 1916, Austria-Hungary defeated and occupied Montenegro. King Nicholas fled. In 1918, after the Allies liberated Montenegro along with the rest of the Balkans, a Montenegrin assembly convened and voted to ban the king from returning, and to unite the country with Serbia. In the Christmas Uprising, a significant number of Montenegrins (called the "Greens") rebelled against the decision to unite with Serbia and fought the unification forces (the "Whites"). At the Paris Peace Conference, despite the rising spirit of self-determination pressed by President Wilson and others, Montenegro was denied a place at the table; her pleas ignored.

In 1922, Montenegro was absorbed into the Kingdom of the Serbs, Croats, and Slovenes. Alexander I, grandson of Nicholas I, became king of the new kingdom, which was renamed Yugoslavia in 1925. In 1929, the Kingdom of Yugoslavia was reorganized into nine provinces, the borders of which were drawn with the aim of diluting historic loyalties and identities. Montenegro, included in a province with parts of Serbia, Croatia, and Bosnia, endured cultural mistreatment that included abolition of the Montenegrin Orthodox Church as well as its distinctive name. The interwar years were marked by three concerns: disputes regarding the organization and extent of centralization of the state, ethnic tensions, and increasing socio-economic problems. The 1938 German-Austria Anschluss, or union, gave Germany a frontier with Yugoslavia, which, along with Italy's apparent readiness to interfere in Balkan affairs, aroused concerns.

In 1941, Benito Mussolini's Italian forces attacked and occupied Montenegro. A 1941 uprising in Montenegro against Italian occupation attracted communist and ex-Royal Yugoslav Army support. Italian forces succeeded in gaining control of the towns, but the communist-led forces remained in control of the rural areas. Montenegro, along with the rest of Yugoslavia, was quickly involved not only in the resist of the occupiers but also in a bitter civil war between Tito's communist Partisans and the royalist Chetniks. After Italy surrendered and withdrew from Yugoslavia in 1943, Germany occupied the country. By the end of 1944, Montenegro, as part of Yugoslavia, was liberated by the Yugoslav Partisans with the assistance of the British—led by, among others, Josip Broz Tito and Montenegrin Milovan Djilas.

After the war, Tito made Montenegro one of the six Republics comprising the new Federal People's Republic of Yugoslavia—comprised of "five nationalities, four languages, three religions, two alphabets, and one Tito." Tito's approach to communism and the Cold War, which differed from Moscow's, led to his estrangement from Stalin and the Soviet Union. Tito's death in 1980 foreshadowed the unraveling of the multi-factional political conglomerate. The drawing to an end of the Communist era, in Montenegro as in other parts of Yugoslavia, escalated the long-suppressed national tensions. The economic crisis across Europe, the collapse of Communism in the Soviet Union, and the lingering effects of the death of Tito fatally undermined what Yugoslavia's one-party regime put together with its cornerstone of "brotherhood and unity." In 1992, four republics declared independence: Slovenia, Croatia, Bosnia and Herzegovina, and Macedonia.

In a 1992 Montenegro referendum, with 66 percent of Montenegrins participating, 96 percent voted for a federation with Serbia. Federation propaganda and pressure and the consequent boycott of Muslim, Albanian, and Roman Catholic minorities, as well as other Montenegrins favoring independence, leveraged the result. Thus, Montenegro, along with Serbia, remained part of the dismembered Federal Republic of Yugoslavia.

In the years 1991–1995, Montenegrin troops joined with Serbian ones in attacking neighboring Kosovo, Bosnia and Herzegovina, and Croatia—acts that led to the NATO bombing of Montenegro.

In 1996, the Montenegrin Government, led by Milo Djukanović, severed ties with the Serbian regime, pursuing its own economic policies and using the German Deutsche Mark as currency—later the euro. In 1989, Milo Djukanović became one of Europe's youngest leaders; 30 years later, at only 57 years old, he is still in power—and has been as president, prime minister, or boss of the ruling-party.

In 2006, of the 86 percent of the electorate in Montenegro that voted, 55.5 percent voted for independence, but they barely passed the 55-percent threshold needed to validate the election; 44.5 percent of the electorate were against. In June 2006, the Montenegrin parliament declared the country's independence, and the United Nations accepted it as its 192nd member. Montenegro has a president—prime minister—parliament form of governance.[41] The country is divided into 23 municipalities, some of which retain their walled core centers.

Montenegro's economic limitations include roads and other infrastructure that have not kept up with the remarkable growth of tourism. Aluminum and steel production and agricultural processing constitute most of the industrial output. The scenic mountains, seascapes, beaches, and walled towns have attracted increasing hordes of visitors over the recent decades.[42] Damage from the bombing in the 1990s conflict only briefly interrupted investment in tourism facilities and residences. Tourists are attracted mainly from Europe and, increasingly, from East Asia, and, before, Montenegro clearly showed its changed opinion on Russia by joining the North Atlantic Treaty Organization (NATO). Until then, many Russians came and invested in Budva, one of Montenegro's coastal walled towns, but no longer.

The major countries to which Montenegro exports are Serbia, Croatia, and Slovenia. It imports mainly from Serbia (29 percent). The GNP in the 2010s was $7.5 billion; the per capita GNP was $12,000. The national budget is $1.8 billion of which $52 million is spent on defense. An indication of

---

41. Filip Vujanović has been president since 2008. Milo Djukanović has served as prime minister since 2012.
42. I can attest to this, having visited there in 1959, 1976, and 2019.

its financial woes is that, as *The Economist* has pointed out, the cost of a motorway represents over one-quarter of Montenegro's annual GNP, making it one of "eight countries at high risk of debt distress thanks to BRI-related lending"—as China's Belt and Road Initiative (BRI) has very long arms.

The country belongs not only to the UN and affiliate organizations such as the FAO, ILO, and WHO but also the OSCE and WTO as an observer. Montenegro maintains missions in Brussels serving the EU and a few neighboring countries and in Washington, D.C., serving both the American Government and the United Nations. It belongs to NATO and expects to join the European Union soon. Plus, it already uses the euro as its currency.[43]

Governance faces the challenge of its identity issue and friction that has critically affected its history—highlighted in the referendums on whether to remain part of Serbia. In the census of 2003, only 270,000, or 43 percent of the nation's citizens, declared themselves as ethnic Montenegrins; 200,000, or 32 percent, preferred self-designation as Serbs. The others were Bosnian (8 percent), Albanian (5 percent), and the rest Muslim, Croatian, and Roma. All speak a dialect of the same language—the official one—that is variously designated as Serbian, Serb-Croatian, or Montenegrin, and is written in Montenegro in modified forms (either in the Cyrillic or the Latin alphabet). The split between the over 70 percent who are Orthodox and the less than 20 percent who are Muslim is critical; the 3 percent who are Roman Catholic live mainly along the coast and add to the volatility. Elizabeth Roberts, in her history of Montenegro, is optimistic: "one clear lesson . . . is the need to bridge the deep division at the heart of Montenegrin society. Prosperity has a way of healing divisions and here the European Union and international financial institutions can help." For the sake of Montenegro and the Balkans, let us hope her optimism is warranted.

---

43. Its neighbor, Croatia, in contrast, belongs to the EU but does not use the euro.

12. AFRICA: ALONG THE COASTS

"In the aftermath of [the capture of] Ceuta [on Africa's Mediterranean coast opposite Spain] . . . Henry the Navigator began to sponsor expeditions down the coast of Africa in search of slaves, gold, and spices. Year by year, headland by headland, Portuguese ships worked their way down . . . [to the] bulge of West Africa . . . Under [Henry's] direction, exploration, raiding, and trading went hand-in-hand with . . . mapping. "

—Roger Crowley

Map 12.1 African Micro-Countries

Source: World Map by M. Ruskin Co. LLC

SEARCHING FOR SPICES, TREASURE, AND PRESTER JOHN

North Africa has long been an integral part of the Mediterranean community. The Saharan Desert cut it off from the "Dark Continent." In the late 1400s, Europeans (initially the Portuguese) fashioned a path from the Atlantic Ocean to western Africa and the Indian Ocean to access eastern Africa. Africa aroused European curiosity and discovered that finding a route around it would open a way to the spices and riches as well as the legendary Christian patriarch, presbyter, and king: Prester John.

The Portuguese Prince Henry the Navigator, after distinguishing himself in the capture of Ceuta—located on the northwest African Mediterranean coast, which Spain still governs as part of the Cadiz province—organized voyages as far as Cape Verde on which Portuguese colonists settled in 1462. After the death of Prince Henry, his brother, King Afonso V assigned the responsibility for exploring the African coast to Fernão Gomes with the condition that "he explored 100 leagues further every year." Captains sailing under Gomes' direction reached the Ivory Coast, the Gold Coast, the Slave Coast, and the Bight of Benin alongside present-day Nigeria.

In 1471, one of Gomes' lieutenants, Fernando Póo, discovered an island, later named after him. In 1472, other captains discovered uninhabited islands that they named São Tomé and Príncipe. In either 1475 or 1476, another of Gomes' captains, Rui de Sequeira, reached a cape which he named for Saint Catherine. By the mid-1470s, the Portuguese were shipping Africans back to Madeira or Portugal as slaves.

As the demand for slaves in the Americas developed, the British, Spanish, French, and Dutch profited in trans-Atlantic slave trade, building factory-fortress-trading posts along the Atlantic coast. São Tomé and Príncipe, further down the Atlantic coast were occupied in 1471. The island of Fernando Póo, part of present-day Equatorial Guinea, was occupied by Portugal in the late 1400s but was later traded to Spain for land in the Americas, a trade that gave Spain a source of slaves for its American territories.

Arab sultans held general sway over much of the African Indian Ocean east coast long before Europeans rounded the Cape of Good Hope and explored this coast and further east. The French formally annexed the Seychelles in 1768, but the British navy seized the islands in 1799 during the Reign of Terror under the Directorate. The French took possession of Mayotte in 1843 and placed the sultans of the adjacent Comoros Islands under its protection in 1886. The French gained control of the Djibouti enclave between 1862 and 1900. The Berlin Conference of 1885 facilitated the division of Africa (along lines that generally ignored tribal and other considerations) among the several European countries including the United Kingdom, France, Germany, Italy, and Belgium.

The per capita GDPs of the African micro-countries range from $1,300 in Comoros, $2,300 in São Tomé and Príncipe, $2,700 in Djibouti, and $4,000 in Cape Verde to $19,000 in oil-rich and corrupt Equatorial Guinea, and $25,000 in tourist attractive and autocratic Seychelles. While Equatorial Guinea has an oil-dependent national budget exceeding $8 billion, the others have budgets ranging from $743 million (Cape Verde), $473 million (Djibouti), and $360 million (Seychelles) to $139 million (Comoros) and $152 (São Tomé and Príncipe). The Comoros, São Tomé and Príncipe, and the Seychelles are dependent on fishing and tourism as their major resources.

Africa has a maze of overlapping regional organizations. The largest is the African Union (AU), initially the Organization of African Unity (OAU), with all 54 African countries members, including the 6 African micro-countries. On Africa's east side, the Comoros and the Seychelles, along with Madagascar, Mauritius, and Réunion, belong to the Indian Ocean Commission (COI). Cabo Verde belongs to ECOWAS, a regional organization of 14 northwest African countries.

While African micro-countries have improved their economies, infrastructure deficits cripple them. Chinese companies are spending some $50 billion on new highways, seaports, airports, and other significant projects (and the military base at Djibouti) as a part of the Belt and Road (or One Belt, One Road) Initiative. The political as well as economic significance of these investments is presented forcefully in Howard French's book *China's Second Continent: How a Million Migrants are Building a New Empire in Africa.* Some African countries are viewed as becoming economic vassals of China.

Of the six African micro-countries' GDPpcs, four rank among the poorest in the world (except Equatorial Guinea with oil and the Seychelles with tourism). Improvement appears destined to remain forlorn. A Ugandan central bank executive said, "Corruption and institutional inefficiency is the biggest problem—extending throughout Africa. If we could alleviate this problem, the others ones would be easier to address."

REPUBLIC OF CABO VERDE (FORMALLY CAPE VERDE)

"[Cabo Verde is] a small nation, possessing limited resources, whose very survival depends on external aid."

—Deirdre Meintel

The archipelago of volcanic islands that constitute the Republic of Cabo Verde lies about 1,500 miles southwest of Gibraltar and less than 100 miles off the western coast of Africa. In 2013, the government announced that it would no longer be known as Cape Verde but as Cabo Verde. Its nearest neighbors are Mauritania, Senegal, Gambia, and Guinea-Bissau.

Diogo Gomes and António de Noli, Portuguese captains in the service of Prince Henry the Navigator, visited a few of these islands. In 1462, Portugal established a settlement at Ribeiro (now Cicada Vela) on the then uninhabited island of São Tiago, reputed to be the first European tropical settlement. The Portuguese settlers brought slaves from the nearby African coasts to work their plantations—and later for shipment across the Atlantic. Jews and convicts exiled from Portugal to the islands in the late 1400s provided a wave of Portuguese settlers. In Inquisition-related efforts, many Jewish children were separated from their parents, transported to Cape Verde, and converted. While statistics do not list any with a Jewish religious affiliation in Cabo Verde, the number of those with Jewish names signals this heritage.

Situated on the trade routes connecting Europe, Africa, the Americas (especially the Portuguese colony of Brazil), and, later, Asia, Cape Verde prospered by supplying ships serving the slave trade and smuggling. This prosperity attracted the attention of pirates and privateers (crown-licensed pirates). Dutch, French, and English (including Drake) attacked from the 1500s.

Beginning in the 1700s frequent droughts, deforestation, erosion of the land, volcanic disasters, and epidemic diseases that killed over 100,000 people impoverished the islands. Portugal neglected the situation. The late 1800s further crippled Cape Verde. Corruption and maladministration added to the islands' plight. The late 1800s witnessed emigration from these islands to São Tomé and Príncipe and other Portuguese colonies. Cohen has observed that, "in the decades after slavery ended in Cape Verde, . . . its own drought-stricken peasants, mostly former slaves and their descendants, were recruited, sometimes under pressure, as labor for the islands of São Tomé and Príncipe." Later, literate Cape Verdeans migrated to fill posts in other Portuguese colonies as teachers, missionaries, bureaucrats, and small businessmen.

The advent of steamships in the late 1800s provided some relief for Cape Verde's economy. The islands' strategic location on shipping lanes facilitated their use for supplying ships with water, coal, and food. American whaling ships also landed there, secured supplies, and recruited crews, some of whom used this opportunity as a means to migrate to the United States—principally to Boston where a Cape Verdean community remains and continues to attract Cabo Verdeans. Remittances from Cabo Verdean migrants are a source of income to stay-behind relatives.

British developed the harbor at Mindelo for the purpose of ship supply, and later placed a coaling and submarine cable station there. With the drastic decline of shipping in World War II, the Cape Verde economy collapsed. The British coal industry decline in the 1980s was an additional final blow to the Cape Verde economy. Portugal, like other European countries, considered their African colonies as opportunities for short-term economic exploitation. Under the military dictatorship of António Carmona in 1926, and then of António de Oliveira Salazar in 1932, Portuguese rule was especially oppressive. But, while Portugal treated the Cape Verdeans less than humanely, Cape Verde fared better than Portugal's other African colonies; for example, they were able to build a secondary school there. By the time of their independence, 25 percent of Cape Verdeans could read, compared to only 5 percent in neighboring Guinea-Bissau. Consequently, Cape Verdeans were recruited as skilled laborers and teachers for São Tomé and Príncipe and other Portuguese African colonies.

Portugal took few steps to introduce limited autonomy to Cabo Verde, including the creation of a partially elected council. As literate Cape Verdeans became aware of the postwar "winds of change" pressures for independence mounting throughout the colonial world, they increasingly articulated their resentment of the severity of Portuguese rule. In 1956, Amílcar Cabral and others founded the initially clandestine network of the

African Party for the Independence of Guinea and Cape Verde (PAIGC). While most non-Portuguese African colonies gained their independence from 1957 through 1964, the liberation movement struggled in Portuguese colonies, including in the overseas province of Portuguese Guinea, which comprised the mainland (now Guinea-Bissau) and the Cape Verde islands 100 miles away.

The liberation movement in several Portuguese colonies was expedited by a covert operation undertaken by the World Council of Churches (WCC) in 1961 (told to the author by Kimbell Jones, one of the participants):

> When the Portuguese dictator Salazar confiscated the visas of students from several Portuguese colonies studying in Lisbon—frustrating their chances to return home to participate in the liberation movement—the WCC organized a smuggling operation to remove the students from Portugal. Many of these students returned home to become active in independence movements: eight became prime ministers, including Pedro Pires of Cape Verde (as it was then [called]); others became cabinet ministers.

A coup in the mainland portion of Portuguese Guinea led to it declaring independence in 1973 and being granted independence in 1974. Abandoning hopes for preserving its unity with Guinea-Bissau, Cape Verdeans formed the African Party for the Independence of Cape Verde (PAICV). Wide unrest forced the Portuguese government, now headed by Marcelo Caetano, who succeeded Salazar when he died in 1968, to negotiate with the PAICV, leading to Cabo Verde gaining its independence in 1975. The PAICV established a one-party state, elected Aristides Pereira president, and ruled until 1990. Responding to pressure PAICV called an emergency congress to discuss constitutional changes. The opposition formed the Movement for Democracy (MpD), which abolished the one-party state and won the first elections. Its candidate, António Mascarenhas Monteiro, convincingly defeated Pereira for president in 1991. In 1995, MpD increased its legislative majority and, in 1996, returned President Monteiro to office. Pedro Pires won close elections in 2001 and 2006. Jorge Carlos Fonseca won as president in 2011 and 2016.

The economy in Cabo Verde is deplorable. The low per capita GDP of $4,200 stems from its harsh heritage, lack of resources, and bleak eroded landscape. Its agriculture has suffered from a severely unequal land distribution, a series of droughts since 1968, and erosion that reduced the tillable land. The economy was primarily dependent on subsistence farming and fishing. The government has focused attention on improving the infrastructure and alleviating the basic needs of the population. This was undertaken by marketing the food and other goods provided through international aid and using the proceeds to fund development programs, followed by a series of five-year development plans, which included land reform plans to increase rural employment. Yet, 40 years after independence, subsistence farming (including livestock, sugarcane, coffee, bananas) and fishing remain the basic employment and underemployment.

Cabo Verde's colonial heritage has affected not only its ethnic composition and its language but also its cultural identity. *Mulattoes* (*mestiços* in Portuguese) comprise 71 percent of the population, Africans 28 percent, and Europeans 1 percent. The official language is Portuguese; widely spoken is a Creole-Portuguese mix with West African vernaculars. But, remarkably, many Cabo Verdeans feel they relate more in terms of values and lifestyles to those in Portugal than to those in other African countries.

Cabo Verde belongs to the African Union (AU) and joined the 15-member Economic Community of West Africa (ECOWAS) in 1999. The country maintains a mission in Brussels to the EU, Belgium, and several nearby countries—and in Washington, D.C., accredited not only to the United States but also the United Nations.

Foreign assistance continues to help develop industry such as food processing, shoe and garment making, and ship repair. Tourism development has had limited success. Cabo Verde's $902 million in imports are principally from Portugal and the Netherlands; its $184 million in exports go mainly to Spain. The wide import-export gap challenges Cabo Verde.

REPUBLIC OF EQUATORIAL GUINEA

"Abundant oil resources and immature governing institutions have cursed Equatorial Guinea with a monolithic economy, endemic corruption, pernicious class inequality, factional dissension, and tyrannical repression."

—Michael Kitti

The Republic of Equatorial Guinea, the former Portuguese Guinea, has a turbulent history of exploitation and terror. The country consists of two distinctly different parts: a group of islands and a mainland much larger in area. Most of the country's population lives on two islands, Bioko (once Fernando Póo) and smaller Annobón; they lie off the West African coast, west of Cameroon and south of Nigeria. Area-wise, the much larger coastal mainland, Río Muni, includes three small estuary islands. The populous main islands and geographically larger mainland are significantly different in history, culture, ethnicity, and economic development.

Bubi and Fang (Bantu tribes) migrants from the neighboring mainland were the original inhabitants of the offshore islands. Portuguese and Spanish colonization brought not only a few Europeans but also slaves freed from slave ships captured by the British as well as former slaves from Sierra Leone and Jamaica. Portuguese Guinea, before the discovery of oil, consisted mainly of a scattering of trading posts on estuaries of four rivers. The country is 85 percent populated by the Bubi peoples and 7 percent by the Fang. Many are mixed-race descendants of the Bubi and Fang peoples with Europeans (called "Fernandinos" after the initial name for the capital). Equatorial Guinea is the only country in Africa where Spanish is widely spoken. Spanish and French are the official languages; Bubi and Fang are also spoken.

The checkered colonial history of what is today Equatorial Guinea began with a landing, probably in 1472, by the Portuguese explorer Fernando Póo on the island once named after him. This landing, and a subsequent one led by Ruy de Sequeira that landed on Annobón in circa 1474, were among those sponsored by the Portuguese Prince Henry the Navigator.

When the Pope awarded what is now Brazil to the Portuguese in 1778, the Spanish were granted the islands and the nearby coastal enclave, which they called Spanish Guinea, in an arrangement trade that gave Spain its own source of slaves.

A late 1770s epidemic of yellow fever forced the shutdown of the Spanish station on the islands. The British abolition of the slave trade in 1807 led to their leasing of the island of Fernando Póo (not until later was it renamed Bioko) as a base for suppressing the slave trade and assuming the administration of the island. In 1843, when Portugal rejected the British offer to purchase Fernando Póo, the British withdrew their base to Sierra Leone.

The British withdrawal led Spain to begin efforts to reoccupy the islands and colonize Fernando Póo. After Spain lost Cuba and the Philippines in the Spanish-American War, Spanish Guinea remained its only significant overseas colony. Starting in 1926, Spain developed cacao and coffee plantations and logging concessions designed to enhance and complement the interests of the Spanish economy. Since the native Bubi people refused to work the plantations and logging concessions, workers were recruited from Nigeria, Liberia, and Cameroon.

In 1959, postwar national and international (including UN) pressures for self-determination led Spain to begin partial decolonization. Following the French practice of extending its mainland system of internal governance to its overseas possessions, Spain changed the jurisdictional status of Spanish Guinea from that of a territory into two overseas provinces: mainland Río Muni with Bata as the capital and the island province of Fernando Póo with its capital of Santa Isabel (since then renamed Malabo). At the same time the citizens, including the Africans, were granted the same rights as those enjoyed by Spanish citizens. A 1963 plebiscite approved a measure granting the country economic and administrative autonomy and changing the country's name to the Republic of Equatorial Guinea. Continuing pressure for full independence and support of the Organization for African Unity led Spain to grant autonomy and independence in October 1968, followed by admittance to the United Nations the same year.

A turbulent post-colonial history followed, dominated first by the tyrannical terror of its president for life, Francisco Macías Nguema (also known as Masie Nguema Biyogo Ñegue Ndong). The closed and repressive nature of the government allowed limited outside attention. Riots in 1969 generated by long-term economic development disparities between relatively developed islands, and the significantly less developed mainland Río Muni highlighted the start of the newly independent country.

Riots provided impetus for the newly elected Francisco Macías Nguema to assert absolute control. With the assistance of a youth organization and the National Guard, he used confiscation of private property, intimidation, and murder to ensure that he was not challenged. The extent of terror drove many Equatorial Guineans to flee the country. The government banned all fishing as a means of preventing flight from the country.

In 1979, Teodoro Obiang Nguema Mbasogo led a bloody coup in which the president, his uncle, was assassinated. Obiang installed himself as president and has won seriously flawed multi-party elections in 1996, 2002, 2009, and 2016. He has liquidated many of his predecessor's supporters, including many of his own relatives, and placed his loyal family members in the key government posts. That he has continued the government's practice of terror was confirmed by a 2012 Human Rights Watch report that reaffirmed that the Obiang regime regularly tortures dissidents.

In 1985, Equatorial Guinea adopted the French-backed CFU currency and aligned itself with former French colonies in the Organisation internationale de la Francophonie, the French equivalent of the (British) Commonwealth. Nominally, multi-party elections held in 1996, 2002, 2009, and 2013 have aroused considerable protest.

The economy once relied on timber, coffee, and cocoa. The discovery of oil enabled a limited investment in roads, and a rapid rise in the GDP to almost $25 billion enabled an increaser of GDP per person of $32,300. But most of Equatorial Guinea's people live on less than $2 per day. Wealth is concentrated on President Obiang, the most influential of his 42 recognized children, and cronies. Government spending on education and health lags far behind the sub-Saharan average. Declining oil production, coupled with lower prices, has led Equatorial Guinea's economy to shrivel since 2013.

Its imports (worth $6.4 billion) are mainly from the United States, Spain, and China; its exports (worth $5.4 billion) are primarily to China, Japan, and the United States. Poverty remains widespread. The national budget is $6.3 billion. The country is considered by Transparency International to be the 14th most corrupt state in a world. Following flawed multi-party elections in 2004, an attempted coup failed, despite alleged help by the American CIA, British MI6, and Spanish government.

In 2011, a new constitution merged the offices of president and prime minister, combined the legislative houses, and stipulated a two seven-year term limit for the president. President Obiang has stipulated that the two-term provision is not retroactive, meaning the term limit would not apply to him until 2030. Obiang won again in the 2016 election when Teodoro Nguema Obiang Mangue, a son of the president—who, while aspiring to succeed his father, posts photos of himself luxuriating as a playboy—was pronounced vice president.

The country is divided into seven local government districts: Bioko Island consisting of two districts, Annobón of one, and mainland Río Muni of four (each one larger in area, much smaller in population, and with a less advantaged economy). The country participates in the United Nations affiliates FAO, IBRO, ILO, IMF, and WHO and observes at WTO. It belongs to the African Union (AU).

The presence of oil and gas has tripled, cursing Equatorial Guinea. It has not developed other industries. It has supported autocracy, nepotism, kleptomania, monopolization of wealth, and lack of investment in infrastructure. The country has been left unprepared for the year when the oil and gas resources are inevitably exhausted.

THE DEMOCRATIC REPUBLIC OF SÃO TOMÉ AND PRÍNCIPE

"During Portugal's half-millennium of rule over the islands, a very little amount of infrastructure was built . . . beyond what was necessary to service the colonial administration and the Portuguese business interests."

—L. M. Denny and Donald I. Ray

The Democratic Republic of São Tomé and Príncipe lies off the Atlantic coast of Africa, west of Gabon and south of Equatorial Guinea. The mountainous country shares a common heritage with Cabo Verde and Equatorial Guinea of Portuguese imperialism and slavery. But, in São Tomé and Príncipe, the impact of Portugal's dictatorial colonial rule was even more degrading and demoralizing, and the efforts of the initial post-colonial, one-party Marxist leadership was less constructive and more corrupt. Subsequent governance has done little better.

The islands of São Tomé and Príncipe were uninhabited before the 1471 arrival of the Portuguese explorers. The first settlers arrived in 1486. In 1493–94, political exiles and 2,000 Jewish children were forcibly converted to Christianity after being taken from their families who were settled there. Settlement on Príncipe began in 1500. The labor-intensive cultivation of sugarcane, for which the islands volcanic soil was well suited, required the importation of slaves. The Portuguese imported slaves from the Guinea coast, Angola, and other parts of the nearby African mainland not only to use locally but also to re-export to the Americas. Their descendants constitute most of the population. The few of European descent are mainly Portuguese. In 1522–23, Portugal's government took over the direct administration of the islands from those who had been given land grants.

In the mid-1600s, the Netherlands, engaged in a protracted revolt against Spain that had included Portugal, attacked and, for a few decades, controlled Brazil and São Tomé, which became an intermediate, a supply stop, and transshipment station for slaves for Spain-controlled Brazil and the Spanish Main at that time.

By the mid-1700s, São Tomé and Príncipe became Africa's leading exporter of sugar. Over the next century, the growth of sugar production created a demand for labor for the sugar-producing plantations. Captives from the Guinea coast were imported as slave labor. Later, São Tomé and Príncipe developed as a deportation station for the slave trade from Africa to the Americas. In the 1800s, coffee and cocoa were successively and successfully introduced. Extensive plantations (*roças*), owned by Portuguese companies and absentee landlords, soon occupied almost all the arable farmland. By the early 1900s, the country became the world's largest producer of cocoa, which remains its primary crop.

The harshly managed *roças* supported an abusive slave system. Even after Portugal officially abolished slavery in 1876, the plantations used forced and maltreated laborers, many of whom were recruited and smuggled from other West African countries such as Cape Verde and Angola where work opportunities were minimal. The labor was obtained, Lord Hailey, whose book *African Survey* remains highly regarded, has written, by methods not very different from those used by slave traders earlier. Denny and Ray have described their work as "the continuation of slavery under the name of contract labor . . . The British (and Quaker) company Cadbury's found itself implicated in the forced labor practices . . . and led an international boycott of cocoa from São Tomé and Príncipe in 1909." That action brought an end to the islands' role as the largest producer of cocoa. Labor unrest continued in the 1900s leading to riots in 1953 in which hundreds of African laborers were killed in clashes with the Portuguese government.

In the late 1950s, as other colonies across Africa were agitating for independence, a small group founded the Committee for the Liberation of São Tomé and Príncipe (CLSTP), which initially established its base in Ghana, and then in Brazzaville, Congo—then in Equatorial Guinea and in Libreville in neighboring Gabon. Later, the group changed its name to the Movement for the Liberation of São Tomé and Príncipe (MLSTP).

São Tomé and Príncipe ceased to be a colony, technically, in 1951 when it was formally granted the status of an overseas province. In 1973, Marcelo Caetano, who had succeeded the long-time dictator António de Oliveira Salazar, instituted cosmetic reforms by which the islands received limited autonomy—one headed by an appointed governor with a legislative assembly composed of six homeland Portuguese and nine elected islanders.

The Cold War impacted the independence movements in many African colonies by introducing an element of the communist revolutionary message to spur national liberation movements maximizing opportunities for Marxists to introduce people to Communist ideas. However, their hopes were stymied by the collapse of the economy and the lack of São Toméans with the skills and the political and managerial experience and lack of education of most MLSTP members. The fact that most Portuguese as well as São Tomé and Príncipe people of Cape Verdean origin left São Tomé after independence and that little effort had been expended in educating São Toméans contributed to the situation. The collapse of the Soviet Union and fading of the Cold War reduced opportunities for São Tomé and Príncipe (and many other third world countries) to secure grants, which further frustrated efforts to improve the governance, the economy, and the standard of living.

The overthrow in 1974 of Portugal's right wing was led by the successive regimes of Salazar and Caetano. Unlike its predecessor, the new regime did not want to retain the colonies, met with the MLSTP in Algiers, and agreed upon a plan for the transfer of sovereignty. After a period of "transitional" government, led by a Portuguese-appointed governor and a half-elected council, São Tomé and Príncipe achieved independence in July 1975. An elected constituent assembly drew up the independence constitution and then transformed itself into the National People's Assembly. All were members of the MLSTP.

Manuel Pinto da Costa, the MLSTP secretary-general, became the first president, heading the government of this ideologically Marxist one-party country for 15 years. The movement developed grass-roots organizations in the tradition of democratic centralism. Agitation led to constitutional revision in 1990—providing for a president elected by universal suffrage for five-year terms with a limit of two terms, a prime minister selected by the president, a cabinet chosen by the prime minister, a 55-member legislature elected for four-year terms, an independent Supreme Court, and legalization of a multi-party system.

Subsequent elections demonstrated the vitality and viciousness of the multi-party system. In the 1991 elections Miguel Trovoada, a former prime minister, in exile since 1986, returned and ran successfully as an independent candidate for president while the new Party of Democratic Convergence (PDC) won a majority of seats in the National Assembly. In 1994 legislative elections, the MLSTP gained a plurality of seats in the Assembly; in 1998 the MLSTP won a majority. Trovoada was reelected president in 1996. A candidate of the Independent Democratic Action Party, cocoa exporter Fradique de Menezes, won in the 2001 presidential elections. For the next four years, a series of short-lived minority governments succeeded one another. In 2003, the army seized power for one week, complaining of corruption and the prejudiced division of projected oil revenue. In 2006, a Menezes-supportive coalition gained sufficient seats to form a government; in 2011, Manuel Pinto da Costa, the former president, was elected president. In the 2014 presidential run-off, da Costa dropped out alleging fraud in the first round; Evaristo Carvalho of the ADI was elected president.

At the time of São Tomé and Príncipe's independence, Portuguese-owned plantations occupied 90 percent of cultivated land. After independence, these plantations came under the control of various state enterprises. Today, the main crop, accounting for 95 percent of export revenue, remains cocoa. Other crops include copra, palm kernels, and coffee. Other economic activities include fishing and producing textiles, soap, and beer. Of the other 41 micro-countries, only the Comoros has a lower GDP per person. Efforts are underway, with grant support, to promote tourism. Despite heavily assisted foreign aid, the GDP per person of this 189,000-citizen country is only $2,400.

THE UNION OF THE COMOROS

"The process of decolonization in the Comoros . . . has to be considered a failure, in both political and economic terms. The very fact of smallness has produced a situation of considerable political and economic dependence on the former colonial power, France."

—Robin Cohen

The Union of the Comoros comprises three islands, Grande Comore, Anjouan, and Mohéli (also known as Mwali), located in the channel between Madagascar, Tanzania, and Mozambique. A fourth Comoro Island, Mayotte, has remained separate from the Union. The Comoros' 800,000 population includes many whose ancestors have migrated to its shores over many centuries. No later than the 6th-century Bantu-speaking Africans from the east coast of Africa, Austronesians from the Malay Archipelago and Madagascar, and Arabs from the Arabian Peninsula and the Persian Gulf settled in the Comoro Islands. This early influx, with the later settling of a few Europeans and large-scale importation of African slaves, generated the present demography of the islands. Proximity to Arab-populated Madagascar and what is now Mozambique led to the immigration of Arabs, the introduction of Islam and Arab culture, the rule of Arab sultans and Malagasy kings, and the trafficking of Arab ships between Africa, India, and the Indies, trading ivory, gold, and slaves.

Portuguese navigators, after rounding the Cape of Good Hope, first visited the Comoros Archipelago in 1505. Throughout the 1500s, the Portuguese used the islands to supply their fort in neighboring Mozambique. The collapse of the East African sultanates led to a powerful Oman (on the Arabian Peninsula) sultan extending his rule to the Comoros. His successor increased Omani Arab influence in the islands and moved the court of his independent sultanate to Zanzibar.

The Comoro Islands long continued serving as a way station for merchants sailing from Europe to India and on to the Far East until the opening of the Suez Canal provided a shortcut. In 1886, the French secured the consent of the then ruling sultan of Mohéli to place his island under

French protection. In the same year, the sultan, ruling part of the Grande Comore Island, fraudulently agreed to place the whole island under French protection in return for French support of his claim to the whole island. Later, the French obtained its cession from the Malagasy king of Mozambique and established a settlement and a colonial government. Sugar plantations were developed and slaves were imported to work them; the other islands followed suit; by 1865, it is estimated that 40 percent of the islanders were of African descent. In 1909, the sultan of Anjouan, the third of the islands now comprising the Union of the Comoros, abdicated in favor of the French. In 1908, the three protectorates had been placed under the Madagascar colonial governor, and they—along with nearby Mayotte— were administratively unified with Madagascar.

In 1973, France reached a tentative agreement with the four islands to gain independence in 1978. In the referenda that followed, three Comoro Islands voted for the proposed union and independence. But Mayotte, the more developed and prosperous fourth island, voted overwhelmingly against joining with the other three.[44]

Following independence, the Comoros endured 20 attempted and successful coups that facilitated a rapid turnover of dictatorial presidents, which punctuated the politically tumultuous decades that followed independence. In 1975, President Ahmed Abdallah was removed in a coup. His replacement was ousted in 1976 by Minister of Defense Ali Soilih who withstood seven coup attempts and initiated socialist and isolationist programs that strained relations with France. French, Rhodesian, and South African troops overthrew Soilih and reinstated Abdallah, who, in 1989, was killed in his office.[45] Said Mohamed Djohar, Soilih's older half-brother, became president. In 1995, Denard returned to attempt another coup; France intervened, forced surrender, removed Djohar, and backed Mohamed Taki Abdoulkarim in his successful presidential election. He died in 1998 and was succeeded by interim President Tadjidine Ben Said

---

44. Nevertheless, the Union of the Comoros has continued to claim sovereignty over all four islands. Mayotte, in 2011, became a French overseas *departement*. Today, of its population of 256,500, more than half are Africans who have illegally migrated into what is now an integral part of France. Officials are rounding up suspected illegal immigrants; they have become scapegoats for a chaotic and rapidly deteriorating situation.
45. Allegedly, the assassination was a plot instigated by Bob Denard, whom French paratroopers had evacuated to South Africa.

Massounde. The next year, Colonel Azali Assoumani, the army chief of staff, seized power.

Concerned by the by the coup and the instability it represented, the Organization of African Unity (the predecessor of the African Union), under the leadership of South Africa's president, Thabo Mbeki, negotiated a settlement. A new system of governance gave each island its own "president," ministers, and legislature and weakened the common Union government. Subsequently, international pressure prompted constitutional changes that enabled new elections. Ahmed Abdallah Mohamed Sambi, a Sunni cleric, won the 2006 elections. A two-day revolt, led by Mohamed Boca in Anjouan, was suppressed by the Union government in 2007. In 2010, former Vice President Ikililou Dhoinine won the presidency.

In an effort to cut down the complicated, cumbersome bureaucracy and swollen budget (80 percent of which is expended on bloated payrolls), a 2009 constitutional change reduced the autonomy of the three islands, converting the three "presidents" and their "ministers" to governors and councilors (department heads), respectively, with the presidency of the Union rotating among the islands. Since President Dhoinine was from Mohéli, the presidents from Grand Comore and Anjouan were expected to follow with five-year terms. President Azali Assoumani was elected in 2016.

A 2018 referendum to amend the constitution extended the presidential limit from one to two five-year terms and eliminated the rotating presidency, thus extending President Assoumani's rule. The opposition regards the referendum as a power grab. One of Assoumani's two vice presidents has turned against him, writing a letter signed by the governors of Grande Comore and Anjouan saying that the referendum risks plunging the country into chaos. The African Union tried to mediate, but President Assoumani was stubborn. To develop foreign support, he sought out Russia and Saudi Arabia. In late 2018, the Comoros Parliament refused to consider a bill to allow him to rule by decree. Following an hour-plus speech threatening to rule without Parliament, he gave in.

The succession of African, Arab, Portuguese, and French influences on the Comoros is reflected unevenly in its ethnicity, languages, religions, and legal system. Most people remain conscious of a tribal affiliation. Islam is the faith of 98 percent of its people. French and Arabic are the official languages; Shikomori (a blend of Swahili and Arabic) is widely spoken. The Comoros' legal system combines Islamic law, the French legal (Napoleonic) code, and customary law, generally administered by village elders (*kadis*) and civilian courts. A Supreme Court oversees the system and is the final court of appeal. In addition to the UN, the Comoros belongs to the African Union (the successor of the Organization of African Unity [OAU]), the Arab League, and the Indian Ocean Commission—linking the Comoros, the Seychelles, Mauritius, and Madagascar—as well as several dependencies and a few intergovernmental organizations.

The Union of the Comoros, whose mercantile trade resembles that of Cabo Verde and São Tomé and Príncipe, has managed to achieve only a precarious, poverty-stricken, and dependent quasi-independence. The Comoros remains heavily dependent upon France, whose colony it last was. The Comoros' GDP per person of $1,300 is the lowest of any of the micro-countries. The public service payroll remains bloated. Subsistence farming, fishing, and forestry constitute 80 percent of the labor force. Inadequate infrastructure limits the fledgling $30 million tourist industry. The GDP, raised by these activities and by aid in the 2010s was only $911 million. The mid-2010s budget was only $167 million. Imports of $208 million from mainly Pakistan, France, and the UAE far exceeded exports of only $20 million (over 58 percent to the Netherlands). Whatever level of productive activity once existed—aside from the maritime traffic, slavery trade, and military garrisons—was typically dependent on a plantation economy worked by imported slave labor. The end of the maritime traffic, slavery, and garrisons has left the Comoro Islands unable to support their increased population without substantial assistance.

Emigration once alleviated the Comoros' rapid population growth, but tightening immigration restrictions elsewhere have limited emigration opportunities. The Comoros, Newitt noted in 1984, is "not alone in being small, desperately poor, and only nominally independent. The problem of combining security and development with a genuine degree of independence may, in the end, prove to be without solution, but the only chance of achieving it will be through cooperation with other states of like size." Thirty years of experience confirms Newitt's cautious assessment.

THE REPUBLIC OF SEYCHELLES

"Unfortunately, the best alternative for the Seychelles might well be the tenuous kind of arrangement it has at present, with foreign influence being confined largely to behind-the-scenes maneuvers that do nothing more than contribute to the mystery and romance and ultimately to the despoliation of this most beautiful of places."

—Marcus Franda

The Republic of Seychelles lies off Africa's Indian Ocean coast, east of Kenya, northeast of the Comoros and Madagascar, and north of Mauritius. The first European sighting of the Seychelles, like that of so many other countries on and off the African coasts, was by Portuguese explorers, including Vasco da Gama in 1502. The Seychelles' strategic location on the sea lanes between Africa, Arabia, and India (before and after the completion of the Suez Canal in 1869), has significantly affected the mixed demography of its people, which includes African, French, Indian, Chinese, and Arab. Seychellois Creole, French, and English are the country's official languages.

Most of the Seychelles' inhabitants are descendants of migrants (both forced and voluntary) who have fused a multi-racial mixture, blending Africans, Arabs, Indians, Chinese, and Europeans, creating a vibrant Seychelles creole ethnic mix and culture. Marcus Franda reports, "To a greater extent than is true in almost any other part of the world . . . slaves or the children of slaves in the Seychelles have intermarried with the non-slave populations, producing a variety of racial types." A few old white settlers (*grands blancs*) continue to speak French because of its snob appeal. Though most speak a Seychelles creole which has 18th-century French roots mixed with Kafir, Malagassy, Bantu, and Indian elements. The predominant faith (82 percent) is Roman Catholicism. The Seychelles' heritage of conflict and coups has not inhibited its post-independence drive to develop its economy, principally by successfully promoting its tourist industry. The Seychelles, comprising of 115 islands, is the least populous of the African micro-countries with 90,846 residents and the most prosperous with a per capita GDP of $9,000.

A few Dutch settled on some apparently uninhabited Seychelles islands in the 1600s. In 1609, an English East India Company (EIC) ship is recorded as landing in one of the islands during its fourth voyage to the East. The Seychelles' location on trade routes attracted pirates, using the islands as bases in the 1600s and early 1700s. In 1756, a French naval captain claimed the Seychelles for France. In 1768, the French intendant (i.e., governor) of Mauritius promoted settlements on the Seychelles islands with the intent of establishing spice plantations.

The outset of the French Revolution led Seychelles' settlers to attempt to secure independence for their islands from France and, consequently, from subordination to the French colony of Mauritius. The revolt was promptly suppressed, first by the Seychelles' French commandant, Jean Baptiste Philogène de Malavois, who then introduced direct rule from Paris and began an orderly system of land tenure. He was succeeded by Jean-Baptiste Quéau de Quinssey who, during the Napoleonic Wars, to avoid an attack, adopted an interesting strategy. Raphael Kaplinsky reports, "When English ships entered the harbor, they were met by Mr. Quincey running up the Union Jack . . . when French ships came, they were met by de Quinssey in French uniform, running up the Tricolour."

The British fought to control Mauritius and the Seychelles because of the islands' strategic location between India and the Cape of Good Hope—and because French corsairs and pirates menaced British East India Company vessels returning from India. Finally, de Quinssey (spelled Quincey when he was dealing with the British), faced with a British squadron of five warships, surrendered to the British. British possession of the Seychelles was confirmed following the war.

Quincey built a thriving shipbuilding and shipping industry and a diversified plantation system. African labor importation continued even after the British prohibited slavery in 1835. Traditional fishing, turtling, and growing of local crops and livestock supplemented the economy. The cutting of timber, principally for shipbuilding, and growth of plantations stripped the Seychelles of their original woodland. Until he died in 1827,

Quincey remained the Seychelles' administrator, subordinate to the Mauritius governor.

In 1898, Britain appointed a governor responsible for separating the Seychelles from Mauritius and, in 1899, established an appointed executive and a legislative council that included only a few Seychellois citizens. The Seychelles became a crown colony separate from Mauritius in 1903. Elected representation was introduced in 1948. On the basis of literacy and property qualifications, about 2,000 local people became eligible to vote. With this limited franchise, the Seychelles Taxpayers and Producers' Association (STPA) easily won all 4 elected seats on the 12-member council and won all 12 seats later when all were elected.

In 1963, two major political parties replaced the STPA: the Seychelles Democratic Party (SDP) led by James Mancham and the Seychelles People's United Party (SPUP) led by France-Albert René. In the 1967 elections, the first following the introduction of universal adult suffrage, the Mancham-led SDP won four of eight seats, René's SPUP won three, and an independent won one. Both parties pressed for a new constitution, which was granted in 1970. In the elections that year, the Mancham-led SDP won 10 seats while René's SPUP won five; Mancham became the first chief minister of the Seychelles. The two party leaders differed dramatically in their orientations. Mancham was comfortable with capitalism, the affluent, and being a playboy. Marcus Franda writes, "René emphasized immediate independence, 'socialism,' and other 'progressive' measures . . . to deal with problems of poverty and British neglect."

Independence was granted in 1976 as a republic within the Commonwealth. Mancham defeated René decisively in the 1976 election, but brought his defeated rival into the government as prime minister. A year later, Mancham was ejected in a coup led by René's SPUP, renamed the Seychelles People's Progressive Front (SPPF). President René ruled as a dictator in the one-party communist state that he established. In 1981, Mike Hoare led a team of South African mercenaries, masquerading as holiday rugby players, in a coup attempt. After a gun battle, many participants escaped via a hijacked Air India plane. More coups failed in the 1980s. In 1986, an attempted coup by the Seychelles' minister of defense, Ogilvy Berlouis, prompted President René to ask India for assistance, which was granted via an Indian naval vessel arriving in the Port Victoria harbor. In

1993, President René was forced to introduce a multi-party system. In 2004, he resigned to be succeeded as president by his vice president, James Michel, who won the 2006 election and—after changing the name of the SPPF to the People's Party (PP)—was reelected in 2011 and 2016.

During the pre-independence era, subsistence-level agriculture employed about one-third of the work force. The principal exports were cinnamon, vanilla, and copra. Another one-fifth of the work force worked in the public sector. Now, agriculture (e.g., sweet potatoes, vanilla, coconuts, and cinnamon) employs only 3 percent of the labor force. GDP per capita has increased to $9,000—significantly more prosperous than other small African countries. The economy relies principally on fishing and tourism. To protect fishing's future, the country made a deal with The Nature Conservatory, an American NGO, which has promised to protect its fishing resource by closing off half of its offshore waters to fishing. The Indian Ocean Tracking Station, built in 1963 by the United States, was discontinued in 1996 when the Seychelles government demanded the rent increase to $10 million. The Seychelles belongs to the United Nations (and the FAO, IBRD, ILO, IMF, and WHO), the African Union (AU), and the Indian Ocean Commission (IOC).

The opening of the Seychelles International Airport in 1971 and the advent of cheaper air travel have facilitated the growth of the tourism. The beach resorts now employ about 30 percent of the work force. The socialist government has developed a pervasive presence in the economic sector with public enterprises active in petroleum product distribution, banking, import of basic products, and telecommunications as well as small-scale processing of frozen and canned fish, copra, and vanilla for export.

THE REPUBLIC OF DJIBOUTI

"Apparently some of them imagine that after they have chosen separation and become in theory a sovereign state, France would continue to contribute to their expenditures and other needs and would use her troops to prevent invasions by neighboring countries. It is necessary to dispel this dangerous illusion. France will certainly not employ her means or men to uphold a pointless façade of statehood which would not be practically viable for many reasons."

—President Charles de Gaulle

Djibouti lies strategically facing the Gulf of Aden, which connects with the Red Sea via the strait of Bab-el-Mendeb (only 16 miles wide at its narrowest width), the strategic link between the Mediterranean and the Indian Ocean, linking Europe and Asia. Northwest of Djibouti, along the Red Sea coast, lies authoritarian, war-ravaged, and destitute Eritrea, which, after a 30-year guerilla struggle, gained its independence from Ethiopia in 1993, continued fighting with Ethiopia until 2018, and invaded Djibouti in 2008. Southwest, along the Gulf of Aden coast, is Somaliland, which in 1991 declared its independence from Somalia and extends further southeast along the coast.[46] To Djibouti's west is land-locked Ethiopia, whose recent double-digit growth economy depends upon Djibouti's port for access to the sea. Across the strait in the southwest corner of the Arabian Peninsula lies Yemen—long engaged in a civil war supported by a Saudi Arabian-led alliance backed with arms supplied by the United States.

In the 800s CE, frequent Arab contact with the indigenous Somali and Afar ethnic groups led to the embrace of Islam and the advent of Arab sultans. The sultanate of Ifat, founded in 1285, had consolidated the area by the 1300s.[47] From 1862 to 1894, Somali and Afar sultans ruled what

---

46. The breakaway government has little credibility and has not been recognized as independent by the United Nations yet displays the trappings of independence—including a flag, a currency, and a few diplomatic missions. It depends on a force of foreign soldiers under a joint United Nations and African Union mandate to keep order.

47. Around the same time, Yekuno Amlak united the monastery-influenced lands in the neighboring Ethiopian highlands.

is now Djibouti. This heritage has led to the country having two official languages: French and Arabic. But most Djiboutians speak Somali or Afar. Djiboutians are 94 percent Muslim; only 6 percent are Christian.

The 1859–69 construction of the Suez Canal transformed the route from Europe to Asia into one of the most strategic and busiest in the world, increasing the interest of Britain, Germany, Italy, and Russia. The Canal led France from 1883 to 1887 to sign several treaties with the local sultans, ceding French control of the area. In 1884, a 24-year-old French officer, Léonce Lagarde, established permanent French administration in Djibouti and named the territory French Somaliland. A governor ruled French Somaliland autocratically until 1945 when a 20-member council was created; 10 were French and 10 were natives. Six from each group were selected by local ethnic electoral groupings to join the council. In 1950, the council was enlarged to 25 members, with 13 seats reserved for natives (elected by 10 ethnic-based electoral groups).

In 1958, with neighboring country Somalia scheduled to gain independence in 1960, Djibouti held a referendum on whether to join Somalia. While the Issas (the Somali local ethnic branch) voted strongly for joining with Somalia, resident Europeans joined with the Afar clans (Ethiopian and Eritrean related ethnic groups) in voting to retain their ties with France. Amid charges of French vote-rigging, the majority voted to remain with France.

A second referendum in 1967 held to determine the future status of French Somaliland was again marked by balloting manipulation and voting along ethnic lines. The Somalis' vote against independence was motivated by the desire to merge with now independent Somaliland. The results supported a continuing but looser relationship with France. Following the referendum, the enclave was renamed "The French Territory of the Afars and the Issas." In a third referendum in 1977, 98.8 percent of the electorate voted for separation from France, which officially sanctioned the independence of Djibouti and the naming of the country as the Republic of Djibouti. A Somali, Hassan Gouled Aptidon, who led the campaign for independence, became the country's first president, serving until 1999. Immediately, Djibouti joined the United Nations, the Organisation internationale de la Francophonie, the OAU (now the African Union), and the Arab League.

The dominant People's Rally for Progress (RPP) has controlled the legislature and executive since its founding in 1979. Early 1990s tensions led to armed conflict between Djibouti's ruling RPP and its opposition: the Front for the Restoration of Unity and Democracy (FRUD). The clash ended with a power-sharing agreement between RPP and a moderate faction of FRUD. Ismaïl Omar Guelleh, the pre-eminent person in Djiboutian politics, became president. Two FRUD members joined the cabinet.

Citing government control of the media and repression the major opposition party boycotted the 2005 election, which RPP and Guelleh then won with 100 percent of the vote. RPP also won the 2008 legislative elections. In 2011, Guelleh won a third term with 80 percent of the vote and a fourth in 2016. In the 2013 elections, the RPP—as a part of a larger coalition—won all the legislative seats. The party and government are dominated by the Somali Issa Dir clan—not the Isse Du clan of the Somalis—with the support of related Somali clans. In a presidential system like that of France, the president shares executive power with his appointed prime minister and presides over a cabinet responsible to the 65-member National Assembly elected for five-year terms.

Despite its lack of resources, Djibouti has maintained an annual growth rate of over 4 percent, essentially due to its strategic port location between Africa and Asia. The port, now connected with a railway to Addis Ababa, serves as the trade access to the world for its over 100 million land-locked Ethiopians that Eritrea relies on Djibouti for 95 percent of their trade.

Even more importantly, several countries maintain military bases on Djibouti. These include America's only permanent military base in Africa (leased for $60 million per year), France's largest military presence abroad, and Japan's only foreign base anywhere. Additionally, Spanish and German soldiers from the EU's anti-piracy force, Saudi Arabia, and China have stakes in building a base there. The location is useful for countries not only to protect shipping against pirates but also as a strategic base for African-Asian operations. Djibouti, in July 2018, opened the first phase of its new free-trade zone.

Chinese companies with Chinese-loaned funds indicated China's strategy of doling out aid throughout Africa—not only for constructing railroads, schools, stadiums, and cultural programs including scholarships to study in China, but also for opening a military base. China has boosted

Djibouti's desire to be more than a superpowers' playground by its investments. With China's help, railway has been built from Djibouti to Addis Ababa, Ethiopia's capital, which enables what once required a three-day truck trip to be transported in 12 hours by rail—significantly enhancing use of the Djibouti port for the Ethiopian imports and exports. The track was constructed by China's assertiveness is increasing evident in an area that the Soviet and the Western powers long contested. China has invested heavily in Djibouti: in the past two years China has loaned over $1.3 billion, more than 70 percent of its GNP, to the micro-country. Debts of this magnitude impose dependency. Among the supranational organizations Djibouti belongs are the FAO, IBRD, ILO, IMF, and WHO, the Arab League, and the African Union. It maintains offices in Washington, D.C. for the U.S. and the UN, and at Brussels for the EU and a few European countries.

The majority (60 percent) of those living in Djibouti are Somalis and Issus who have emigrated in the last few decades from Somalia; most of the others are Ifars (35 percent). Europeans (mainly French, some Italian), Arabs, and Asians (Indians and Chinese) comprise the rest. Virginia Thompson and Richard Adloff, authors of *Djibouti and the Horn of Africa* and numerous other books on African countries, write the following:

> The presence of so many heterogeneous ethnic groups gives Djibouti cosmopolitan atmosphere not found to the same degree in any other Somali coastal towns. The various ethnic groups live compartmentalized lives in their own "quarters." As a consequence, Djibouti did not become a melting pot, and tribalism was reinforced.

Those not of Arab, Asian, or European descent are mainly engaged in subsistence herding or farming or are unemployed, a condition affecting about 60 percent of the population, which explains the country's per capita GDP of only $2,700—one of the lowest in the world.

The service sector accounts for about 80 percent of Djibouti's GDP, followed by industry at 17 percent and agriculture at 3 percent. The Port of Djibouti container terminal serves as a major refueling center (undermining the role once played by ports such as those in the Comoros and the Seychelles before the opening of the Suez Canal) and handles the bulk of the nation's trade. About 70 percent of the port's activity serves neighboring Ethiopia. Efforts to develop the economy include constructing an additional port, a telecommunication infrastructure, a thermal power plant, and a large-scale desalinization project as well as supporting small businesses. These projects are supported by massive international investment, including substantial sums from China. The port provides the base for the burgeoning commercial, service-oriented economy, most of whose companies are foreign-owned—and most of whose senior employees are Europeans, Arabs, and Asians.

13. ASIA: ALONG SOUTH ASIA'S FRINGES

"[The Portuguese King] Manuel believed he had inherited the mantle of his granduncle Henrique, 'the Navigator.' Since the fall of Constantinople, Christian Europe had felt itself increasingly hemmed in. [Its ambitions were] to outflank Islam, link up with Prester John and the rumored Christian communities of India, seize control of the spice trade, and destroy the wealth that empowered the Mamluk sultans in Cairo . . . The project was at the same time imperial, religious, and economic. . . . Manuel started to assemble the expedition to reach the Indies."

—Roger Crowley

Map 13.1. South Asian Micro-countries

Source: World Map by M. Ruskin Co.

CONTINUING THE SEARCH FURTHER EAST–
TO INDIA AND THE INDIES

The three widely dispersed Asian micro-countries differ significantly in geography, demography, economics, and political dynamics. On Indian Ocean fringe is the Maldives; in the Himalayas is Bhutan, squeezed between China and India; and on the South China Sea fringe is Brunei, an enclave almost surrounded by Malaysia, lying on the north coast of Borneo. Brunei is geographically closer to the Oceania micro-countries than the other South Asian ones.

Portuguese voyages to South Asia, following the successful rounding of the Cape of Good Hope, continued the search further east—to India and the Indies.[48] In 1497, Portuguese navigator Vasco da Gama sailed around Africa to India and traded with Indians and other Asians running South Asian merchant networks.

In the 1500s, the Portuguese were the first Europeans to control the Maldives. In the 1600s, the newly independent Dutch briefly dominated the Maldives; not until 1887 did the Maldives, threatened by war, become a British protectorate. The country became independent in 1965. Long a sultanate, the country became a republic in 1968.

Bhutan, an isolated mountain enclave wedged between China and India, long had its international interests protected by Britain. India assumed that role upon its becoming independent in 1947. Bhutan maintains it has always been independent. In 1971, Bhutan became a United Nations member.

Brunei, an oil-rich sultanate, which once ruled Borneo, became a protectorate of Britain before attaining independence in 1984. The sultan of Brunei was the most important prince the Spanish encountered and perhaps the most difficult the British endured. The present sultan is the 29th in a succession dating back to 1368.

The per capita GDP of the Maldives is $8,800. Bhutan's is $6,000. In contrast, oil-rich Brunei's GDP per capita is $50,000, but this figure disguises the fact that wealth is concentrated in the sultan, his family, and

---

48. The word "Indies," according to Samuel Morrison, was then used in Europe to include China, Japan, the Ryukyu Islands, the Spice Islands, Indonesia, Thailand, and everything between them and India proper.

their cronies. Brunei's national budget is $5.1 billion; Bhutan and the Maldives have national budgets of $73 and $758 million, respectively. Brunei depends on oil; Bhutan and the Maldives depend primarily upon tourism.

Both the Maldives and Bhutan belong to the South Asian Association for Regional Cooperation whose other members are Afghanistan, Bangladesh, India, Nepal, Pakistan, and Sri Lanka. Bhutan also belongs to the Bay of Bengal Initiative for Multi-Sectoral Technical and Economic Cooperation (BIMSTEC) whose other members are Bangladesh, India, Myanmar, Nepal, Sri Lanka, and Thailand.

THE REPUBLIC OF THE MALDIVES

"The Government of the Maldives is the government of a small, isolated, non-industrial society, presiding over a population that expects little in the way of public services."

—Theodore Stoddard and others

The Republic of the Maldives is a chain of islands in the Indian Ocean about 400 miles southwest of Sri Lanka. Its total population of about 400,000 lives on 92 of the 1,192 islands in the archipelago. With an average ground level of less than five feet above sea level, the islands are extremely vulnerable to the rising sea levels and periodic inundations, which threaten the islands' and, therefore, the country's existence.

Based on oral cultural traditions, the earliest settlers of the Maldives were probably Dravidian fishermen from India's southwest coast and Sinhalese from the west coast of Sri Lanka. Buddhism began replacing Hinduism in the Maldives at the time of the vulnerable Emperor Ashoka's expansion of his Indian realm in the 200s BCE. Buddhists from Ceylon (modern day Sri Lanka) probably settled in the Maldives by the 500s BCE. From the 1100s CE, contact with the Arab traders introduced Islam, sultans, and the Dhivehi language into South Asia. The sultanate government, founded in the 1200s, lasted until 1968. Its impact has been pervasive. The Dhivehi language, an Indo-Aryan language, is principally Arabic but incorporates Persian, Hindu, Portuguese, French, and English and is the official as well as the most widely used language in the Maldives. Sunni Muslim is the official faith. The ethnic groups reflect the Maldives' multiple heritages: Dravidian (from South India), Sinhalese (from Sri Lanka), and Arab.

From 1558 to 1573, the Portuguese controlled the main islands as way stations on the route around Africa to India and beyond. In the early 1600s, the Dutch, who were winning the Hundred Years' War against Spain, as a *de facto* independent country were already developing trade and overseas colonies in Ceylon and other Asian sites, dominated the Maldives for 15 years. Following Britain's displacement of the Dutch from Ceylon from 1802 during the Napoleonic wars, they took possession of the Maldives.

In 1887, the sultan of the Maldives signed an agreement with the British governor of Ceylon, making the islands a protectorate.

While still a protectorate, the Maldives faced objections to the autocratic nature of sultanate rule that led to the sultan being briefly replaced by Mohamed Amin Didi as president in 1953. Within seven months, however, the president's unpopular policies led to his deposition. Upon his departure, the former sultan, Ibrahim Nasir Rannabandeyri Kilegefan, was reinstated. In 1959, objections to the centralism of Mohamed Amin Didi led to the three southernmost atolls briefly seceding.

The shift from protectorate to independent status in 1965 occurred with little tension. Having already presided over the independence of neighboring India in 1947 and Ceylon in 1948 (and many others, including Ghana in 1958, Nigeria in 1960, and Kenya in 1963), the British recognized that the age of empires and colonies was rapidly fading. Furthermore, "indirect rule," as practiced by the British protectorate in its relations with the Maldivian sultanate, had not been sufficiently intrusive to generate the degree of anti-colonial spirit that had arisen in many British colonies. It was the sultanate against whoever aroused the most public opposition. The Republic of the Maldives was one of the first colonies with under 1 million inhabitants to win their independence after World War II, along with the island countries of Cyprus in 1959, Samoa in 1962, and Malta in 1964.

In anticipation of independence, an amended constitution provided that the sultan, as head of state, be elected by a special convention in 1964 and that the prime minister, as head of government, be appointed by the sultan on the advice of the legislative body—the Majlis, or council. All legislation was subject to ratification by the sultan. In the words of Theodore Stoddard in his co-authored book, *Area Handbook for Indian Ocean Territories* (1971), "The modern republican form of government [of the Maldives] is not based on any democratic traditions but is superimposed on an Islamic society. Politics, like internal security, is based on face-to-face relationships and personal recognizance." In 1967, parliament voted for replacing the 853-year-old sultanate state with a republic, and, in 1968, a referendum was approved. The former sultan, Ibrahim Nasir, was elected its first president.

Political infighting from 1975 led to the exile of President Nasir, who fled to Singapore in 1978 along with millions from the treasury. Maumoon Abdul Gayoom began a 30-year tenure as president; his authoritarian rule facilitated his winning six consecutive elections. The constitutionally provided concentration of power enabled Gayoom's promotion of economic development, especially tourism. His autocratic style, however, proved highly controversial. Nasir supporters attempted coups in 1980, 1983, and, more critically, 1988. The 1988 coup began when an 80-person Tamil Indian mercenary force seized the airport. Gayoom fled. The intervention of 1,600 airlifted Indian military restored Gayoom. The Indian intervention appeared to be an act maintaining its regional dominance.

President Gayoom's authoritarian style was facilitated by the terms of the 1997 constitution that concentrated power in the executive presidency, which combined not only the roles of Head of State and Head of Government but also Commander-in-Chief of the armed forces and the police. His authoritarian style spurred opposition that challenged his Maldivian People's Party (MPP) whose demands for political reforms led, in 2008, to a new constitution providing for a separation of powers. In 2009, the first contested presidential elections, Mohamed Nasheed, leader of the Maldivian Democratic Party (MDC) defeated the Gayoom "dictatorship," winning presidency—even though he won two fewer parliamentary seats than Gayoom's MPP. Under the 2008 constitution, the president is elected for five-year terms, the unicameral 77-member parliament is elected, and the judiciary is separated from the president.

President Nasheed, upon taking office, confronted many issues, including huge debts, the tsunami-induced economic downturn, unemployment, corruption, and drug use. Among the steps the new president took to meet the challenges was the introduction of taxation on goods. He also announced his intention to create a fund to purchase land elsewhere for Maldivian people to relocate should rising sea levels inundate their country. This was spurred by a 2004 tsunami with waves as high as 14 feet (about 4 meters); 57 islands were seriously hit, 14 evacuated, 6 totally destroyed, and another 21 resort islands closed.

In December 2011, the opposition held a mass meeting in the capital (Malé), ostensibly to protest the president's inadequate protection of the role of Islam, the Maldives' official religion (the open practice of any other faith is prohibited and subject to prosecution). The protest ignited continuing social unrest. In February, President Nasheed ordered the army and the police to subdue the protests—and ordered the arrest of Abdulla Mohamed, chief judge of the criminal court. Instead, the police protested against the government. Four days later, President Nasheed resigned, later telling media that he had been evicted by a military coup led by his vice president, Mohammed Waheed Hassan, who was immediately sworn in as president. In 2013, Abdulla Yameen Abdul Gayoom, half-brother of long-term President Maumoon Abdul Gayoom, was elected president.

In 2015, President Yameen (he prefers ignoring the "Abdul Gayoom" part of his name to distinguish himself from his half-brother, the former president) claimed there was an aborted assassination attempt upon him instigated by his vice president, Ahmed Adeeb, who was immediately taken into custody. Later, the president ordered the arrest of former President Nasheed on the grounds that he had unconstitutionally arrested a criminal court chief judge. Continuing public protests demanded the resignation of President Yameen and the release of former President Nasheed from custody. Protests led to clashes with the police and the arrest of three opposition leaders and 191 others, which escalated tensions between the two major parties.

Since 2015, Yameen continued to use the courts to attack his political opponents. The opposition criticized the closed-door trials. Yameen, in just three years in office, dismissed two vice presidents and his first defense minister. His newest adversary was his half-brother Maumoon Abdul Gayoom, the country's strongman from 1978 to 2008. In 2017, former President Mohamed Nasheed, who now lives in exile in London, announced that he met with members of his party, the MDC, in Colombo, Sri Lanka, and that he planned to run in the party's internal primaries with the hope of becoming a presidential candidate in the elections next year. As MPs supporting Yameen's ruling party defected to the opposition, Yameen clung to power by ejecting opposition MPs from parliament and using the army to prevent their return.

In 2018, the Maldives' Supreme Court provoked Yameen's ire by over-turning the convictions of the president's opponents. Yameen reacted by declaring a state of emergency and arresting two Supreme Court justices and former President Gayoom. He behaves like a dictator who feels his opposition threatening a coup. Hassan Moosa and Jeffrey Gettleman in *The New York Times* have opined that Yameen, having cultivated ties with China and Saudi Arabia, believed he could survive the crisis.

But he did not. Despite extraordinary efforts to rig the September 2018 election, among which were the enforced absence of Nasheed, government-backed thugs intimidating opposition members, vote buying, the opening of a few new projects including a new China-backed bridge linking the capital to the international airport (according to the September 22 edition of *The Economist*). In the election, in which there was a very high turnout not only in the Maldives but also in countries overseas with large Maldivian expatriate communities, Ibrahim Mohamed Solih, leader of the opposition, defeated President Yameen's bid for a second term.

In the mid-2010s, the GDP was $3.1 billion; the GDP per person was about $9,000. Farming and fishing, long the principal industries in the Maldives, continue to be major factors in the economy. The major crops are coconuts, corn, and sweet potatoes. Tourism, prompted by sand, surf, and scuba diving, has developed as the major source of income, generating $1.9 billion. Though, these sources of income have not been sufficient to reduce unemployment below 11 percent. Imports cost $1.4 billion; the major countries from which the Maldives import goods are Singapore, the UAE, and India. Exports amount to $283 million; most exports are to France and Thailand. The budget of about $90 million is insufficient to continue to develop the infrastructure and the economy—and to purchase land abroad, which the government has determined is needed for resettling Maldivians in the event of their islands being inundated by natural disasters.

The Maldives belongs to the United Nations and several related global organizations such as the WTO and WHO. It withdrew from the Commonwealth after being warned that it might be suspended for subverting democracy. The Maldives has missions at the EU in Brussels, whose ambassador is also accredited to Belgium and other European countries, and at the UN, whose ambassador is also accredited to the US. The Maldives maintains 14 missions accredited to several Asian and Pacific countries. Until recently, Sri Lanka was the only country to maintain a mission in Malé, but China has now opened an embassy there, ratified a free-trade agreement, leased an island to one Chinese firm, and awarded major infrastructure contracts to another firm that involve the construction of the Maldives' longest road and a 1.3 mile bridge that will link the airport island with the main one. China also holds 75 percent of the Maldives' debt.

The Maldives is a founding member of the South Asian Association for Regional Cooperation (SAARC) and the Alliance of Small Island States (AOSIS). The principal concern of the AOSIS is that the impact of global warning will inundate their islands, at least the low-lying portions near the coastlines where all or most people concentrate. Ambassador Ahmed Sareer, when asked what his country's biggest concern at the United Nations was, replied firmly:

> Preserve our country's existence!!! To do so, we have led the effort to unite the small island nations to generate support in the United Nations and its members to take steps to minimize the conditions that increase prospects successfully combating of global warming, thus increasing the inevitability of our islands not being inundated.

The Maldives combines a dismal reputation for its autocracy and religious intolerance with prosperous sand, surf, and scuba-diving based economy, and the threat of inundation.

THE KINGDOM OF BHUTAN

"Bhutan undoubtedly was one of the last nations in the world to embrace modernity and launch the process of modernization. Bhutan's modern era effectively begins only in the middle of the 20th century . . . Bhutan's culture, art, governance, and worldview were primarily informed by a spiritual ethos, and religion played an important part in state administration and ordinary lives."

—Karma Phuntsho

Bhutan, a Himalayan mountain country bounded by China's "autonomous region" of Tibet on the north and the Indian states of Assam and West Bengal—which lie immediately north of Bangladesh—on the south. On the east is the Indian border state of Sikkim, which was independent until 1975. East of Sikkim is Nepal, an independent country with a population of over 20 million inhabitants, which, like Bhutan, is sandwiched between Tibet and India. Bhutan's existence has been threatened repeatedly throughout its recorded history. After India gained independence in 1948, the British government recognized Bhutan's autonomy and passed on to the newly established Indian government the working relationship with the Bhutan monarchy. Bhutan's mid-2010s population of about 784,000 residents presents a steep decline from the mid-1990s population of 1,950,000—a decrease that contrasts with the population increases recorded in many emerging countries.

By the 7th century, the Tibetan Empire had extended into Bhutan and into neighboring India, driving out the Indian population and introducing Vajrayana Buddhism. Various factions emerged around competing monastery-based schools of Buddhism that were patronized by warlords. In the 1600s, a Buddhist lama (monk) and military leader, Ngawang Namgyal, after fleeing Tibet, built a network of fortresses (*dzongs*).[49] He unified the country, bringing the local lords and religious factions under central control and creating a quasi-theocratic Tsa Yig code of law. He combined

---

49. Many are still used as centers of religious and district administration.

heading the political, religious, military, and cultural communities, adopting the title of Zhabdrung.

His leadership in founding the country was developed so thoroughly that the title was bestowed upon his successors and applied to the ethos of the Bhutan court administration. His role was so crucial in the development of the kingdom's traditions and institutions that, when he died in 1651 without a credible heir, his passing was kept secret for 54 years. It was the challenge of preserving the first Zhabdrung's status as the divine royal ancestor that, when faced with a lack of male heirs, the practice was adopted of selecting heirs through reincarnation. Until 1907, the Buddhist theocracy governing the country was a tandem of the Zhabdrung and a secular leader who controlled the district heads.

The first recorded Europeans to visit Bhutan were two Portuguese Jesuits, Estêvão Cacella and João Cabral, who, in 1627, presented Namgyal with firearms and gunpowder and offered their services in the war against Tibet; the offer was rejected. In the 18th century, Bhutan invaded the neighboring Kingdom of Cooch Behar (the frontier Indian state on Bhutan's southern border that has been absorbed into West Bengal), an action that led Cooch Behar in 1774 to appeal to the British East India Company to oust the Bhutanese and later to attack Bhutan itself. Envoys sent by Lord Hastings arranged a peace treaty in which Bhutan agreed to withdraw to earlier borders. Frontier clashes threatening Bhutan continued for the next 100 years, skirmishes that culminated in the Bhutan War, or Duar War (1864–65), which ended with neighboring Duar land in Bengal being ceded to British India.

From 1870 to 1885, power struggles between valley communities lit a Bhutanese civil war in which the Trongsa Penlop (governor), Ugyen Wangchuck, from his more densely populated central power base eventually united the country. In 1907, an assembly of the leading Buddhist monks, government officials, and heads of leading families unanimously chose him as the hereditary king of the country; the British government promptly recognized the new monarchy. The new king was a product of and a facilitator of a new political era in Bhutan. Although the Zhabdrung religious institutions remained large in the civic consciousness, its potent political role had faded. The new political system in which the keystone of the state administration was the hereditary king took off with relative ease enabling the new king to assert his leadership.

In the early 20th century, increasing Chinese pressure on Tibet and a consequent threat to the Tibet-Bhutan frontier led Bhutan to take steps to resist interference and insist on protecting its own identity and interests. Shortly after the installation of the new Bhutan king, Bhutan and Britain signed the Treaty of Punakha (1910), which acknowledged Britain's oversight of Bhutan's foreign affairs. The ambiguous treaty served Britain's hegemonic South Asian interests and Bhutan's concerns with protecting and preserving its porous border with China, which embraced Tibet.

Karma Phuntsho notes, "Generations of Bhutanese would deny that Bhutan was ever under the British Empire; the British, on their part, have also remained equally ill-informed about Bhutan's historical relations with its neighbors. In their official administration, the British never considered Bhutan to be under their rule nor gave Bhutan the kind of benefits received by princely states in their region. Bhutan has not chosen to belong to the Commonwealth of Nations." When India was negotiating its independence from the United Kingdom in 1947, Bhutan forcefully presented the case that it had always been independent and should not be absorbed into India along with the princely states.

When India declared its independence, Bhutan was one of the first to recognize its independence. In 1949, India and Bhutan signed an India-Bhutan Friendship Treaty that replaced the Treaty of Punakha. The new treaty stipulates that Bhutan be guided by India in its external relations. Bhutan asserts this pertains only to matters concerning India's interests; the Indian interpretation is broader, extending to maintaining troops in Bhutan close to the contested Himalayan border with China.

In 1953, King Jigme Dorji Wangchuck established Bhutan's first legislature, a 130-member National Assembly. He abolished serfdom in 1958, created the first Royal Advisory Council in 1965, and founded the first Cabinet in 1968. With India's support, Bhutan became an observer at the UN in 1968 and a member in 1971. A treaty in 2007 significantly revised the 1949 India-Bhutan treaty, stating that each "shall cooperate closely with each other on issues relating to their national interests." In 1972, 17-year-old Jigme Singye Wangchuck succeeded to the throne upon the death of his father. In 1973, the "Dragon King" (as he was called) introduced a civil service. In the 1990s, during a period of ethnic cleansing, the king expelled more than 100,000 of the southern Bhutan population, mainly ethnic Lhotshampa minorities (an ethnic group of Nepalese origin) who had not conformed to the king's demands for conformity in religion, language, and dress. Many fled to Nepalese refugee camps.[50] In 1999, the government lifted a ban on TV and the internet, making Bhutan one of the last countries in the world to welcome television.

Following King Jigme Singye Wangchuck's pledge for modernization in 2001, a new constitution was adopted in 2005 and a prime minister was appointed. In 2006, the healthy 50-year-old king abdicated the throne in favor of his son, Jigme Khesar Namgyel Wangchuck. Parliamentary elections with universal suffrage followed in 2007–8. The major parties contesting the first elections were the Bhutan Peace and Prosperity Party (DPT) led by Jigme Thinley and the People's Democratic Party (PDP) led by Sangay Ngedup. The DPT won the first election and Thinley, once Bhutan's permanent representative to the United Nations, became prime minister and won the 2013 elections. In the first round of the 2018 elections, voters selected two parties to nominate candidates for the second round. Surprisingly, the ruling party lost; an upstart party, the Bhutan United Party (DNT) won the most votes; the People's Democratic Party came second.

The 4,000-kilometer long Himalayan border that separates China's Tibet from India, Bhutan, and Nepal continues to be a source of friction. A 1962 bloody border conflict, provoked by China building a road across what India considered to be disputed territory, humiliated India. Again, in 2017, the presence of Chinese road-making equipment in the disputed Dolam Plateau led to the military forces of China confronting those of India and Bhutan. To support its entrenched role defending this border kingdom, India maintains a garrison (with a training academy, hospital, and golf course) only 13 miles from the disputed Himalayan frontier. India's army builds and maintains Bhutan's roads. India claims it is protecting Bhutan interests, but its actions arouse concern about being drawn into India-China clashes.

---

50. After nearly a decade, 40,000 Bhutanese refugees have settled in the United States.

Bhutan's mountainous terrain, its setting between Tibet and India, and its potent religious heritage have impacted not only its history but also ethnicity, linguistics, and occupational specialization (lama, subsistence farmer, and other) separatism. The numerous lamas (Buddhist monks) tend to live in communities separate from the working population. The country's rugged mountains long divided its people into distinct isolated districts, speaking 19 different languages—languages that share Sino-Tibetan and Burmese roots but are mutually unintelligible, meaning the speakers of one language cannot understand the others. Until the mid-1900s, Tibetan was the language of communication and business.

Only in modern times has the hegemony of the capital governing class led to Dzongkha, or Bhutanese, the language of the capital area, becoming widespread, emerging as the common language (and official language) after it was developed as a written language in the 20th century. English, the only foreign language taught in schools, has become the dominant medium of official correspondence and written communication. The transition to the rise of English for official correspondence reflects the shift in social, cultural, and political life to more international connections.

While 75 percent of the Bhutanese people are Lamaistic Buddhists, many Hindus live on the plains adjoining India. The labor force is 44 percent devoted to subsistence agriculture (rice, corn, and citrus on the 2.8 percent of the land that is arable). Light manufacturing (wood products, processed fruit, and cement) and service industries (principally tourism) account for 56 percent of the labor force. There may be as many lamas as agricultural workers. The GDP is $5 billion. GDPpc is $6,800. India accounts for more than 75 percent of its exports and imports. Nearly 60 percent of India's foreign aid goes to Bhutan.

Bhutan not only belongs to the United Nations and several of its affiliate organizations but also to the South Asian Association for Regional Cooperation (SAARC) whose other members are Afghanistan, Bangladesh, India, Nepal, Pakistan, Sri Lanka, and micro-country Maldives. It also belongs to the Bay of Bengal Initiative for Multi-Sectoral Technical and Economic Cooperation (BIMSTEC) whose other members are Bangladesh, India, Sri Lanka, Thailand, and Myanmar. Its 16,000-troop standing army is trained by the Indian Army and relies on the Indian Air Force for air assistance. Bhutan has diplomatic relations with 52 countries and the EU and maintains missions in India, Bangladesh, Thailand, Kuwait, and two at the UN (one in New York and one in Geneva). Only India and Bangladesh maintain missions in Thimphu, the Bhutanese capital. Other countries maintain diplomatic contact through their embassies in New Delhi, India and Dhaka, Bangladesh.

Bhutan's mountainous people's ingrained suspicion of foreign ways and forces presents an uncertain future. While resisting outside influence and pressures from two directions, it faces the challenges of developing its tourist industry as a viable source of revenue, improving its physical and social infrastructure, and its representative governance institutions.

BRUNEI DARUSSALAM

"Sir Hassanal [Bolkiah] is reckoned to be the 29th sultan in a royal line that, in the opinion of some scholars, stretches further back in time than those of the world's other remaining monarchs . . . [the sultan possesses] the world's largest residential palace [with 1,778 rooms] . . . [he is] often considered to be the world's richest."

—David Leake, Jr.

Brunei lies on the north coast of Borneo, bordering the South China Sea. It is surrounded and divided by a prong of the Malaysian state of Sarawak. Evidence exists that Arabian, Indian, and Chinese traders have been coming to Borneo for centuries forming trading polities. The present sultan of Brunei is the 29th—a direct descendant of the first.

The opening of a sea route between Europe and India in the first years of the 16th century brought European power to the gates of southeast Asia. When Magellan anchored there in 1521, the then sultan controlled not only almost all of Borneo but also the adjacent Sulu islands and a southern part of what is now the Philippines.

Bruneians were converted to Islam in the 14th or 15th century, following conversion of the first or third sultan (sources differ). Their conversion was part of a sweep of Islam throughout Southeast Asia. By the 16th century, Islam was firmly rooted in Brunei. By the 17th century, the power of the Bruneian Empire was at its peak, extending over most of Borneo and present-day southern Philippines. Beginning in 1521, when Magellan's world-girdling voyage had landed in Brunei Bay (without its leader who had already been killed), a series of European countries and their trading companies expedited their efforts to open up trade in the East Indies, efforts that gradually led to an end of Brunei as a regional power.

European commercial pressure and Brunei's internal strife over sultanate succession led to the decline of the power and prestige of its sultanate. In collaboration with two of the leading Bruneian nobles who hoped to become sultan, the Spanish invaded Brunei in 1578. But pestilence that caused high fatalities forced their withdrawal a few months later to their

Spanish base in Manila.[51] Brunei suffered a civil war from 1660 to 1673, further weakening its ability to withstand European efforts to develop trade hegemony.

In the first of several interventions in the affairs of Brunei, the British, in 1846, intervened to settle internal conflicts regarding the sultanate succession. In the 1880s, as the decline of Brunei continued, the sultan granted land within the sultanate to James Brooke, who had assisted him in subduing a rebellion. Initially owing allegiance to the sultanate, over time, Brooke and his nephews who succeeded him created a local dynasty called the White Rajahs. Their polity, Sarawak, became autonomous and annexed more land.

Threatened by the Brookes, European powers, and piracy, Sultan Hashim Ahmad negotiated a treaty with the British in 1888, granting Brunei British protection and British control over Brunei's foreign affairs but no right to interfere with internal affairs. In 1890, despite the treaty, when the Kingdom of Sarawak annexed the Brunei's Limbang district, which severed the major part of Brunei from a district across the bay, the British did not object. In 1888, despite the sultan's objections, the British imposed a protectorate status upon Brunei and posted a "resident" who "advised" the sultan regarding how he should govern his sultanate—acting as the true ruler. In 1906, faced with continuing decline, debts, and threats, the sultan reluctantly signed a new protectorate treaty, the terms of which confirmed the British assignment of a "resident" who advised the native chief—who remained as the nominal head of state. This resident system continued until Brunei regained internal self-governance in 1959; the protectorate status ended in 1984 with the formal granting of independence.

The discovery of oil in 1929 transformed the Brunei economy from rags to riches, providing the basis of Brunei's development, wealth, and role in local and international politics. The Japanese invasion of Southeast Asia engulfed Brunei shortly after the 1941 attack on Pearl Harbor. Under Japanese occupation, Sultan Ahmad retained his throne, albeit nominally; a secretary to the former British resident was appointed chief administrative officer under the Japanese governor.[52] In 1944 the Allies began bombing

---

51. Brunei mythology considers this Castilian War a heroic victory.
52. The former British resident declined the invitation to retain his post and was interned with other British nationals in a prison camp.

the occupying Japanese, destroying much of the Brunei forces. In June 1945, Australian land forces, supported by American air and naval units, recaptured Brunei.

After the Pacific War ended in August 1945 and the Japanese troops formally surrendered in September 1945, the Brunei State Council was revived, and a British Military Administration—consisting principally of Australian servicemen—charged with restoring the Bruneian economy took over from the Japanese and remained until July 1946. In 1946, a political movement formed, with a Malay orientation; but government action limiting participation led to its 1948 demise.

In 1953, prompted by the rising tide of nationalism and anti-colonialism, the 28th sultan of Brunei, Omar Ali Saifuddien III created a committee to ascertain public views regarding a new constitution, a step that led to drafting a constitution, accepted by the United Kingdom in 1959, declaring Brunei self-governing, replacing the "resident" with a High Commissioner (ambassador) and retaining Britain's protectorate role in international affairs. While the new constitution created an Executive Council headed by the sultan and Legislative Council, the members of both bodies were appointed and granted only limited power.

Beginning in 1953, Sir Saifuddien III initiated a series of national development plans. The first promoted the development of oil and gas fields, new roads, and public education. The second development plan, beginning in 1962, not only continued the development of oil production, but also supported husbandry and fishing, expanded the provision of electric power, and improved public sanitation and disease control. In 1956, Sheikh A. M. Azahari formed the first Brunei political party, the socialist Brunei People's Party (PRB), which appealed to Malay nationalism by advocating an independent Malay country embracing the whole Malay Archipelago. In 1962, there were District Council elections, seen as a test between the sultan and PRB, in which the PRB took 32 of the contested 55 seats. In 1965, the sultan agreed to hold elections for the Legislative Council, composed of 10 elected and 11 appointed members. He also agreed to introduce a ministerial form of government and promised to eventually enlarge the Legislative Council to 20 members. The democratization progress proceeded slowly.

In 1963, Britain began negotiations regarding federating with the neighboring Malay states. In 1967, a short Brunei rebellion, prompted by Brunei's reluctance to share the oil wealth and the sultan's reluctance to share power, was suppressed with the assistance of UK Gurkha forces. The rebellion and the reluctance to share wealth and power contributed to the failure of two successive federation proposals. First, the sultan decided against a North Borneo Federation with Sarawak and North Borneo (Sabah) by proposing a federation with the Malayan states. Later, spurred by disagreement regarding his status, he opted out of the larger Malaysian federation with the Straits polities and Singapore.

In 1967, resisting the British pressures to democratize, Sir Saifuddien III precipitously abdicated in favor of his oldest son, Hassanal Bolkiah, bringing him home before completing his studies at the Royal Military Academy at Sandhurst. He then argued that the democratization steps should be delayed while the young ruler learned the job. In fact, Sir Saifuddien III initially continued to run the government while his son took the opportunity to enjoy a playboy lifestyle and acquire an additional wife. Only when the young Sultan Bolkiah developed an interest in governing did disruptive tensions develop with his father. In January 1979, a new treaty was negotiated with Britain that turned over international responsibilities to the Brunei government and signaled the intent to grant it independence. In 1983, it was announced that the date of independence would be January 1, 1984, but the actual celebration was delayed until the completion of the new, gigantic palace in February.

Under Brunei's constitution, Sultan Bolkiah is both head of state and head of government with full executive power; he is his prime minister, finance minister, and defense minister. Most of the other ministries in Brunei are headed by family members. The family has a venerated status that is reflected in the retention of traditional Malayan and other Islamic customs. The country maintains a military force of 7,000 soldiers that includes Gurkha troops (Nepalese who for decades have served in the armed forces of Britain, India, and other former British-ruled countries).

Brunei's traditional ties to Britain led to it becoming the 49th member of the Commonwealth on the day of its independence in 1984. One week later, it became the sixth member of the Association of Southeast Asian Nations (ASEAN). Shortly afterwards, it became a member of what has become the Organisation of Islamic Cooperation and a member of the UN. In 1989, it joined the Asia-Pacific Economic Cooperation (APEC) and in

1995, the WTO. Brunei's firm support of ASEAN—plus the money that it derived from oil since the 1920s—gave Brunei clout that belies its size.

Oil has gushed since the 1920s. Crude oil and natural gas production account for about 90 percent of Brunei's GDP of $30.2 billion in 2016, providing a per capita GDP of $73,200. The national budget is $6.3 billion—of which $573 million is spent on defense. In an effort to keep the populace content, the government budget supports subsidized food and an inflated payroll that employs close to 15 percent of the population.

Even as the oil reserves dwindle, David Leake, Jr. observes, "Brunei's foreign reserves are so large, in fact, that investments earning alone could cover the government's [present] annual budget." Brunei depends heavily on imports such as agricultural products, cars, and electrical products—principally from Singapore, China, the United Kingdom, Malaysia, and the United States. It exports mainly to Japan (45 percent), South Korea, Australia, India, Indonesia, and Vietnam. Brunei belongs to APEC, ASEAN, and Pacific Islands Forum, thus straddling East Asian and Pacific communities.

Tumbling oil prices,—and the projected prospect that the wells will run dry by 2040—increasing commodity prices, and the need for increasing its heretofore desultory investment in roads, bridges, and power stations to attract Japanese and other investors are forcing the contraction of Brunei's economy.

More than 75 percent of the 400,000-plus residents in Brunei live in urban areas, and almost 67 percent live in Brunei-Muara District (the compact area where the capital is located). The other three districts are sparsely populated. Over 66 percent of its people are Malay; 11 percent are Chinese, who dominate commercial life, and the others are British, Australian, Filipino, Thai, and Cambodian. The industry employs 63 percent of the labor (a significant percent of whom are foreign), service 33 percent, and agriculture 4 percent.

Migrants make up almost one-half of the population. Multinationals, most notably Brunei Shell Petroleum and its numerous contractors, recruit overseas; a large expatriate *orang puteh* (white men) community dominates the upper-administrative level. Brunei's a very stratified society. But Brunei's hereditary autocratic government, with its hereditary nobles who hold the highest posts, has become ingrained. The extent of its commitment to conservative Islam, autocracy, and determination to preserve an Islamic culture and way of life was reflected in its 2019 passage of a harsh new criminal law prescribing death by stoning for sex between men and for adultery, and amputation of limbs for theft.

The Malay spoken in Brunei differs sufficiently from standard Malay and other Malay dialects that the tongues are mostly mutually unintelligible. English is widely used and is a language of instruction in the schools. The Chinese, many of whom are shopkeepers, use a variety of Chinese dialects. The culture of Brunei is predominately Malay with heavy input from its animist, Hindu, Islamic, and Western heritage. Islam's strong influence is evident in its ideology, philosophy, and faith. Islam, Brunei's state religion, is the faith shared by 79 percent of its people, most of whom are Malay. Many Chinese practice Buddhism; some Chinese and non-Malays are Christian.

The well-entrenched degree of sultanate control and the well-appreciated, oil-based standard of living are threatened. The erosion of oil revenue—along with the prospect of royal family dissension—and the increasing pressure for democratization both challenge Brunei's economy and its political stability.

14. THE AMERICAS: THE EXTENDED CARIBBEAN

"The Greater Caribbean was the first part of the Americas to experience the shock waves of European colonialism . . . While the Greater Caribbean was a dynamic place of movement and change, the Spanish conquest transformed the region and its peoples with lightning speed and on an unprecedented scale."

—Matthew Mulcahy

MAP 14.1 American Micro-countries

Source: World Map by M. Ruskin Co. LLC;

EXPLORING AND EXPLOITING THE SPANISH MAIN

While the initial European transcontinental voyages sought spices and gold in the mystic East by sailing south around the Cape of Good Hope at the foot of Africa and then east, another series of adventurers sailed west in search of the Indies in 1492. Instead, they landed on islands off the coast of unknown continents, blocking the way to India and its surrounding countries. The Spanish undertook most of the initial explorations—exploiting and colonizing the Americas. French, Dutch, Danish, and English navigators soon joined in the competitive empire-building race. Over time, the British, with their naval power, developed an American empire extending from the North American coast to the northwest coast of South America.

The Caribbean Sea extends east for 1,500 miles from Central America to the Lesser Antilles and north from South America to the Greater Antilles. Seven micro-countries lie in the Lesser Antilles (Saint Kitts and Nevis, Antigua and Barbuda, Dominica, Saint Lucia, Saint Vincent and the Grenadines, Grenada, and Barbados), and there are four along or on the coast of the American continents (the Bahamas, Belize, Guyana, and Suriname). The Caribbean neighbors are British, Dutch, and U.S. dependencies, French local governments, and independent countries. All the American micro-countries received their independence from Britain except Suriname, which was a former Dutch possession.

The colonial history of the Caribbean and adjacent American lands began with Christopher Columbus. In his search for a passage to Asia, he landed in Hispaniola in 1492. Before Columbus' arrival, Arawak tribes— the name applied by the Spanish to Amerindian tribes they encountered in the West Indies and in northeast South America—appeared to have been driven out of the Lesser Antilles by the Caribs, the other major group of Amerindians in the region that gets its name from the Caribbean. Spanish adventurers in search of quick fortunes in gold and silver led to Spanish conquest and settlements initially centered on Hispaniola, then on other islands, then on the adjacent mainlands of Mexico and Central America, and then on South America. Columbus and his successors' efforts developed the "Spanish Main," an area providing a lucrative source of income for the Spanish crown. The importance of the Spanish Main as a source and as a route for the shipment of treasure led to maritime plundering by the British, French, Dutch, and Danish.

Between 1595 and 1620, the English, French, Danish, and Dutch made unsuccessful efforts to settle in the Caribbean. The Lesser Antilles, having been ignored by the Spanish because of a lack of self-evident treasure, attracted British, French, Dutch, and Danish explorers, entrepreneurs, and settlers. Cultivation of sugar made the Caribbean islands financially attractive and a highly contested scene. The British captured Jamaica in 1655, and the French captured western Hispaniola in 1697.

Throughout the long British-French duel from the late 1600s to 1815, war on the European continent was accompanied by war in the colonies. The North American phases of the European wars included the following: Queen Anne's War, part of the War of Spanish Succession (1702–13), the War of Jenkins' Ear (1739–45), King George's War, also known as the War of Austrian Succession (1743–8), the French and Indian War, also known as the Seven Years' War (1755–63), the American Revolution (1775–83) that engaged French and Dutch support, and the War of 1812 (1812–15)—considered the last phase of the Napoleonic Wars (1792–1815). The wars provided the British the opportunity to seize colonies earlier claimed and settled by rival European powers.

The cultivation of sugar generated the need for cheap labor: American-Indians, indentured whites, Africans, and Asian-Indians. By the 1700s, African slaves significantly outnumbered the whites, a free mulatto population was growing, and a merchant and professional class was emerging. The successive imports of labor of different ethnic backgrounds to the Caribbean dependencies led to an almost complete displacement of the original people and culture and the lack of cohesion among the replacements—a condition that continues to characterize Caribbean countries. Charles C. Mann, in *1493: Uncovering the New World Columbus Created*, observed that fewer blacks succumbed to the prevalence of malaria than the plantation native laborers (killing 6,000 in Barbados alone), which strengthened the African slave trade.

V. S. Naipaul has noted the following:

These Caribbean territories are not like those in Africa and Asia, with their own internal reverences that have been returned to themselves after a period of colonial rule. They are manufactured societies, labor camps, creations of empires, and for long they were dependent on empires for law, language, institutions, culture, even officials. Nothing was generated locally, [so] dependence became a habit.

Despite the lack of cohesion—the absence of pre-colonial traditions and the enduring tensions—these tropical colonies became so profitable that they were once considered more valuable than the 13 British North American colonies.

Post-World War II Britain's Colonial Office considered it "clearly impossible in the modern world for the present separate (West Indian) communities, small and isolated as most of them are, to achieve and maintain full self-government on their own." In 1958, driven by an urge to achieve dominion status, independence within the (British) Commonwealth, and increased bargaining power, especially in sugar quotas and prices, the West Indies Federation was formed, which absorbed the previously formed Leeward Islands Federation.

The West Indies Federation survived for four years but dissolved in 1962 on the same date on which it was originally planned that it should become an independent country, succumbing to a few countries' desires for separate independence. Upon its demise, several Lesser Antilles colonies agreed to seek a new federation, the West Indies Associated States. But, beginning in 1966, several of these island dependencies negotiated their independence: Barbados in 1966, Saint Vincent and the Grenadines in 1970, Grenada in 1970, Saint Lucia in 1978, Dominica in 1981, Antigua and Barbuda in 1981, and Saint Kitts and Nevis in 1983. Six Lesser Antilles islands, not Barbados, and three associate members—the dependencies Anguilla, Montserrat, and the British Virgin Islands—then joined together in the Organization of East Caribbean States (OECS).[53]

OECS works closely with its affiliate, the East Caribbean Central Bank (ECCB), which manages the East Caribbean dollar. Michael Baptiste, a former minority leader in the Grenada legislature and now the macroeco-

---

53. The 2005-founded Caribbean Court of Justice is the ultimate court of appeal to which most Caribbean micro-countries and a few neighbors belong; only a few independent countries continue to use the Privy Council in London as the final court of appeal.

nomic adviser to Grenada's Ministry of Finance, Planning, Economy, and Energy explains that "the fact that it is tied to the United States dollar accounts for its strength and stability." Baptiste continued to say, "Without being tied like the Trinidad dollar, it would be worthless." All eleven micro-countries in the extended West Indies belong to the Caribbean Community and Common Market (CARICOM).[54]

"Small countries suffer from a lack of significant financial or human capital organizations," AJ Mediratta, a Greylock Capital partner, points out, "so pooling resources to form a Central Bank and currency union provided a measure of stability. However, this stability has often been cited as coming at the price of economic growth, as exports and tourism have lagged behind international peers in recent years. Until recently, the strength of the U.S. dollar, high energy prices, and high reliance on imported goods, combined with the currency peg (the exchange value of the local currency with the U.S. dollar) hurt the OECS' common currency."

Saint Kitts and Nevis, Antigua and Barbuda, and Dominica have fewer than 70,000 inhabitants; more populous countries in the Americas include Grenada with 90,000 inhabitants, Saint Vincent and the Grenadines with 108,000, Saint Lucia with 166,000, and Barbados with 279,000. More populous, still, are Belize with 280,000 inhabitants, the Bahamas with 302,000, Suriname with 438,000, and Guyana with 765,000. The seven Lesser Antilles countries' GDP per person range from about $14,000 in Dominica and Grenada to $25,000 in Barbados. Helped by proximity to and tourists from the United States, the GDPpc in the 2010s of the Bahamas was $31,500; but that of Guyana was $7,800—Belize's was $8,400 and Suriname's $9,600.

The economy of the Caribbean micro-countries will be affected by the prospects for the sugar and tourist industries, the continued development of infrastructure and utilities, and the weather. The sugar industry declined for 200 years, beginning with the abolition of slavery and accelerating with competition from mechanized sugar operations in Brazil and Australia, and the abandonment of sugar for more profitable activities. An

---

54. CARICOM also includes Haiti, Jamaica, Trinidad and Tobago, Montserrat, Anguilla, Bermuda, British Virgin Islands, Cayman Islands, and Turks and Caicos Islands as associate members.

EU withdrawal of protective quotas would be devastating. Tourism already accounts for a larger share of the Caribbean's GDP than any other region of the world according to the World Travel and Tourism Council. Even less tourist-attractive Antigua and Barbuda depends on tourism for 60 percent of its GDP, and Dominica depends on tourism for 35 percent GDP.

Even before 2017, striking differences existed among the East Caribbean countries in the sun-sand-surf-scenery that attracts tourists and related features including their economies, infrastructure, the extent to which their island is populated with part-time expatriate residents with ostentatious houses, the extent to which the commercial areas include store fronts displaying world-fashionable brands, and prosperity. Post hurricane, the differences are even more striking. The sites with the better beaches and ports (sea and air) recovered faster. The more "attractive" countries in the Caribbean received more aid and investment. Urban areas received more than the rural regions.

Several Caribbean micro-countries, like fortunate ones elsewhere, have taken advantage of good climate and beaches, a location proximate to tourists and convenient connections, and with relatively good infrastructure and utilities to develop tourist or finance and light industry. But global economics, overstaffing, deficits, debts, and failing foreign reserves have debased investors' interest. The need to invest clashes the need to reduce the debt in order to attract investment.

Hurricanes Irma, Jose, and Maria hit the Caribbean in 2017. Press and TV's focus on Texas, Florida, the British and U.S. Virgin Islands, the Bahamas, Cuba, and Puerto Rico drew attention from category four or five hurricanes that brought winds, rain, ruin, and devastation to Caribbean micro-countries. Prime Minister Roosevelt Skerrit of Dominica, before the island became incommunicado, reported that his country was completely flattened. Barbuda was completely evacuated. The number of hurricanes in the North Atlantic has been increasing, reports NASA and the National Oceanic and Atmospheric Administration (NOAA) in a recent study. The Caribbean may expect many more storms.

DeLisle Worrell, a former governor of the Barbados central bank, has published some severe criticisms of and strong recommendations for the

Barbados fiscal system, noting the lax control of expenditures and citing specific reforms (briefly summarized in following section on Barbados). He has asserted, in a response to an inquiry to the author, "the lessons are of a more general applicability," not just in the Caribbean.

BARBADOS

"Adventurers came, full of greed and mysticism, to probe, and wonder at, and desecrate . . . Fortunately for the islands, they came and saw— and passed on; seeking the greater glories of the [Spanish] Main . . . Nearly a century passed before the English found their way to this gem of the so-called Indies. They, too, brought in their wake much evil-doing and sordid discord. Time, it is true, has done much for the healing of these wounds, although a scar remains in the form of a complex racial problem."

—Vincent T. Harlow

Barbados, the eastern-most Caribbean island, was sometimes a first port of call for ships crossing from Europe or Africa to the Spanish Main. Cristóbal Colón (Columbus) is credited with "discovering" Barbados at the end of the 1400s. Arawaks had then populated it. Spanish landed in 1518 and took Arawaks as slaves for Hispaniola, depopulating Barbados. Two English ships landed in 1625; the captain of the second, John Powell, on his return to England convinced his employer, Sir William Courten, to fund and found a Powell-led settlement on the island in 162–8. From the arrival of the first settlers until 1966, the British ruled Barbados—more than three centuries. Barbados was the only island that did not undergo a change of rulers in the almost continuous warfare among the Spanish, Portuguese, Dutch, and British for control of the Caribbean and its resources.

Over time, it was the British, with its navy, which ruled most of the North American coast until 1776–83 and the Caribbean islands until the 1960s. The Commonwealth Caribbean islands comprise a major share of the hundreds of Caribbean islands, forming an arc from Florida in the north to Venezuela in the south with Barbados as a projection at the far eastern end of the arc. Barbados played a major role extending these communities and commonalities. Matthew Mulcahy, author of *Hubs of Empire: The Southeastern Lowcountry and British Caribbean* (2014), points out the following:

It was in Barbados that the deadly combination of sugar, slavery, and large landholdings first took hold within the British Empire during the 1640s, and it was the migrants from Barbados searching for new opportunities who carried this plantation complex to Jamaica and the Leeward Islands during the 1650s and 1660s, and eventually to the South Carolina Low Country during the 1670s.

The colonial plantocracy emphasis on slavery has significantly impacted its present demography. Close to 93 percent of the almost 300,000 Barbadians (also called Bajans) are of Afro-Caribbean or mixed descent. The rest include those of European, North American, Asian,—principally Chinese but also South Asian—and immigrants from Guyana. At least 75 percent of Bajans are Christian: 63 percent are Protestant (mainly Anglican), 7 percent Catholic, and 4 percent other Christian. English is the official language, used for commerce and administration throughout the island. A few Barbadians speak a Bajan creole variant of English.

Barbados' colonial start was hardly promising. In 1629, an earlier grant of the islands to the Earl of Carlisle was declared to apply to Barbados. The strife between rival claimants seriously retarded its development. From 1641 to 1650, during the English Civil War between king and parliament, there followed a period of ambiguity during which Vincent Harlow has noted, "The close ties, both economic and political, with which the Mother Country had endeavored to bind the colonies to herself in order to build up a centralized, self-contained empire, were now quickly broken." Assemblies of freeholders had been held even before 1639, when Governor Hawley established the first Barbadian assembly, which was composed of elected burgesses. In 1641, the governor changed the status of this assembly from an advisory body into one with the right of initiating legislation. The presence and pressure of local representative bodies continued to irritate the colonial authorities for more than three centuries.

In the initial years of the colony, most of the population was white and male who, with a few African slaves and indentured European workers, undertook the cultivation of tobacco, cotton, ginger, and indigo. But the initial English yeoman farming economy soon faced a severe crisis; it could not compete with the mid-Atlantic North American colonies' larger-scale sugar cane production. The development of a slave-dependent plutocracy system from the 1640s dramatically increased the number of slaves on the islands.

As the plantation economy developed, land was consolidated in the hands of a few elite white families. Consequently, between 1650 and 1680, about 30,000 landless Barbadians left the island for other Caribbean islands and North American colonies. The vast systematic import of slaves increased the black population to far exceed that of the whites. By 1680, sugar had transformed Barbados into the richest colony in English America. But Barbados' economy faltered in the 1700s when the price of sugar fell sharply and European-based wars and the American Revolution interfered with trade.

The abolition of slavery in Britain in 1772 led to the British abolition of the slave trade in 1807 and the government's energetic efforts to discourage other countries from engaging in the practice. The cessation of the slave trade foreclosed the plantations' ability to augment their labor force. Knight points out that "the year 1810 marked the apogee of the system . . . About 60 percent of all Africans who arrived as slaves in the New World came between 1700 and 1810, the period during which Jamaica, Barbados, and the Leeward Islands peaked as sugar producers." Already, by the early 1800s, the expansion of plutocracy and slavery had multiplied the black population of Barbados—and in all the Caribbean except Trinidad—to over 90 percent. Slavery was converted to an "apprentice" system in 1834 and fully abolished in 1838, but, as the ex-slaves remained on the island as laborers, the country's demography remained essentially the same.

Barbados continued to successfully resist successive British efforts in the 1800s to abolish its elected House of Assembly. The British had found the local assemblies argumentative and cumbersome to manage and preferred to install a crown colony government (with a strong crown-appointed governor) as they had in the other Caribbean possessions. While the Barbadian elite managed to retain its House of Assembly, which functioned alongside the governor's Legislative Council, for almost 300 years Barbados remained in the hands of a small, white, propertied minority elite who maintained a limited white male electoral franchise.

Not until after World War I did reform come. Returning veterans inspired the founding of the Democratic League, which espoused franchise reform, old-age pensions, compulsory education, scholarships, and trade union organization, and the election of a few representatives to the House of Assembly. A riot in 1937 inspired Grantley Adams (later Sir Grantley) to found the Barbados Labour Party (BLP) in 1938; he also became president-general of the Barbados Workers' Union (BWU) in 1941.

Under terms of constitutional reforms introducing a semi-autonomous form of government and progressive liberalization of the franchise, Adams became leader of the government in 1946. In 1951, in the first election conducted under universal adult suffrage, the Adams-led BLP gained a majority of seats in the House. Under a new ministerial system of governance introduced in 1954, Adams became premier (the pre-independence title for prime minister) and resigned as president of the BWU. Meanwhile, Errol Barrow, a newly elected member of the House and a nephew of the founder of the Democratic League, emerged as leader of the discontented left-wing BLP and formed the Democratic Labour Party (DLP). Despite the defection, the Adams-led BLP won the 1956 election.

Plans for a West Indies Federation had been negotiated in 1953, and elections for a federative assembly were held in 1958; Adams was elected premier of Barbados. The Federation dissolved in 1962 when Jamaica and Trinidad opted for independence. Adams' successor as Barbadian premier was unable to hold the party together in the face of persistent unemployment exceeding 20 percent. Consequently, the DLP won the 1961 elections, and Barrow became premier.

Between 1961 and 1966, anticipating the securing of independence in 1966, Barrow expanded free education, instituted a program of industrialization, and replaced the governor's Legislative Council with a Senate appointed by the governor. The possibility of forming a federation of Lesser Antilles islands was explored unsuccessfully. Upon achieving independence in 1966, Barbados became a Westminster-like constitutional monarchy, with a governor-general representing a queen as head of state and a prime minister elected by and may be dismissed by the House, which was initially comprised of 24 members and was later increased to 30.

Since independence, Barbados has functioned with a two-party system. The Barrow-led DLP won reelection in 1966, and Barrow became the first prime minister. The Barrow-headed DLP won again in 1971. The BLP, led by G. M. Tom Adams, son of Sir Grantley Adams, returned to power in 1976 and again in 1981. Tom Adams died in 1985; his successor was defeated by the Barrow-led DLP in 1986. When Barrow died in 1987, Lloyd Erskine Sandiford succeeded him. In 1993, 1988, and 2003 the opposing BLP won consecutive terms in office. In 2008, the DLP returned to power; its leader and prime minister, David Thompson, died in 2010; Freundel Stuart succeeded him. The DLP, with Stuart as prime minister, was reelected in 2013

In the 2018 election, the opposing BLP won a stunning victory, capturing three-quarters of the popular vote and all 30 of the Assembly seats. *The Economist* said, "The electorate simply seemed to have tired of a mucky status quo." The new prime minister, Mia Mottley, a self-identified lesbian, faces an unusual constitutional dilemma: with no official opposition in the legislature, who performs the opposition's constitutional duties including nominating two seats in the upper house and consulting on judicial appointments? Motley proposes to restructure the public sector and attract foreign investment as well as seek a constitutional amendment to resolve the constitutional question.

In the 1990s, at the suggestion of Trinidad and Tobago's Patrick Manning, Barbados began negotiating a political union with Guyana and Trinidad and Tobago. The project stalled after Sandiford became ill, and his BLP lost the next election. Adams pursued regional economic integration, promoting CARICOM, the Caribbean Free Trade Association (CARIFTA), and the Caribbean Basin Initiative (CBI). Barbados also belongs to the UN, several of its affiliate organizations, the Commonwealth, and the Organization of American States (OAS). But, by a closely contested vote in the legislature, the country did not join the Organization of East Caribbean States. It has its own central bank, which manages the Barbadian dollar.

In 2016, the Barbadian GDP was $4.5 billion; the per capita GDP was $16,200. Barbados had been dependent on sugar cane, but, in the late 1900s, the economy diversified into tourism and light manufacturing. In 2016, 75 percent of the labor force was in the service industry, 15 percent in industry, and, because of the shriveling of the sugar industry, only 10 percent remained in agriculture. Unemployment was 12 percent. Since World War II, tourism has grown to become the major source of revenue with the advent of airlines and cruise ships and second homers, accounting for almost $1 billion—30 percent from the U.K., 20 percent from the U.S., 20 percent from Canada, 20 percent from the Caribbean, and 10 percent from mainland Europe and South America. In 2016, most of the Barbadian $1.7 billion of imports were from Trinidad and Tobago, the U.S., and Canada. Its exports of $775 million came from the same neighbors. Its annual budget was $1.5 billion—defense accounted for $8 million.

In the early 2010s, the debt-to-GDP ratio was 117 percent and rising, sufficiently high enough to cause continuing concern to investors (private as well as public). DeLisle Worrell has pointed out that an overstaffed government, repetitive deficit budgets, a growing debt, and an economic malaise has led to an increasing reluctance of investors to invest—to fund construction in infrastructure supporting the economy and its growth. The challenges exceed the prospects of the frequent use of Barbados' independent monetary system's flexible exchange rates to alleviate the stress by allowing the currency to fluctuate.

Worrell has cryptically assessed Barbados' prospects as follows:

> Barbados has a very competitive economy, with high-end tourist facilities, highly regarded international finance and business services, world renowned rum, and entrepreneurs that are making a name internationally. However, the bad news is that our government is overspending, and is failing to deliver public services efficiently. This has stifled the economy, growth, and eroded the foreign reserves of the Central Bank.

He has set forth recommendations for revitalizing the economy with the aim of eliminating the government's operating deficit. The recommendations included drastic public sector job cuts, reductions of transfers to government-owned enterprises of 10 percent per year over three years, a divestment of selected public assets, and temporary freezing of all public works (except those funded by foreign finance and projects directly related to tourism).

GRENADA

"The grim events of October 1983 . . . mark a crucial turning point in the history and character of the Caribbean, and especially of the English-speaking Commonwealth Caribbean . . . the murder in cold blood of [Prime Minister] Bishop and his loyal ministers, the imposition by the Revolutionary Military Council of a harsh curfew, and the U.S. invasion . . . took on the character of a Greek play."

—Gordon L. Lewis

The history of colonial Grenada, located south of Saint Vincent and the Grenadines and north of Trinidad has been one of irreconcilable conflict. Before 1492, rival Amerindian tribes fought; from 1500, Amerindians fought Europeans. From 1600, Europeans fought each other. The aggressive Caribs displaced the earlier Amerindian inhabitants, the Arawaks (a Taíno tribe). Columbus visited Grenada in 1498 on his third and last voyage to the Caribbean.

In 1609 and 1638, there were efforts to initiate English settlements; the Caribs drove them off. George Brizan states that although "Grenada was technically a part of the Spanish Empire, both England and France wanted to claim her as their possession." In 1627, King Charles I of England included Grenada in an Eastern Caribbean islands grant of islands to the Earl of Carlisle. In 1636, Cardinal Richelieu formed a joint stock company, which included Grenada as part of its possessions. The Caribbean scene was set for a series of wars between European powers that continued intermittently for more than two centuries.

In 1650, the French governor of Martinique, Jacques Dyel du Parquet, after buying Grenada from the Richelieu-led joint-stock company, made the first successful attempt to colonize the island by erecting a fort and starting a large tobacco plantation. Later, the company turned to sugar cane and indigo. Continuing Carib attacks led to three efforts by French forces to eliminate them. In the 1650s, the French massacred the Carib villages. Many Caribs committed suicide by jumping from a cliff. The few surviving Caribs fled to the forests where they joined escaping slaves. The resulting mix became known as Maroons. A subsequent uprising and ad hoc execution of the governor led to Grenada coming directly under the French Crown.

The British captured Grenada in 1762 at the tail end of the multi-continental Seven Years' War in which France formally ceded Grenada to Britain. Grenada then became part of an administrative entity that included the Grenadines, Saint Vincent, Tobago, and Dominica.[55]

France recaptured Grenada in 1779 during the American Revolution when France came to the aid of the American colonies. (The coalition interrupted Britain's long naval primacy.) At the end of that war in 1783, the island was returned to the British. Resentment of religious discrimination by Grenada's Roman Catholic French residents led to a 1795 French-assisted rebellion which failed to reunite Grenada with France.

The two-stage abolition of slavery in 1834 and 1838, though the 1834–8 "apprentice" system was a thinly veiled transparent form of slavery, forced a hastened transformation of the economy. When labor recruited from India was inadequate to fill the void emancipation had left, many plantations were broken up into smaller farms or let out to the ex-slaves for sharecropping. Furthermore, the decreasing international competitiveness of sugar led to replacement of work-intensive cultivation of sugar with that cocoa and nutmeg.

Brizan has written "the emancipation of the Blacks in 1838 raised the fundamental issue of the form of government that would develop in the West Indies . . . In Grenada, the political dominance of the plutocracy was prolonged until 1877, when crown colony government was introduced." The imposition of a British-appointed governor, all-appointed Legislative Council, and the termination of the five elected town and parish councils assured that the designation of Grenada as a crown colony was viewed as regressive by all but the elite who were "in" with the governor.

In 1917, Theophilus A. Marryshow founded the Representative Government Association (RGA) to promote more participative government for the Grenadian people. Prodded by the RPA, lobbying the Wood Royal Commission of 1921-2 recommended a constitutional reform modifying

---

55. Jane Harris, in her recently published novel *Sugar Money*, uses a fictional tale to tell a real life one—in this case of a religious order to round up and bring slaves remaining in newly British-occupied Grenada to French-owned Martinique.

the crown colony government by granting 4 percent of Grenadians (the wealthy property owners) the right to elect 5 of the 15 members of the Legislative Council.

In 1950, Eric Gairy founded the Grenada United Labor Party (GULP). In 1951, GULP led a general strike, which aroused severe unrest and destruction. British troops quelled the situation. In the 1951 general elections—the first based on universal adult suffrage recently negotiated—Gairy's party won six of the eight contested seats. In 1959, new constitutional amendments became effective. From 1958 to 1962, Grenada was part of the Federation of the West Indies. The Federation's mid-60s breakdown led to an attempt to federate the so-called "Little Eight islands," which included Grenada; but, when Barbados achieved independence singly without the others, the effort folded. During the 1960s, the multi-island Regional Security System (RSS) and the Caribbean Free Trade Association were formed.

In 1967, Grenada became an associate state in the British Commonwealth, a status short of full independence but one that gave it full autonomy over its internal affairs while the British government retained control of its foreign affairs and defense. Herbert Blaize served a few months as the first premier. Eric Gairy succeeded him as premier, serving from 1967 to 1974. Grenada achieved its independence despite the fact that the opposition with a more radical agenda boycotted the negotiations with Britain. Gairy became the first prime minister, but civil strife developed between the Eric Gairy-led GULP and the opposition parties. Prominent among the anti-Gairy opposition was the New Jewel Movement (NJM) led by Maurice Bishop whose precursor organization, Joint Endeavor for Welfare, Education, and Liberation (JEWEL), had already begun to call for independence and for a whole range of programs in 1973—not only agricultural reform, education, and health but also nationalization of banks and insurance companies—with rhetoric that embraced Marxist-Leninist rhetoric.

Nevertheless, Gairy's GULP won the 1976 elections. Claiming fraud, Bishop's People's Revolutionary Government (PRG), successor to the NJM, overthrew the government, suspended the constitution, and ruled by decree. With the help of Cuban doctors, teachers, and technicians, strides were made in health, literacy, and agriculture. Bishop cooperated with Cuba and the USSR on selected trade and foreign policy issues but projected a "non-aligned" status in the ongoing Cold War. Considering Bishop insufficiently Marxist, the Marxist party members, led by Deputy Prime Minister Bernard Coard, unsuccessfully demanded that Bishop abdicate or share power.

Having failed to bully his way into power, Coard and his American-born wife Phyllis, with the support of the Grenadian Army, led a coup against Bishop in 1983 and placed him under house arrest. When Bishop attempted to resume power, he was captured and executed along with seven others. Coard put Grenada under martial law. U.S. President Ronald Reagan became concerned with what he had believed was a pro-communist government.[56] President Reagan sent United States troops to occupy Grenada, asserting that the action was taken to protect the lives of U.S. citizens (mainly students at the local medical school) despite the lack of any evidence that they needed or requested any protection. Grenadian reaction to the invasion was mixed.

Britain, Canada, and Trinidad and Tobago criticized the invasion; the UN General Assembly condemned it as "a flagrant violation of international law." After the invasion, the pre-revolutionary constitution was restored, and 18 members of the PRG, and the top-ranking military were arrested for the operation that led to the murder of Bishop and his colleagues; 14 were sentenced to death, but the sentences were commuted to prison sentences. Following the withdrawal of U.S. troops from the country, the Grenadian Governor-General Paul Scoon appointed Nicholas Brathwaite of the National Democratic Congress (NDC) as interim prime minister.[57]

The 1984 elections, the first democratic ones since 1976, were won by the Grenada National Party (GNP) led by Herbert Blaize, who served as prime minister until his death in 1989, when he was succeeded by Ben Joseph Jones. The NDC, led by Brathwaite, won the 1990 elections; he served until he resigned in 1995 and was briefly succeeded by George Bri-

---

56. Reagan believed this, in part, because a 10,000-foot airstrip was being built that he believed Cuban and Soviet planes might use despite assurances from the European Economic Community and airfield contactors that these concerns lacked substance.

57. In 2017, the words "Thank you Ronny," painted conspicuously in bold letters on a barn, reminding passersby that their view of the war remained poignant.

zan. In the 1995 elections, the New National Party (NNP), led by Keith Mitchell, won, winning again in the 1999 and 2003 elections. His record 13 years as prime minister ended with the 2008 elections that the NDC won. Then, Keith Mitchell returned to office in 2013 and 2018 with a clean sweep of all 15 seats in the Legislature.

The British monarch is represented by a governor-general as head of state whereas the prime minister is selected by and supported by the majority party in the elected lower house. The legislature consists of an appointed Senate as well as the 15-member House of Representatives. Grenada (which includes the two southern-most Grenadine islands) is divided into six parishes, which are merely electoral districts for the national legislatures.

Grenada's top court is its Supreme Court. From there, one may appeal to the Court of Appeals for the east Caribbean and to the Privy Council based in London. Celia Edwards QC (Queen's Counsel, an honorific) warns, "In theory one can appeal to the Privy Council . . . in fact one must be very wealthy to do so. Such appeals cost several hundred thousand dollars."[58]

Grenada belongs to the United Nations and several other global specialized agencies including the World Trade Organization (WTO). It also belongs to the OAS, CARICOM, OECS and its East Caribbean Central Bank (ECCB), which maintains the East Caribbean dollar. Finally, it belongs to the Commonwealth and to the ACP. It has a mission in Washington, D.C., that serves the United States and the United Nations.

Grenada's slavery heritage has created its dominant ethnic group. Of the 110,000 plus Grenadians in the world, more than 90 percent are descendants of African slaves. Of those 90 percent, 82 percent are considered black, and another 13 percent are considered "mixed." East Indians, descendants of those recruited from India to work on plantations after the abolition of slavery, constitute about 3 percent, and those of French or

---

58. In 2018, Grenada and Antigua and Barbuda voted to replace the Privy Council as their ultimate court of appeal with the Caribbean Court of Justice—to which 13 Caribbean countries belong.

English descent make up the rest.[59] More than 53 percent those living in Grenada are Roman Catholic, 14 percent are Anglican, and 33 percent are "members of various Protestant denominations." The Grenadian heritage is also reflected in everyday speech. While English is the official language, everyday conversation is in either an English-based creole or a French-based creole, which reflects the population's roots.

Grenada's economy has shifted over its history from tobacco, to sugarcane, coffee, cocoa, and nutmeg, which is second only to Indonesia in its production. Major promotion has developed tourism into Grenada's major industry, accounting for $120 million in the mid-2010s. The labor force comprises 69 percent in services, 11 percent in agriculture, and 20 percent in industry. In 2012, 40 percent of Grenada exports ($40.5 million) went primarily to Nigeria and nearby islands. Grenada's 2012 imports ($297 million) were primarily to Trinidad and Tobago. The Grenada GDP was $1.5 billion; the per capita GDP was $13,800. The 2012 national budget was $196 million. St. George's University—with its medical and other programs with 9,000 students from 140 countries—is a distinct economic and prestige boost for the island nation.

A primary challenge facing Grenada is the continued improvement of the tourism industry. The island faces the mutually interdependent tri-fold challenge of developing hotels, air (and sea) ports, and airline scheduling. Global increasing protectionism, especially in the U.S., is a threat. Hurricane Ivan struck Grenada in 2004. The 2017 hurricanes missed Grenada; recognition of the continuing threat is a continuing challenge.

---

59. Emigration abroad to the United States, Britain, Canada, and even Australia has attracted so many that as few as one-half of those born in Grenada continue to live there.

SAINT VINCENT AND THE GRENADINES

"The 1975 Lomé Convention had outlined special trade provisions for the African, Caribbean, and Pacific states (ACP) to help develop their economies, though [the 2012] change in banana tariffs will cause the price of Latin America bananas to drop, eroding the advantage enjoyed by fruit growers in Dominica, Saint Vincent and other Windward Islands, and potentially ending their livelihood as they are priced out of the market. "

—Carrie Gibson

Saint Vincent and the Grenadines is a micro-country consisting of a main island and several smaller ones west of Barbados and north of Grenada. It was "discovered" by Columbus in 1498, colonized by French in 1719, ceded to the British in 1763, became a British Associated State in 1969, and gained its independence in 1979.

The warlike Caribs fended off European settlement for over two centuries until 1719, when French from the island of Martinique settled on the Leeward side of Saint Vincent Island. During the Seven Years' War in 1756–1763 (called the French and Indian War in North America), the British occupied the island; the island was ceded by France to Britain after the war.

Friction between the British and the Caribs led to the First Carib War. The French gained control of Saint Vincent in 1779, but the British regained the island by the treaty after the war. Increasingly resentful of British rule, the Caribs revolted in 1795; with French support, they gained control of the island. The British regained control the following year and deported most of the Caribs to British Honduras (present day Belize).

By the 1800s, sugar plantations were developed, and African slaves were imported to work them. The abolition of slavery in 1834 throughout the British colonies—and the subsequent termination of the "apprentice" period in 1838—led to labor shortages and the recruitment of indentured workers from Madeira, Portugal, and India.

A representative assembly was initiated in 1776, a crown colony was established in 1877, a legislative council was created in 1925, universal adult suffrage was introduced in 1951, and "associate state" status was introduced in 1969. Through this status, Saint Vincent and the Grenadines controlled internal affairs, and the British continued to manage external affairs. Saint Vincent and the Grenadines gained independence in 1979. Saint Vincent was grouped in the West Indies Federation, which disbanded in 1992.

The first political party to develop mass support was the People's Political Party (PPP), founded in 1952 by Ebenezer Joshua. The second party to emerge was the Saint Vincent Labour Party (SVLP) founded in 1955 by R. Milton Cato, whose conservative platform stressed law and order, a pro-Western foreign policy, and a mixed economy; the black middle class were its main support. Another contender was the United People's Movement (UPM) formed in 1979 by Ralph Gonsalves. Drawing much of its support from trade union members and at the forefront of national policy making prior to independence, the pro-Western PPP won elections in 1957, 1961, and 1966. Largely because of the emergence of a more conservative middle class, the PPP waned in strength after the late 1960s. It formed a coalition government in 1971 and relinquished the post of premier to James Mitchell who had earlier resigned from the SVLP. The coalition collapsed in 1974, leading to a PPP election win. But the PPP lost decisively to the SVLP in 1979. The SVLP, in the 1984 elections, suffered a humiliating defeat at the hands of the New Democratic Party (NDP) founded by Mitchell in 1975. The PPP disbanded in 1984 and reorganized in 1987 as the People's National Movement (PNM). The Gonsalves-led Unity Labour Party won in 2001 and subsequent elections, serving as prime minister of Saint Vincent since 2001.

A governor-general represents the British monarch as head of state; the prime minister is head of government and leader of the majority, or plurality, party and, with his cabinet, is responsible to a unicameral legislature. The Assembly is composed of 15 members elected from single member constituencies and 6 appointed by the governor-general (4 on the prime minister's advice and 2 on the leader of the opposition's advice): "although [this] political system was one of the most stable of Britain's former colonies . . . the leadership of the political parties was erratic; founders of one party frequently emerged as leaders of another party only a few years later." The country is divided into six administrative divisions, five on Saint Vincent and one covering the Grenadine islands. The independent judiciary is divided into district courts, appeals may go to the Eastern Caribbean Supreme Court (a joint institution of the six Organization of East Caribbean States, which is headquartered in Saint Kitts) and then to the Privy Council in London, but the latter is rarely used. Saint Vincent and the Grenadines, like the other Lesser Antilles islands, belongs not only to the United Nations and several affiliates but also to the Commonwealth, the OAS, CARICOM, and OECS and its affiliate the East Caribbean Central Bank. The country maintains diplomatic missions at the EU and the UN, where its mission to the United States is co-located.

In 2013, the ethnic composition of the 100,000 plus population of Saint Vincent and the Grenadines consisted of 66 percent African descent (most Vincentians are descendants of Africans imported as slaves), 19 percent mixed descent, 8 percent east Indian, 4 percent European, and 2 percent Carib Amerindian. While the official language is English, which is used in government, business, education, and formal occasions, informally, most Vincentians speak Vincentian creole, a dialect that reflects their French and English heritage. Three-quarters of the people are Protestant (47 percent are Anglican and 28 percent Methodist); 13 percent are Roman Catholic; the others are Seventh-day Adventist and Hindu.

In the mid-2010s, Saint Vincent and the Grenadines' GDP was $1.3 billion; its GDP per capita was $12,100. Agriculture, long dominated by banana production since the fading of the sugar plantocracy, remains the staple (albeit threatened) of this lower middle-income economy. A growing tourism industry is the other major industry. The government has been less successful in promoting manufacturing. Agriculture accounts for 26 percent of the labor force, tourism and other services 57 percent, and industry only 17 percent. High unemployment continues to be a major challenge with rates at 20 percent. Most of the exports (66 percent) go to nearby Lesser Antilles countries. Most imports ($301.5 million) come from Singapore (28 percent), Trinidad and Tobago (23 percent), and the United States (19 percent). The national budget in the mid-2010s was around $185 million.

The country's continued dependence on bananas is a major challenge. Tropical storms have wiped out substantial portions of crops for years. Employment and the standard of living will be difficult for the nation to maintain.

SAINT LUCIA

"Saint Lucia, however, was not to enjoy a lengthy period of peace. Military conflicts among the Dutch, British, Spanish, and French, both on the European continent and in the colonies, resulted in Saint Lucia's falling alternately under the control of France and Britain 14 different times in the 18th and early 19th centuries."

—John F. Hornbeck

Saint Lucia, located between Martinique and Saint Vincent and the Grenadines, has, like other Caribbean islands, a mixed cultural heritage including Amerindian, African, Asian Indian, and European backgrounds. Carib Amerindians inhabited Saint Lucia when a Spanish ship (many believe captained by Columbus) first sighted the island in 1502. It was not until almost a century later that Europeans made their first efforts to settle on the island. The Dutch made the first brief attempt in 1600. The English attempts to plant settlements in 1605 and 1638–9 were met with disaster at the hands of the Caribs. Not until after French King Louis XIII ceded the island to the French West India Company in 1642 and the Caribs were appeased with a treaty in 1660 did the French succeed in planting a settlement. In 1674, the French settlements were annexed to the French dependency of Martinique.

During the 17th- and 18th-century European wars and the corollary-American colonial wars, the British and French fought for control of Saint Lucia. Its deep-water harbors and strategic location provided a base for monitoring maritime activity in the Caribbean. After the sugar industry developed, increasing the prosperity of Saint Lucia, the British and the French invested much military might in contesting the island. There were 14 changes in control of Saint Lucia in the turbulent late 1700s! Half of these turnovers occurred within the two decades before and during the Napoleonic Wars. The turnovers and consequent turmoil enabled many slaves to escape. In 1794, the French governor declared all slaves were free, but the decree was unevenly carried out. The French government ceded Saint Lucia to the British in 1815.

When the British finally regained control of the island, many more slaves escaped into the forests; there, they mixed with the Amerindians already there, forming Maroon (mixed ethnic) communities. Slavery on the island continued for a short time, but the rise of anti-slavery sentiment in Britain led to the abolition of the slave trade in 1807, the abolition of slavery in 1834, and the end of its maligned "apprenticeship" extension in 1838. By that time, people of African descent far outnumbered those of other descents.

The continued presence of the French settlers on the island after the British secured it led the British administration to resist allowing popularly elected local assemblies—as were allowed, or tolerated, in other British Caribbean possessions. Instead, the British imposed a crown colony government on Saint Lucia by which the British governor ruled in conjunction with an appointed Legislative Council. Not until 1924 did a constitution introduce representative government. For the next 34 years, though, there was only incremental progress toward locally controlled governance.

In 1958, Saint Lucia, encouraged by the British government, joined the short-lived West Indies Federation, which dissolved in 1962. Following its dissolution, Saint Lucia became an Associated State in which the island nation gained control of local domestic affairs while Britain retained control of defense and foreign affairs; it became a member of the West Indies Associated States.

In 1979, Saint Lucia gained independence and joined the Organization of East Caribbean States (OECS). The new constitution, copied from the British one, provided for the British monarch (represented by a governor-general) and a prime minister selected by and responsible to the assembly. The assembly was comprised of 17 elected members (each from a single member electoral district) and an appointed 11-member senate. Independent Saint Lucia's first prime minister was Sir John Compton, the leader of the conservative (compared to its opposition) United Workers Party (UWP). He served until 1996, when he was succeeded by Vaughan Lewis. During these first two decades of independence the opposition parties included the Saint Lucia Labour Party (SLP), the Progressive Labour Party (PLP), and briefly a relatively new party formed from a disaffected split from the SLP in 1982. The SLP, led by Dr. Kenny Davis Anthony, won the election in 1997 and was reelected, serving until 2006. In 2006, the PLP, again led by Compton, won. Compton served until he suffered a series of strokes, died, and was succeeded by Stephenson King in 2007. The Anthony-led SLP returned to office in 2011.

The multiple heritage of Saint Lucia is reflected in its present-day institutions and demography. Its slavery legacy is evident in the fact that more than 85 percent of the more than 163,000 inhabitants of Saint Lucia are of black or African descent, and another 11 percent are mixed. The rest include those of Asian Indian (24 percent) and European and North American descent. The population continues to grow despite significant emigration, primarily to Britain and North America. While English is the official language, almost everyone speaks Patois, an Antillean Creole that emerged during the early period of French colonization, based upon French and west African languages, with some Carib words thrown in.

The French colonial legacy accounts for the fact that 62 percent of Saint Lucians are Roman Catholic—only 26 percent are Protestant, including Seventh-day Adventism (10 percent), Pentecostalism (9 percent), and Baptist and Anglicanism. The culture and food reflects Amerindian, African, and Asian Indian, as well as French and British traditions. Saint Lucia is unusual in that belongs not only to the United Kingdom-led Commonwealth but also to the France-led La Francophonie. Saint Lucia belongs to the United Nations and affiliated agencies, the OAS, CARICOM, and the Organization of East Caribbean States (OECS).[60]

In the mid-2010s Saint Lucia had a $2.2 billion GDP and a $13,100 per capita GDP, supported by an economy based principally upon bananas, tourism, and offshore banking. Its labor force was 22 percent agriculture, 25 percent industry, and 54 percent services including tourism and government. The extent on Saint Lucia's development was an incentive to the rapid repair of the 2017 hurricane damage.

With its educated workforce and improvements in infrastructure, the island nation has attracted foreign investment, especially in offshore banking, which is the island's main source of revenue. Its imports (totaling $593 million) in the mid-2010s were primarily from Brazil. Its exports (totaling

---

60. Its principal town and capital, Castries, serves as the seat of the OECS Commission, the governing body of the OECS.

$207 million) were scattered throughout the U.S., UK, Peru, France, and a few nearby islands. The 2011 estimated budget was $222 million.

Major concerns include the continued diversification of the economy, the continued development of tourism that at present is dependent largely on cruise ships, development of light industry, and the revitalization and privatization of the banana industry. Of particular concern is the change in the EU import preference system and the accompanying increased competition from Latin American bananas.

THE COMMONWEALTH OF DOMINICA

"The early establishment of a group of free black inhabitants, . . . many of whom later owned small estates and slaves, . . . [led to a] unique mix of slave plantations owned by Europeans and Africans, existing alongside small garden plots and farms cultivated by escaped slaves, freed slaves, and Carib Indians, charted a markedly different colonial course for Dominica compared to that of the sugar colonies of Barbados and Jamaica."

—Atherton Martin

Dominica, a mountainous Caribbean island inhabited by almost 75,000 people, lies between the French islands of Martinique and Guadeloupe.[61] Dominica was settled successively by Arawak and Carib Indians, "discovered" by the Spanish, colonized by the French, then by the British. Eventually, Dominica secured independence. Its rugged terrain facilitated the early Carib Indians' resistance against European colonizers, later provided a haven for escaped slaves and facilitated French resistance to British aggression.

Dominica's enduring multiple heritages reflect its multi-ethnic, or eclectic, culture. Like Saint Lucia, Dominica provided significantly more French resistance to British influence than the other islands, which the continued prevalence of the markedly French-based Patois language demonstrates. For more than two centuries of British administration Dominica has undergone administrative combinations with other islands, before being grouped from 1958 to 1962 in the more inclusive 10-member West Indies Federation. Then, after gaining Associated State status, it was grouped in the West Indies Associated States. Upon achieving independence in 1978, it became a founding member of the Organization of East Caribbean States (OECS).

As happened elsewhere in the Caribbean, the more aggressive Carib Indians displaced the Arawak Indians around 900 CE. Six centuries later,

---

61. Dominica is not to be confused with the Dominican Republic, a Spanish-speaking country with more than 10 million inhabitants.

Columbus first sighted the island in 1493 on his second voyage to the Americas. As the Spanish extended its empire in the Americas in the 1500s, its use of Dominica was limited to the securing of timber and fresh water; the aggressive behavior of the Caribs posed too hostile a threat to potential settlers. In 1632, the French Compagnie des îsles d'Amérique claimed Dominique and other Lesser Antilles for France. By the late 1600s, French woodcutters from Martinique and Guadeloupe had gradually become permanent settlers. They were augmented by white smallholder farmers from Martinique and Guadeloupe.

As these settlers developed their farms and plantations, they met their growing labor needs by importing African slaves. The Dominican practice of allowing slaves to tend their own gardens, sell the produce at the local markets, and purchase their own freedom led to many African freedmen owning property and slaves alongside the French settlers.

The continued evolution of this hybrid French-African socio-economic community was interrupted by the Seven Years' War, during which the British, after several battles, occupied Dominica in 1761. In 1763, France formally ceded the island to the British. The British takeover introduced several changes that significantly affected Dominica's governance. By 1773, Dominica had a two-chamber legislature: a limited franchise elected House of Assembly and a governor-appointed Council. Most of those entitled to vote were the British landlords; excluded from suffrage were the free blacks and those who did not repudiate the basic tenets of Catholicism, which effectively limited political participation by the French. While many British purchased estates in Dominica, they tended to become absentee landlords. Dominica's French inhabitants long continued resisting British efforts to "British-ize" the established French-oriented culture of Dominica.

The American War of Independence, by distracting British attention to the North American mainland and disrupting trade between the British North American colonies and Dominica, provided the French the opportunity to reclaim Dominica and other West Indies colonies. But, following the terms of the treaty after the war, Dominica and the other Caribbean islands were returned to the British.

The British abolition of the slave trade in 1807, passage of a bill granting political and social rights to free colored, the abolition of slavery in 1834, and the termination of its internship extension in 1838 fundamentally altered Dominica's socio-political climate. By 1838, the elected House of Assembly, unlike those in other British Caribbean possessions, had a majority of ethnic African and mixed descent inhabitants. Reacting to a perceived non-white threat to their power, the planters appealed to British colonial office for more direct rule. In 1865, the colonial office replaced the wholly elected Assembly with a half-nominated one. In 1871, Dominica was merged into a Leeward Islands administration, which, in 1896, the United Kingdom made a crown colony.

Following World War I, a rise in political consciousness throughout the Caribbean led to the formation of the Representative Government Association, which won one-third of the popularly elected seats in the partially elected assembly in 1924—and one-half in 1936. Shortly thereafter, Britain removed Dominica from the Leeward Islands administration.

Riots throughout the West Indies in the 1930s led to the release of the Report of West India Royal Commission (the Moyne Report) in 1940. The report called for a comprehensive development program that would facilitate investment in infrastructure, education, health facilities, and jobs. Dominica was then governed as part of the Windward Islands until 1958, when it became part of the short-lived West Indies Federation. The federation, which folded in 1962, left Dominica to be governed alone.

Since the 1950s, demands for better conditions drove social and political activism and the emergence of trade unions and political parties representing the interests of workers and small farmers on one side and business interests on the other. This activism facilitated the foundation of the Dominica Labor Party (DLP) led by Edward Oliver LeBlanc defeated the Dominica United People's Party in the 1961 election and subsequently led the DLP to victory again in the 1965 and the 1970 elections. Atherton Martin writes the following:

> In 1968 the LeBlanc government responded to incipient signs of social unrest . . . [by] pushing a bill through the House of Assembly to curb criticism of government officials. This act . . . resulted in the formation of the Dominican Freedom Party (DFP) under the leadership of Mary Eugenia Charles.

In 1967, he negotiated Associated State status for Dominica in which Dominica ran its internal affairs and Britain ran its external affairs. In 1974, LeBlanc resigned as premier (the pre-independence title for the head of the government) to be succeeded by Patrick John, whose party won the 1975 and 1980 elections and who, after independence, served as Dominica's first prime minister. John's political style and prejudices were evident when he "declared war on . . . 'dreads,' intending to make it illegal for them to grow their hair in locks." The Dreads Act passed, Carrie Gibson asserted, partly because "John had linked the dreads and Black power with communism."

In 1978, Dominica achieved full independence. After elections in 1980, the Dominica Freedom Party (DFP) replaced the interim administration led by Prime Minister Patrick John with Eugenia Charles, the first female prime minister in the Caribbean. In 1981, there was a "never-left-the dock" failed attempt allegedly by North American interests to help ex-prime minister John regain power in exchange for gaining control of development. The United Workers' Party (UWP), one of the major parties by the 1980s, won the 1995 election and installed its leader, Edison James, as prime minister. In 2000, the DLP returned to power, but its leader and prime minister, Pierre Charles, died before the scheduled 2005 elections. Roosevelt Skerrit, then 31 years old, led the opposition DLP to victory in the 2005 and subsequent elections.

Like Saint Lucia, but unlike the other Commonwealth Caribbean countries, Dominica opted to become a republic with an elected president—not with the British monarch as head of state. But, in other respects, Dominica is a British-style parliamentary democracy in which the prime minister, elected by the legislature, heads the executive. The unicameral legislature, the 31-member House of Assembly, consists of 21 directly elected members and 9 presidentially appointed senators.

Dominica comprises 10 parishes, ranging in population from that of Saint George (more than 21,000) where the capital Roseau is located, to that of Saint Peter (1,430). Dominica belongs to the UN and several of its affiliates and maintains missions in the EU and Washington, D.C., serving the United Nations and the United States. It belongs to the OAS, CARICOM, and OECS and its OCCB. Like Saint Lucia, with which it shares a strong French culture, Dominica belongs not only to the Commonwealth but also to the Organisation internationale de la Francophonie (OIF), the French equivalent of the Commonwealth.

Dominica, with its population of 73,448, is the 10th least populated country of the world; 87 percent of the population is of African descent, another 9 percent is mixed, 3 percent is Carib Indian (living on land reserved for them by the British in 1903), and 1 percent is European (mainly descendants of French, British, and Irish colonists). While English is the official language, a French-based Antillean Creole that developed under French colonial rule, which has survived through almost two centuries of British administration, remains in everyday use, especially among older inhabitants. More than 60 percent of Dominicans are Roman Catholic, and more than 30 percent are Protestant with one-third being Jehovah's Witnesses.

Bananas and other agricultural products have dominated the vulnerable agricultural economy of Dominica, which, in the mid-2010s, had a GDP of barely $1 billion with a per capita GDP of $14,300. Dominica's reliance on agriculture is reflected in the fact that 40 percent of the labor force is engaged in it compared to 32 percent in industry and 28 percent in services. Even before the devastating hurricanes in 2017, the many challenges facing Dominica included improving its infrastructure, which had been limited to one main road, one port, and two small air fields. The IMF had pointed out other challenges: the further reduction of its public debt, increased financial sector regulation, and further diversification of the economy.

Efforts to develop a niche boutique tourism industry that accents snorkeling and hiking based on the untrammeled typography of its scenic mountainous terrain are underway. Other diverse efforts include developing coffee and exotic fruits as agricultural products. Manufacturing soap was the most significant effort to develop light industry. Dominica's imports (totaling $220 million) were mainly from Japan, the United States, and Trinidad and Tobago. Its exports (totaling $40 million) are principally to Japan, and nearby islands. Dominica's mid-2010s national budget was $185 million.

In 2017, Hurricanes Irma and Maria devastated the island. Prime Minister Roosevelt Skerrit described the damage as "mind-boggling." Over 90 percent of the nation's buildings were damaged or destroyed, power lines were toppled, and landslides and raging rivers tore away roads and bridges. Winds of up to 160 miles per hour left virtually nothing else undamaged; the immediate death toll was at least 31; 50 more were reported missing. A police officer was on duty when the hurricane struck. The roads were so torn up and blocked with debris that he had to walk home, a journey of eight hours. Residents were evacuated to other islands or to civic centers. Eighteen months later, many former residents had not returned. Most still lived in tents or houses with blue tarpaulin roofs, especially in the rural areas.

By late 2018, the roads had been cleared, roofs had replaced the tarpaulins covering the buildings, and the capital was busy, but power was still lacking. Recovery could take years and billions of dollars for the construction of stronger housing, commercial buildings, and the infrastructure upon which its existence depends! International attention focused on the hurricane damage to the American states of Texas and Florida, and the U.S. territory of Puerto Rico. Little to no coverage was given to the smaller countries, and the even more devastated independent Dominica received less.

In the words of Liam Localio, principal of a firm specializing in sovereign debt, the state of the Caribbean was as follows:

> Even before the devastation caused by Hurricane Maria Dominica was a minor actor in the eastern Caribbean. Without meaningful air traffic from the U.S. and Europe, the country's tourism is limited to boutique and eco-tourism. No large hotels and resorts, such as those on many neighboring Caribbean countries, have been developed. Its sub-75,000 population lacks the critical mass for robust domestic industry.

Bananas brought in a third of foreign earnings a generation ago but have dwindled to zero since 2008, when the EU terminated privileged access to its market. Dominica's main manufacturing venture, which made soap and toothpaste, closed after tropical storms. Tropical Storm Erika struck a few years before Maria. The American-owned Ross University School of Medicine, which directly accounted for 8 percent of the GDP, was closed by the storm and relocated to Barbados.

Despite its mountainous beauty, can Dominica attract sufficient funds to be reconstructed sufficiently to resist the devastation of future storms? Can it be reconstructed sufficiently to develop a sustaining economy? The dire environmental and economic outlook limits the hope for such significant aid. Dominica faces a bleak future.

ANTIGUA AND BARBUDA

"As sugar took hold in the Leeward Islands, the plantation complex established in Barbados repeated itself. The islands became increasingly deforested, smaller farms were displaced, an elite group of powerful planters consolidated land into larger estates, all available land was shifted into sugar cultivation, and the number of enslaved Africans increased dramatically. They even experienced their own period of outward expansion as rich planters engrossed the best lands and displaced small planters who sought new opportunities in the smaller nearby islands of Barbuda, Anguilla, Tortola, Virgin Gorda, and Saint Croix."

—Matthew Mulcahy

Antigua and Barbuda are two small flat islands with islets east of Saint Kitts and Nevis in the Leeward Islands. Columbus sighted one of these islands in 1493. Carib hostility and insufficient potable water long discouraged European settlement. In 1632, the English claimed Antigua and Barbuda and colonized them. From the 1670s, the development of plantations made Antigua one of the most profitable of Britain's Caribbean colonies. The two together were granted their independence in 1981. Since 1949, with two interruptions, the politics of Antigua and Barbuda has been dominated by the populist Antigua Labour Party (ALP), whose leadership passed from father to son from the 1940s to 2012.

English colonization of Antigua was only interrupted once. The French occupied it in 1666 during the Anglo-French/Dutch War, but it was returned by treaty after the war. From 1674, Sir Christopher Codrington developed Antigua, establishing sugar plantations and the importing of slaves from Africa to work the fields. The extent of the demographic impact of the establishment of plantations and the importation of slaves has been pointed out by Matthew Mulcahy:

One parish in Antigua provides a microcosm of the social changes accompanying the rise of sugar. There were 53 taxpayers in St. Mary parish in 1688. Sixteen were slaveholders, but only six owned

20 or more slaves. By 1706, the number of taxpayers fell to 36, 30 of whom owned slaves. Sixteen planters owned 20 or more slaves, while four owned more than 100.

From 1685 to 1870, Codrington leased Barbuda from the crown; it was annexed to Antigua in 1860. During its colonial rule, Antigua and Barbuda, like other Caribbean islands, underwent several jurisdictional combinations with other settlements. Antigua and Barbuda were initially included in the Leeward Caribbean Islands' administration. In 1806, the Leeward Caribbean countries were split into separate administrations with Antigua separated from Saint Kitts-Nevis-Anguilla. The Leewards were re-united as a single administrative entity in 1871 with Dominica included in the grouping.

The empire-wide abolition of slavery by the British Parliament in 1834 disrupted the plantation system. But, rather than extend the system with mandated apprenticeships, the Antigua government freed its slaves. Nevertheless, the plantation owners continued to exploit their workers.

In the 1930s, the Great Depression drove a decline in the price of sugar, which, combined with a severe drought, badly damaged the Antiguan sugar industry. The resulting depression further undermined the already poor social conditions and aroused protests that threatened law and order. The British established the Moyne Commission in 1938 to investigate the Caribbean situation. Encouraged by the British Trades Union Congress, workers in Antigua formed the Antigua Trades and Labour Union (ATLU) in 1940. In 1946, the ATLU established a political wing, the Antigua Labour Party (ALP), which ran and elected five parliamentary candidates. One of these, Vere Cornwall Bird, Sr., was selected to serve on the Antiguan Executive Council. In 1961, he was appointed to the newly created post of chief minister (the title given, along with premier, to the head of government in a dependency).

In 1958, the colony became part of the West Indies Federation, which dissolved soon after in 1962. Following a Bird-led delegation to London to negotiate independence, a 1967 constitutional conference determined that Antigua, with Barbuda and Redonda as its dependencies, become an Associated State (i.e., autonomous in domestic affairs but externally continuing its dependence on Britain) and participate in the West Indies Associated States.

During the period of associated statehood (1967–81), Antigua experienced the rise of the following organizations and events: a second labor union and its affiliated political party, the Progressive Labour Movement (PLM); a secessionist movement in Barbuda; and tourism replacing sugar as the major driver of the economy. The ALP continued to dominate the political scene, winning every general election, except PLM's win in 1971. In the 1980 general elections, the ALP campaigned calling for full independence. Its victory, considered a popular mandate, led to another constitutional conference. Overcoming the issue of Barbudan separatism by granting it substantial internal autonomy, Britain granted independence to Antigua and Barbuda in 1981. Bird went on to lead his party and serve as prime minister until 1994. His second son Lester Bryant Bird, who had served as deputy prime minister under his father, succeeded him as prime minister, ending the first Bird's multi-decade career as the ALP leader and head of government.

In 2004, suffering from allegations of corruption on the part of Vere Bird, Jr. (the older son of Vere Bird, Sr.), the ALP lost the general election to the United Progressive Party (UPP). The ALP lost again to the UPP in 2010, causing the ALP, at its party biennial convention in 2012, to elect Gaston Browne to replace Lester Bird as the ALP party leader and to rename the party the Antigua and Barbuda Labour Party (ABLP). The ABLP went on to defeat the UPP in the 2014 general election; its leader, Gaston Browne, became prime minister.

Antigua and Barbuda, like most other former British colonies, has a Westminster-model government; the British monarch, represented by a governor-general, is head of state, and a prime minister is head of government, supported by a cabinet. The government is selected by, and may be dismissed by, the bicameral legislature, which consists of a senate—whose 17 members are nominated by the majority and opposition parties in the lower house and approved by the governor-general—and a House of Representatives—whose 17 members are elected from single-member districts. The country is divided into six parishes and two dependencies (i.e., parishes that retain their historic title). The judiciary is the Eastern Caribbean Supreme Court, based in Saint Lucia, of which one judge resides in

Antigua. While a few other Caribbean countries have abolished the right of appeal to the British Privy Council in favor of the Caribbean Court of Justice, Antigua and Barbuda continue to maintain the Privy Council as a seldom-used and prohibitively expensive court of appeal.

The Antigua and Barbuda government participates in several international organizations: at the world level, the UN, its affiliates, and the World Trade Organization. Along with most of the other former British colonies, and many remaining dependencies, it belongs to the Commonwealth, several regional organizations including the OAS, CARICOM, and, most important, the OECS and its central bank (OCCB). It maintains missions at the EU and in Washington, D.C. (the latter is accredited to the United States and the United Nations).

Of the 90,000 inhabitants of Antigua and Barbuda, 91 percent are black or mulatto. "Whites" (generally British or Irish) constitute less than 2 percent of the population. The rest are mixed, East Indian and other Asian, Portuguese (Madeiran), Middle Easterners (many of whom are shopkeepers), and a few immigrants from other West Indies countries. And a few thousand American citizens now make these islands their seasonal or full-time homes. As in other Caribbean islands, the history of racial tensions that preceded the late 1900s redefined the role of those of African descent in the islands' political life. The shade of skin color stratifies society. Over 75 percent of the population profess one of many denominations of Protestantism—principally Anglicanism (26 percent), but also Baptists, Presbyterianism, Adventism, Pentecostalism, and Moravianism; 10 percent are Roman Catholic. The official language is English. Informally, many speak an English-based Antiguan Creole.

The country's agricultural production (mainly fruits, vegetables, bananas, coconuts, and livestock) focuses on the domestic market. Constrained by a limited water supply and a labor shortage stemming from the higher wages present in tourism and construction, only 7 percent of the islands' labor force is in agriculture. Manufacturing focuses on assembly for export, the major products being mattresses, electronics, and handicrafts. Its mid-2010s imports totaled $400 million; its exports totaled $40 million. Tourism has long replaced agriculture as the dominant driver of the Antigua-Barbuda economy, accounting for more than half of the mid-2010s $1.6 billion GDP, and the per capita GDP of $16,300. In 2009, the economy supported a $293 million budget. Antigua attracts many cruise ships, possesses many luxurious resorts, and lures many snowbirds and retirees to own or rent homes. Investment banking and financial services contribute significantly to the economy. Resorts and banks, being financially intensive investments and operations, are often foreign-owned multinational operations, which remit their profits abroad.

The growth challenges facing Antigua and Barbuda were magnified in September 2017 when Hurricane Irma struck the northeast Caribbean with battering rain and winds of up to 185 miles an hour. Gaston Browne, Antigua and Barbuda's prime minister, reported that Hurricane Irma destroyed the banana plantations.

Over 95 percent of the structures were destroyed on the smaller island of Barbuda, whose pre-hurricane population was less than 1,800. Most of Barbuda's residents lived in the village of Codrington, which possessed only one gas station. Eighteen months later, most of the buildings were still unrepaired, some had blue tarps for roofs and some were abandoned—so was the sole gas station. Many families lived in tents, while many others left the island, likely with no plans of coming back. Why return to an island with such an unpromising future? What future did they have?

Can the government of Antigua and Barbuda solicit sufficient aid to afford the expensive job of reconstructing an island with so little potential? The banana trees alone will take years to regrow, and the sand and surf do not attract tourists. In what appears to be a more skeptical and less generous world, how will donor countries and international institutions weigh their requests versus those of other "attractive" countries in need?

THE FEDERATION OF SAINT KITTS AND NEVIS

"[A] 'small-island'. . . mindset, which is a general feeling of inferiority suffered by the residents of small islands in relation to the residents of larger islands such as Jamaica and Trinidad and Tobago, [may be blamed] for the failure of the West Indies Federation and even less successful efforts at unification. Others have noted the 'push and pull' on migration from the smaller islands to the larger islands."

—Sandra Meditz and Dennis M. Hanratty

The last wave of Amerindians, the Kalinago (Caribs), settled in these islands more than three centuries before the arrival of the first Europeans. Christopher Columbus "discovered" Saint Kitts—the shortened form of Saint Christopher. In 1538, French Huguenots established a settlement on Saint Kitts; later, the Spanish razed the settlement and deported survivors. In 1623, the English founded a settlement, which was soon followed by French settlers, the island being divided by mutual assent. In 1626, the English and French massacred the Kalinago. In 1628, English from Saint Kitts settled on Nevis.

The long continuing decline of Spain began with the Revolt of the Netherlands (1555–1609), during which the Dutch successfully won their independence from Spain.

The defeat of the Spanish Armada when it attempted to invade England in 1588 undermined Spain pretensions in the Caribbean. During the Thirty Years' War, which coincided with the last thirty years of the Eighty Years' War, Spain sent what was to be its last expedition to enforce its Caribbean claims to destroy the French and English settlements on Saint Kitts in 1629, but their colonists returned without opposition the following year. The French, in 1666 during the Anglo-French-Dutch War, expelled 5,000 British colonists from the island; the French returned the English section of Saint Kitts after the war in 1667. Similar warfare continued in the Nine Years' War (1688–1697) and the War of Spanish Succession (1701–1714). Britain gained full possession of Saint Kitts and Nevis with the treaty ending the War of Spanish Succession. Full British control was recognized by the treaty that ended the American War of Independence.

The Saint Kitts plantations, owned by well-to-do (often absentee) British, initially focused on indigo and tobacco, but, by the mid-1600s, sugarcane was the principal crop; slavery had replaced indentured servants. African slaves were imported to handle the laborious and prosperous sugarcane. The abolition of slavery in 1834 (and the follow-up "internship" variation of slavery in 1838) led to the curtailment of sugarcane cultivation by plantation owners and the growth of smallholdings.

Worker discontent brought efforts to seek greater control of their working conditions and the governing of it. A sugar price collapse triggered the birth of the Workers' League (WL) organized by Robert Bradshaw in 1932 and sparked the labor riots of 1935–6. The Workers' League renamed itself the Saint Kitts and Nevis Trades and Labour Union in 1940 and established the Saint Kitts and Nevis Labour Party. It secured Bradshaw a seat on the Legislative Council in 1946. It also dominated politics in the two-island polity for more than 30 years. Saint Kitts belonged to the West Indies Federation from 1958 to 1962.

During its reign, Bradshaw's Labour Party moved steadily toward more government control of economic development, culminating in 1975 when it took over all the sugarcane fields and, a year later, the central sugar factory. Resentment of the heavy-handedness of party's rule, especially acute in Nevis, contributed to the party's decline and the growing tension between Saint Kitts and Nevis. Bradshaw's death in 1978 marked the decline of the Labour Party. Only a year later, the death of his successor as premier, C. Paul Southwell, precipitated a party leadership crisis that curtailed the ability of the new leader, Lee Moore, to cope with the growing opposition. Saint Kitts-Nevis-Anguilla became an Associated State with full internal autonomy in 1967 and became a member of the West Indies Associated States.

The opposition coalesced around two parties founded earlier. On Saint Kitts was the People's Action Movement (PAM), initially formed to protest electric rate increases and, later, advocated economic diversification from sugar, the promotion of tourism, and more autonomy for Nevis. The opposing party on Nevis was the Nevis Reformation Party (NRP), which focused on separating from Saint Kitts. Following the 1980 elections, PAM and NRP formed a coalition with Kennedy Simmonds (PAM's founder) as premier and Simeon Daniel (founding member of NRP) as minister of Finance and Nevis Affairs. Formation of the PAM-NRP coalition was facilitated by reducing the demand for Nevisian secession, which, in turn, secured the independence of Saint Kitts and Nevis.

In June 1984, the PAM-NRP coalition won the first election in the independent country, and Simmonds became its first prime minister. The Labour Party defeated the PAM-NRP coalition in the 1995 elections and installed as prime minister its leader, Denzil Douglas. He led the party to victory in the 2000, 2005, and 2010 elections, serving as prime minister until 2015. The combined Unity Party, as PAM-NRP was known, won the 2015 elections and installed its leader, Dr. Timothy Harris, as prime minister.

The constitution Saint Kitts and Nevis adopted in 1983 providing for a Westminster-modeled government, with the British monarch (represented by a governor-general) as chief of state and a prime minister as head of government, heading a cabinet. Its unicameral National Assembly consists of 14 members; 11 are elected, and 3 are appointed by the governor-general. There are 14 parishes—9 on Saint Kitts and 5 on Nevis. The country belongs to the UN and affiliated organizations, the Commonwealth, the OAS, CARICOM, and OECS. Its mission in Washington, D.C., is accredited to both the United States and the United Nations.

In the mid-2010s, 75 percent of the 51,000 inhabitants were of predominantly African descent, 12 percent were Afro-European, 8 percent were South Asian, and 5 percent were other—mainly British and Portuguese. Continuing, but declining, migration from the islands, especially to the U.S., has slowed the country's growth. The country's major religions are Anglicanism, other Protestant denominations, and Roman Catholicism. English, the official language, is in common use.

Tourism, agriculture, and light industry drive the country's $962-million economy with its per capita GDP of $16,300. Tourism has been expanding since 1978. Rising production costs, low world market prices of sugar, and the government's efforts to reduce dependence upon sugar have led to a growing diversification of the agricultural sector to include vegetables, rice, and bananas. In 2005, the government closed the state-owned sugar company whose deficits were a major factor accounting for the country's deficits and debt. The debt-to-GDP ratio was 153 percent in the mid-2010s—higher than several other Caribbean islands with debts exceeding 100 percent GDP.

The challenges confronting Saint Kitts and Nevis include continuing to increase tourism, diversifying agriculture, and developing more export-oriented light manufacturing. These goals are ones they share with—and compete against—countries across the world, which make the challenge especially difficult. Making headway will require investment by a country with a frighteningly high debt-to-GDP ratio.

THE COMMONWEALTH OF THE BAHAMAS

"The Bahamas stands out among the Commonwealth Caribbean nations because of its relative wealth and prosperity, political stability, and close proximity to the United States."

—Mark Sullivan

The Bahamas, a chain of more than 700 islands—about 30 of which are inhabited—lies east of the U.S. state of Florida and north of Cuba. In 1492, one of the islands upon which Christopher Columbus landed was one of those in the Bahamas' chain. Between 1492 and 1508, Spanish depopulated the islands of the native Arawak by shipping them as slaves to Hispaniola.

While there may have been attempts to settle the islands by groups from Spain, France, and Britain, essentially, the islands remained deserted until Puritans from Bermuda settled there in 1649. In 1666, other English settlers established Charles Town, later renamed Nassau, near the end of the 17th century. A Spanish corsair raided the capital in 1684. A joint Franco-Spanish expedition briefly occupied the capital in 1703 during the War of Spanish Succession (1701–1714).

As early as 1664, Sir Thomas Modyford, governor of the Bahamas, had orders to rein in the pirates and privateers. Instead, he profited by allying with the buccaneers by authorizing missions against the Dutch and Spanish ships and securing a percentage of the captured goods. Craton and Saunders report, "The structure of law and order had completely broken down by 1704 . . . Hardly a proprietary governor escaped censure as a harbourer or encourager of pirates, even if he was not engaged in piratical activities or behavior himself." Pirates such as Blackbeard (Edward Teach) and Henry Morgan left trails of terror and destruction in the Spanish Main, making reputations, which have provided characters, plots, and settings for countless legends, books, and movies.

Wartime prosperity and interwar slumps during the 1700s set the pattern of "boom and bust" of subsequent Bahamian history. The prosperity brought by the war between Spain and England that began in 1739 with the War of Jenkins' Ear in the Caribbean and continued from 1743 in the War of the Austrian Succession (known as King George's War by the North American colonies) ended in 1748. The economy revived during the Seven Years' War, but the progress languished in subsequent decades.

In 1776, during the American War of Independence, American naval forces briefly occupied Nassau, the capital and largest city of the Bahamas. In 1782, following the British surrender at Yorktown (that effectively won the Americans' War of Independence), Nassau capitulated to the threat of a Spanish fleet. At the conclusion of the American War with the Treaty of Versailles, the British regained the Bahamas. The following year, the British resettled over 7,000 American loyalists—with their slaves—in the Bahamas, granting land to the planters to help compensate for their losses resulting from the divisive American War. The loyalist plantation owners became the major political force on the islands.

The Royal Navy attempted to enforce this prohibition by intercepting slave ships at sea, resettling in the Bahamas thousands of African people freed from the slave ships. The British intercepted not only slave ships coming from the African coast but also American ships engaged in the domestic coastal slave trade, which led to the War of 1812. The British abolition of slavery in 1834 intensified the effort to eliminate the slave trade. During the 1820s, the Seminole Wars in Florida led hundreds of American slaves and "Black Seminoles" (mixed African-Native American) to escape from Florida to the Bahamas, further escalating the preponderance of blacks and mixed blacks in the island chain. The British Slavery Abolition Act, which came into effect in 1834, challenged the Bahamian lifestyle. In the immediate pre-emancipation era, the Bahamian population included 26 percent whites, 23 percent free mixed-coloreds and blacks, and 50 percent about-to-be-emancipated slaves.

During the Civil War in the United States, Nassau prospered as a base for blockade runners supplying the Confederate States. For the next five decades after the war, the economy dragged. It revived with the coming of the prohibition in the United States; the islands served as a base for supplying the runners. The advent of commercial aircraft in the 1930s enabled the Bahamas to begin developing tourism as the backbone of its economy—a development that helped mitigate the impact of the 1933 repeal of prohibition in the United States in 1933.

Following World War II, anti-colonial frustrations inspired the forming of political parties, steps toward self-governance, and then independence. For decades prior to the achievement of self-governance, the Bahamas' political and economic systems were dominated by a small elite referred to as the "Bay Street Boys," so named because most of their business and economic activities were concentrated on Bay Street. In 1953, the first Bahamian political party, the Progressive Liberal Party (PLP) was formed by the blacks who were discontented with the policies of the governing elite; the PLP's popular success forced the elite to form a party of its own in 1958, the United Bahamian Party (UBP). The 1958 strike of the Bahamas Federation of Labour strengthened the PLP's image as champion of the working masses as well as of black Bahamian pride.

Responding to Bahamian as well as worldwide decolonization pressures, constitutional changes were negotiated at a conference in London in 1964. A new constitution established a self-governing government with a premier (the pre-independence title for the chief minister) and a cabinet. Sir Roland Symonette of the United Bahamian Party became the first premier. In 1967, a bicameral legislature was established, and the first independent government was elected. The PLP defeated the "old guard" UBP, installing Lynden Pindling (Sir Lynden from 1983) as the first black premier of the preponderantly black colony. In 1973, the colony achieved full independence, retaining membership in the Commonwealth and Queen Elizabeth II as head of state with a governor-general as her representative—and becoming a United Nations member.

The Bahamas is a constitutional monarchy modeled on that of the United Kingdom. The prime minister heads the government that is responsible to the legislature. The bicameral legislature consists of a 38-member House of Assembly elected from single-member constituencies and a 16-member Senate with members appointed by the governor-general (9 on the advice of the prime minister, 4 on the advice of the opposition leader, and 3 on the advice of the prime minister after consultation with the opposition leader).

The Bahamas has a two-party system with the UBP on the center left and the Free National Movement (FNM) on the center right. In 2017, the FNM defeated the UBP and the FNM's leader, Hubert Minnis, replaced Perry Christie as prime minister. There are 32 local government districts,

31 of which have elected councils; the exception is the island in which the capital is located, which contains 70 percent of the country's population and is governed directly by the national government. In addition to its UN membership and affiliated agencies and the Commonwealth, the Bahamas belongs to the OAS, CARICOM, and OECS and its Organization of Caribbean Central Banks. Among its missions are ones at the European Union, the United Kingdom, the United States, which also serves the United Nations, and the Dominican Republic.

The 319,000-resident population of the Bahamas (2015) consists of 85 percent blacks (including mixed). Intermarriage and interracial unions over the centuries account for skin color ranging from light to brown to dark. The whites, which comprise 12 percent of the population, are mainly the descendants of the English Puritans and American loyalists that arrived in 1649 and 1783 respectively. Hispanics account for 3 percent of the Bahamas' population. 68 percent of the population is Protestant (35 percent Baptist and 15 percent Anglican); 14 percent is Roman Catholic, and 15 percent is considered "other Christian." English is the official and commonly used language, but many residents speak a Bahamian Creole, which has a largely English-based vocabulary. Haitian refugees, whose influx is a troubling issue for the island nation, use a French-based Creole.

The Bahamian mid-2010s GDP is $11.2 billion; its GDP per capita is $31,900, the third highest in the Western Hemisphere. The economy relies principally on tourism (totaling 50 percent of its GDP), and 50 percent of the labor force is employed by it. With its proximity to the U.S., convenient travel arrangements, and excellent resort and hotel accommodations, it has a promising future. Financial services including offshore banking are the second-most important sector, along with associated business services, which account for 35 percent of the economy; the Bahamas is resisting efforts of the OECD to have it comply with the Common Reporting Standards (CRS) aimed at speeding up data-swapping among countries.

The country's national budget is $1.8 billion, $57 million of which constitutes the mid-2010s defense budget that supports about 67,000 troops. Most Bahamian exports ($750 million) go to Singapore and the United States, and 13 go to the Dominican Republic. The imports (totaling $2.9

billion) were principally from the United States and India. The Bahamas have experienced a slight decline in population and GDP in recent years.

Agriculture and manufacturing account for only 10 percent of the economy. Banking is important. Tourism is the mainstay of the economy. Proximity to the United States provides the major support for the Bahamian economy. Drugs and narcotics traffic to the U.S., a Caribbean-wide concern, is a continuing challenge for the Bahamas.

Hurricanes also pose a major threat. In September 2019, Hurricane Dorian hit the Bahamas with catastrophic category-five force winds (up to 185 mph) striking the islands. In the direct path of the storm were Grand Bahama Island with 52,000 inhabitants and the Abaco Islands with 17,000. The full force of the storm brought with it a storm-wave surge that raised water levels as much as 26 feet above normal heights, not counting the wave surges.[62] These islands are populated sparsely by fishermen and manual laborers as well as many migrants from Haiti living in shantytowns. Many homes were still awaiting repair from Hurricane Matthew, which hit three years earlier. Given the extent of devastation and prospect of future storms, there is little reason to anticipate the reconstruction of the devastated area—all of the Abaco Islands and one-half of Grand Bahama. Fortunately, these islands are not the center of the tourist industry on which the Bahama economy depends.

The rest of the Bahamas is considered by many to be well prepared to confront most storms, having updated its building codes in the early 2000s with some of the strictest standards in the region. Its attractions, accommodations, and airlines bring visitors from Britain and Canada as well as nearby United States.

---

62. At its highest point, Grand Bahama is only 30 feet above sea level.

BELIZE

"Two themes dominate the history of Belize: the outward struggle to establish and maintain an English-speaking nation in an area dominated by Hispanic peoples and culture, and the inward interaction between groups of different races and cultural backgrounds. Understanding contemporary social relations and the politics of Belize depends on understanding these diverse groups and their interpretations of past events."

—Nigel Bolland

Belize lies along the Caribbean coast of Central America. Bordering on the north is Mexico—on the west and south is Guatemala, with whom there have long been disputes regarding the common border. Belize maintains a cultural identity and political and economic institutions similar to that of the English-speaking Caribbean, despite its coastal Central American location. This English-speaking enclave in Latin America embraces a heritage, which includes the pre-Columbian Mayan civilization, the Spanish conquest in the mid-16th century, pirates that came to plunder, German and Dutch Mennonites who came to worship, English who came to colonize in the 18th and early 19th centuries, and Africans who were brought, enslaved, to work on its plantations. The country, once called British Honduras, changed its name to Belize in 1974 and acquired independence in 1981.

The Mayan civilization in its prime—250 BCE, when many of the ceremonial centers were built—covered what is now southern Mexico, Belize, Guatemala, El Salvador, and northern Honduras. After the 9th-century Mayan civilization collapse, populations shrived, and religious centers were abandoned along the Honduras Bay (now Belize). When the Spanish invaded Central America in the 16th century, they initially largely ignored this apparently resourceless coast. In the early 1700s, Spain allowed Britain to cut timber in the northern half of what is now Belize. The timber cutting moved southward over the ongoing decades.

In the early 1800s, when Spain, confronted with revolts, withdrew from Latin America, "Britain formally took over the entire territory, naming it British Honduras. The new state claimed it had 'inherited' the region from Spain. Guatemala gave up its claim in 1859 in exchange for Britain building a road from Guatemala City to the Caribbean." But Britain did not build the road, and Guatemala declared the treaty void. Guatemala's claim was pressed for most of the 20th century. After years of stalled talks, in 2008 the two countries agreed to submit the issue to the International Court of Justice (ICJ) but only if voters in each country approved, in a referendum, of submitting the issue to the Court. After a decade, Guatemala's voters have agreed to submit the issue to the ICJ, but Belize (as of mid-2018) had not scheduled a referendum. Most people considered the long-simmering issue irrelevant, except as a distracting political issue in Guatemala's politics.

The need for labor for the profitable hardwood cutting initiated the importation of slaves. By the early 1800s, more than half of the population was slaves. A few settlers owned most of the land and dominated the politics.

Pirates and British privateers (crown-licensed pirates) used the Belize harbor to threaten the Spanish shipments of treasure and other goods to support the war-starved Spanish treasury, further increasing Spanish-British tensions. The attacks led to periodic Spanish attempts by force as well as diplomacy to evict the intruders, culminating in a major naval attack in 1798 at Saint George's Caye in which the British settler-led forces routed the Spanish. Despite the outcome of the battle, the British government was reluctant to assert control over this land (then called Northern Nicaragua and then British Honduras) between the Yucatan and Nicaragua over which the Spanish continued to claim sovereignty. After Guatemala declared independence in 1821, it continued its claims over much of Belize. Despite numerous aborted efforts to resolve these claims, they remain a source of friction.

Local magistrates, the only local officials, were responsible for law and order by default. A "public meeting," elected from a limited suffrage and, therefore, dominated by the wealthy landowners, coordinated the governing efforts. Not until 1862, after continued skirmishes, did Britain proclaim the area a British colony, subordinated to the governor in Jamaica. The new governor replaced the public meeting with an appointed legislature.

By the 20th century, P. A. B. Thomson, author of *Belize: A Concise History*, writes the following:

> [The] development of British Honduras [Belize] had come to resemble the motion of a see-saw . . . constitutional development had inched forward, albeit only slightly, in the years that followed the adoption of Crown Colony government in 1871.

This introduced a council comprised of the chief justice, colonial secretary, officer commanding troops, treasurer, attorney general, and four unofficial appointed members. Local pressure led to the ratio being changed in 1989 to three official members and five unofficial nominated members. Representative government, albeit once again on a minimal basis, returned to British Honduras with the 1935 constitution, which provided for a Legislative Council with limited control over local affairs, of which 5 of its 12 members were elected from a limited franchise of property owners.

The 1960 constitution expanded the legislature to 18 elected members with only 5 nominated ones and 2 ex-officio ones but retained an executive chaired by the British-appointed governor. Prime Minister Harold Macmillan's "Wind of Change" speech on colonial policy in 1960 led to constitutional reforms and the granting of self-government in 1963. The legislature became a bicameral body renamed the National Assembly, with a nominated Senate and a wholly elected 18-member House of Representatives. In 1964, the Governor's Executive Council was replaced by a cabinet of ministers headed by the chief minister. Negotiations with Guatemala regarding the boundary delayed the recognition of Belize's independence until 1981.

The constitution, the latest version of which came into force in 2012, provides for a British-style constitutional monarchy: the Queen is the head of state represented in Belize by the governor-general who is a Belizean proposed by the Belize government. The prime minister, the head of government and leader of the majority party in the House, is appointed by the governor-general and serves at the pleasure of the House of Representatives—composed of 31 elected members. The Senate, composed of 12 members, is nominated by the governor-general. In recent decades, the two major parties, People's United Party (PUP) and United Democratic Party (UDP) have succeeded each other, winning the popular vote and, thus, becoming the majority party in the House and forming a government. The Judicial and Legal Services Commission appoints judges. The justices of the Supreme Court are appointed by the governor-general upon the advice of the same Commission, with the exception of the chief justice whom the governor-general selects with the advice of the prime minister and consultation of the leader of the opposition.

Attorney-General Michael Peyrefitte has proudly stated, "The mixed heritage and the location of Belize has led to its efforts to straddle the Caribbean and Latin American communities, as well as the community of former British colonies. It belongs not only to the OAS, the British Commonwealth, the African, Caribbean, and Pacific Group of States, the Central American Integration System (SICA), and CARICOM"; it also belongs to the United Nations, several of its affiliates, and to the WTO. Among its diplomatic missions are ones in the United States and at the European Union.

Reflecting its mixed heritage, the population of 334,000 includes Mestizos (about 50 percent), Creole or mulatto (25 percent), Maya (about 10 percent), and the Garifuna, another indigenous people (6 percent). Among the languages spoken by Belizeans are English (the official language), Spanish, Kriol (an English-based creole integrating African tribal languages), German, and Mayan dialects. About 40 percent of Belize's population is Roman Catholic, and 35 percent adhere to a variety of Protestant denominations including Pentecostal, Jehovah's Witnesses, Anglican, Mennonite, and Seventh-day Adventist; 10 percent adhere to other religions including Mayan, Hinduism, Islam, Rastafarian, and 15 percent say they are not religious.

Situated in Central America, on the Caribbean coast, in the orbit of North American economy, and a former British colony, Belize belongs to an array of regional organizations through which they cooperate to advocate matters of common interest and join forces in common endeavors. The Caribbean Community and Common Market expands trading opportunities with its single market and maintains the Caribbean Court of Justice. Other regional organizations include the Organization of American States, the Central American Integration System, and the Community of Latin American and Caribbean States (LIMUN). Multi-continental organizations to which Belize belongs include the Alliance of Small island States (AOIS), the African, Caribbean, and Pacific Group of States (ACP) through which the European Union provides assistance, the Commonwealth, and the Forum of Small States which coordinates issues before the United Nation.

The economy, which supports a GNP per person of about $9,000, depends principally upon tourism, garment and food production, and agriculture (principally bananas, cocoa, citrus, and sugar), is growing at a rate of over 5 percent a year. Its exports are directed principally to the United States and the United Kingdom. Its imports are from mainly from the United States and Germany. Belize's challenge is to maintain the growth of its tourism and diversified industry. To what extent will it be able to take advantage of its sun, sea, and surf, its proximity to North American tourists, and its unique position straddling multiple cultural and political cultures to foster West Indies political and economic collaboration?

THE CO-OPERATIVE REPUBLIC OF GUYANA

"In order to confront the enormous problems that confront the country, some sort of broad political alliance is highly desirable, particularly to counteract the divisive influence of ethnic differences. The problem of providing a government with sufficient power to enable it to enact unpopular but necessary legislation without having to be in continuing danger of falling is a difficult one and may easily lead to repression and misery."

—Raymond T. Smith

Guyana, a country of more than 700,000 people located on the northwest coast of South America, calls itself the land of six peoples. As the only English country on the South American continent, it has more in common with the English, Dutch, and French speaking Caribbean islands than with Brazil and Venezuela, its immediate Portuguese and Spanish-speaking South American neighbors. Like Suriname, British Guiana (the name of Guyana before independence) was colonized by both the Netherlands and Britain. While Suriname was initially settled by the English and lost to the Dutch, British Guiana was initially colonized by the Dutch and taken over by the British. Both countries share a pre-independence legacy of slavery, indentured labor, and multiple separatist ethnic communities, followed by a post-independence marked by one-party rule, strong-man politics, racially biased voting patterns, and disenfranchisement of the Guyanese people.

The sharply split ethnic composition of the population has created an environment of racially tinged politics facilitating strikes, riots, and political intimidation. Afro-Guyanese, descendants of the slaves from West Africa brought over by the Dutch, comprise 30 percent of the population; they typically live in the capital and towns. The Indo-Guyanese, descendants of indentured laborers from India brought in by the British, comprise 44 percent of the population; most live in rural areas. Mixed heritage Guyanese constitute 17 percent of the population. Nine percent are Amerindian, most of who live in the forested inland.

English is the official language and is used for media and services. The vast majority speaks an English-based Guyanese creole with slight African and Asian Indian influence. Religious identification reflects heritage and missions: 57 percent are Christian (31 percent of Christians are Protestants ranging from Pentecostal to Methodist), 8 percent are Roman Catholic, 18 percent are "other" Christian denominations, 28 percent are Hindu, and 7 percent are Muslim.

When the first European explorers sighted the northwest South American coast two groups of Amerindians inhabited the area: the Arawak along the coast and the Carib in the interior. Columbus sailed along the coast in 1498 as part of his third and last voyage across the Atlantic but failed to see any prospects of gold and was discouraged by hostile Amerindians. English courtier and navigator, Sir Walter Raleigh's visit in 1595 directed English attention to the region. The Dutchman, Cabelieu, after his 1597 trip along what is now the Guinea coast, asked the Dutch parliament for permission to colonize the Guinea coast.

In 1616, the Dutch founded the first European settlement—a trading post along the Essequibo River in what is now Guyana.[63] In 1621, the Dutch government gave complete control over the Guyana operation to the newly formed Dutch East India Company. In 1627, the company established a second settlement along the Berbice River and founded a third settlement at Demerara, granting 'patroons' large estates (as in New Netherland—later New York). The growth of agriculture and dissatisfaction with the Amerindians' adaptation to plantation labor led to the import of African slaves.

In 1781, when the Netherlands joined Russia, Sweden, and Denmark in the League of Armed Neutrality during the American War of Independence, Britain declared war and occupied Dutch Guinea colonies. A few months later, the French who supported the American colonies took the three settlements and held them until 1783 when, at the conclusion of the American Revolution, the Dutch regained them. The Napoleonic Wars, and the accompanying French occupation of the Netherlands, led to the British recovery of the Netherlands Guinea colonies in 1796. In 1802,

---

63. By the late 1500s, the Netherlands had gained *de facto* independence from Spain; by 1648, the Dutch Republic had formally gained independence and developed as a major naval, trading, and colonizing power.

during a short-lived intermission in the Napoleonic Wars, the three Guinea colonies were returned to the Dutch; when the war resumed, the British once more seized control of the three Dutch colonies. At the 1814 London Conference, the settlements were formally ceded to the United Kingdom and were unified as British Guinea.

Political, economic, and social life in the 1800s was dominated by the plantation owners' close ties with the royally appointed governor. The next level consisted of a small number of freed slaves—many of mixed African-European heritage and a few Portuguese merchants. The lowest level were the African slaves who lived and worked the plantations. In this socio-economic plutocracy, a few whites controlled a vast number of slaves.

The end of slavery not only radically changed Guinea colonial life, but also fashioned the dynamics of its post-colonial ethnic-driven politics. Mounting abolitionist pressure led to the prohibition of the international slave trade in the British Empire in 1807 and, by 1838, total emancipation. An immediate impact of the abolition of slavery was that many former slaves left the plantations; some slaves moved to towns and some joined together to purchase abandoned estates. The emancipation, thus, led to the emergence of a new independent-minded Afro-Guyanese middle class challenging the plantocracy. Initial efforts to recruit Portuguese labor proved unsuccessful; not adapting to fieldwork, many became shopkeepers. Concerned about the shrinking labor force and production, the British, like their Dutch Guiana neighbor, contracted for indentured labor from India, adding another ethnic group to Guyana's demography.

Macdonald asserts, "Planter political power in British Guinea was based in the Court of Policy and two Courts of Justice, established in the late 1700s under Dutch rule." The courts, which combined legislative and executive functions, consisted of the governor, three colonial officials, and four governor-appointed colonists. British constitutional revisions in 1891, incorporating changes demanded by reformers and resisted by the planters, included relaxation of voter qualifications, the enlargement of the Court of Policy to 16 members—of which 8 were elected—the adding of 6 elected members to the Combined Court, and the formation of an Executive Council. In 1897, the secret ballot was introduced. In 1909, the electoral franchise was so broadened that, for the first time, the Afro-Guyanese constituted a majority of the eligible voters. Continuing workers' widespread dissatisfaction manifested itself in the 1905 Ruimveldt Riots that began with a stevedore strike and grew to the point that Britain brought in troops. The riots promoted the later emergence of an organized trade union movement.

World War I expedited social and political change. In 1917, in the face of widespread business opposition, the colony's first trade union, the British Guinea Trade Union (BGTU) was formed. Also in 1917, British and Indian nationalist concerns regarding the oppressive nature of indentured service led to the end of recruiting contract labor from India. The Indo-Guyanese former indentured laborers, over time, became rice farmers and merchants. The Indo-Guyanese, who ended to retain their Asian homeland culture and customs, had become the country's largest ethnic group, pitting them socially and politically against the Afro-Guyanese.

This rivalry presaged post-World War I development of political parties. Many Guyanese who had served in the British armed forces during the war came home with an international perspective and self-confidence that encouraged and empowered social and political activism and leadership. In 1928, a new constitution made British Guinea a crown colony under the tight control of the Colonial Office-appointed governor and, with a half appointed legislature, changes seen as gains for planters and setbacks for the emerging middle classes.

The main economic activities have been sugar, rice, and bauxite mining. Sugar, once pervasive in Guyana, still accounts for 28 percent of export earnings; the plantations, after nationalization, are largely run by the Guyana Sugar Company (GuySuCo), which employs more people than any other industry. Rice cultivation came with the arrival of, and remains dominated by, the Indo-Guyanese. Bauxite mining is run mainly by foreign companies, principally Reynolds Metals (American), the aluminum division of Rio Tinto (British-Australian), and Barama Company (Korean/Malaysian).

Bauxite mining and sugar and rice farming declined during the 1930s Depression, an era of escalating unemployment rates, multiplying political demonstrations, and agitation of emerging classes for more political power. A commission, set up to consider the British Caribbean unrest, reported in 1938. World War II delayed implementation of their recommendation until 1943 when the governor extended the suffrage to women and non-land-owners. Tourism accounts for $80 million of the GDP. New oil production in 2020 (the first in Guyana's history) is expected to booster the economy; its revenues are counted on to support the many new programs launched by President Granger's government.

The $1.3 billion of exports are mainly to the United States and Canada. The $21.1 billion of imports are principally from Trinidad and Tobago and the U.S. The nation's GDP is $6.3 billion; the GDPpc is $8,100, making Guyana South America's second poorest country (after Bolivia). The country's budget in 2014 was $806 million, of which $31 million is for defense.

The postwar period witnessed the founding of political parties: the People's Progressive Party (PPP)—initially the National Democratic Party (NDP)—led by Cheddi Jagan, an Indo-Guyanese, and, later, the People's National Congress (PNC) led by Linden Forbes Burnham, an Afro-Guyanese. In 1950, a British commission recommended a new constitution featuring a ministerial system with a British-appointed chief executive and an elected legislature. In new elections in 1953, the PPP's initial coalition of lower-class Afro-Guyanese and rural Indo-Guyanese workers soundly defeated a NDP-headed coalition. The PPP's appeals to nationalism over-came the opposition's branding as communist. A pro-PPP Guinea Indus-trial Workers' Union (GIWU) strike supported a PPP-proposed Labour Relations Act.

The strike led the British government to suspend the constitution. During the constitutional suspension (1953–57) a conservative interim ad-ministration governed the country. The interim administration constructed a new constitution that was adopted, providing for limited self-government with a 24-member Legislative Council (15 elected members, 6 nominated, and 3 ex-officio). In 1955, Jagan and Burnham formed rival wings of the PPP—largely along ethnic lines—with the Forbes wing shifting ideologi-cally to the right.

The 1957 election was won by Jagan, who supported more land and government posts for Indo-Guyanese, advocated Marxist-Leninist princi-ples, and admired Stalin, Mao, and Castro. Because of Indo-Guyanese con-cerns about becoming an ethnic minority within the British proposed West African Federation, Jagan vetoed British Guinea participation, a stand that resulted in the total loss of Afro-Guyanese support for his PPP party. The West Indies Federation subsequently collapsed.

Burnham developed support for his faction by increasing his appeal beyond the lower class Afro-Guyanese community to include the increasing potent middle class Afro-Guyanese. Before the 1961 election, the British promulgated another new constitution introducing a bicameral legislature (an elected lower house and an appointed upper one) and a legislature-selected chief minister, changes that enhanced self-governance. Constitutional changes again led to Jagan's PPP winning the 1961 elections and becoming chief minister again.

Labor violence increased from 1961 to 1964. Responding to a strike in 1964, the PPP government proposed a Labour Relations Act (similar to the 1953 one). The reaction was such that the governor felt compelled to declare a state of emergency. To quell the turmoil, at the request of the political parties, the British colonial secretary considered constitutional changes and concurred with a PNC-advocated proposal modifying the constitution by introducing a 53-member legislature elected by proportional representation. The change was opposed by the PPP, but the revision was adopted.

In the 1964 election, Burnham's socialist-oriented PNC, together with the less left-leaning United Force (UF) party, won the majority of legislative seats and once again became chief minister—and thus became the first prime minister of Guyana when it won its independence in 1966. In 1968 Burnham won reelection without the support of the UF. In 1970, the country declared itself the Co-operative Republic of Guyana. It abandoned ties to the monarchy, replaced the British-appointed governor-general with a president/head of state. In 1973, the constitution was again amended to terminate legal appeals to the Privy Council in London. Guyana belongs to the United Nations and several affiliates, CARICOM, the (British) Commonwealth, and the OAS, and among the missions it maintains are ones to the European Union (Brussels) and the United States (Washington, D.C.).

In the 1970s, Burnham consolidated his power over the PNC and the country. The PNC won the 1973 elections. Burnham continued as prime minister; Jagan's political power waned. Following the 1978 Jonestown communal murder-suicide, led by Jim Jones' Peoples Temple of the Disciples of Christ (which had had the PNC's support), Burnham's control over Guyana weakened. Violence escalated in 1979. Still another new constitution was promulgated in 1980 (the seventh since 1792 and the fourth since 1953), which combined the roles of head of state and prime minister/head of government into the post of a legislature-elected executive president.

The PNC won the following election, and Burnham became Guyana's first executive president. In the midst of the 1980s economic crisis, President Burnham unexpectedly died in surgery. Vice President Desmond Hoyte became president, won the subsequent 1985 elections (that the opposition boycotted), and lifted curbs on foreign activity and ownership. In 1992, the Indo-Guyana-dominated PPP won a majority of seats; Jagan became president. After his 1997 death, his American-born wife, Janet, was elected president.

Former finance minister Bharrat Jagdeo, an Indo-Guyanan, became president in 1999 and won again in 2001 and 2005. Following a further constitutional revision, the posts of a president and prime minister were separated once again; the PPP won the plurality of seats in the 2011 elections, PPP General Secretary Donald Ramotar won the direct election to the presidency, and Samuel Hinds was confirmed as prime minister.

In the 2015 elections, a fractious coalition, which included the Afro-Guyanese Party for National Unity (successor to the PNC), the People's National Congress-National Reform Party led by David A. Granger (an Afro-Guyanese retired military commander), and a third new multi-ethnic Alliance for Change (AFC) Party. Granger's party narrowly defeated the PPP, which had ruled since 1992. Granger was installed as president with Moses Nagamootoo, AFC leader, as prime minister. President Granger followed through on many of his campaign promises, including increasing funding for the national university, raising teachers' salaries, upgrading the pumping system in Georgetown (the capital), and slashing subsidies for the sugar industry (putting thousands of Indo-Guyanese out of work). But the fragile Granger-led coalition has not cut the bureaucracy that mainly employs Afro-Guyanese. Although Granger promised to fight corruption, his critics accuse him of moving too slowly.

The 2018 discovery of oil by a consortium led by Exxon Mobil off Guyana's Atlantic coast may extract 750,000 barrels a day, worth about $15 billion a year at current prices, four times Guyana's current GDP. This discovery not only spectacularly transforms the future expectations for Guyanese economy, revenue, and opportunities for corruption but also the immediate national electoral dynamics and international threats from Venezuela, who has reasserted an old claim to its coastal waters and two-thirds of its land.

Shortly after this oil discovery, the fragile multi-racial coalition government led by Granger was felled by a vote of no-confidence, triggered by the switch in allegiance of a single member of the National Assembly belonging to the junior party in the coalition. The Indo-Guyanese backbencher that defected gave no advance warning and has fled to Canada. The no-confidence vote set off a general election campaign nearly two years before Granger's term of office was scheduled to end. The next elections, whenever they happen, given the stakes and the deep-seated racial-cum-partisan divide, promise to be even more bitter than earlier ones. Dr. Mohammed Infann Ali (PPP) became president in 2020. (He is the first Moslem president.)[64]

Guyana's problems are grave; they include the world's third highest suicide rate, the highest rate of maternal mortality in South America, and the migration of 80 percent of university graduates from the country. Guyana's future, however glittering, confronts the hazards of dominance by a multinational firm, corruption, Venezuelan threats, and a deeply ingrained, racially-based partisan divide. Like Venezuela, oil riches may exacerbate, rather than alleviate, Guyana's future.

---

64. This is because he was not chosen from a list proposed by the opposition, which is customary.

THE REPUBLIC OF SURINAME

"Consociational democracy means government by elite cartel designed to turn a democracy with a fragmented political structure into a stable democracy."

—Arend Lijphart

Suriname, once known as Dutch Guiana, is one of two micro-countries on the northeastern Guiana coast of South America. It borders French Guiana on the east, Guyana on the west, and Brazil on the south. It is the only country in South America whose prevailing language is Dutch. When the first European explorers sighted the Guiana coast, two groups of Amerindians inhabited the area: the Arawak along the coast and the Carib of the interior.

The English planted a settlement along the coast in 1651, which it traded to the Dutch in 1664 for the North American colony of New Netherland (Manhattan Island and upstream Hudson). Except for certain intervals during the Napoleonic Wars, the Dutch ruled the country for more than 300 years until 1975, when it became independent and named itself Suriname.

In the 16th century, French, Spanish, and English explorers visited the Guiana coast. A century later, Dutch and English colonists established plantation colonies along the rivers leading to the coast. In 1651, the Barbadian colonial governor established a small settlement at Marshall's Creek. The settlement was contested between the Dutch and English until the Second Anglo-Dutch War in 1665, during which the British captured New Amsterdam (now New York) from the Dutch, and the Dutch captured Dutch Guiana from the English. After the war, the Dutch kept the prosperous sugar colonies and gave up New Netherland (now New York City and the Hudson Valley) to the British Dutch

Guiana's sugar plantocracy was initially so dependent on the importation of slaves from Africa that, Charles C. Mann reports—in his book titled *1493: Uncovering the New World Columbus Created*—that for each European, the colony had more than 25 Africans. Many escaped to the forests, mixed with natives, and formed hybrid groups of Maroons. Mann notes that, from the 1670s, guerilla warfare "continued for almost a century . . . [until 1762 when] the Dutch colonial government signed a humiliating peace treaty." But slaves continued to run away. A second guerilla war ensued. The planters begged for help. More than 1,000 Dutch troops arrived; almost all were casualties of sickness or injuries. Eventually, the Dutch and Maroons reached a kind of standoff accommodation.

Slavery was finally abolished in 1863. Most of the emancipated slaves departed the plantation estates, many to the forests. Since plantation agriculture was perceived as the only viable economic option, replacement laborers were required; tens of thousands of contract workers were imported from northwest India, and later Java. Slavery, emancipation, and successive waves of imported labor from other sources set the stage for the ethnic separatism, sequential coups, and military dictatorship tinged with Marxism that have marked Suriname's post-independence ethnic fragmentism, partisan factionalism, and domination.

In 1883, the Dutch West India Company with partners chartered the Society of Suriname to develop, manage, and defend the Dutch Guiana colony. The Dutch planters imported African slaves to cultivate the sugar cane—and coffee, cocoa, and cotton—plantations, the base of the economy. The ill treatment of slaves led many to escape to the adjoining rainforests, where they interbred with Amerindians; their descendants are the Maroons. After protracted negotiation, the Dutch parliament abolished slavery in Suriname in 1863 (shortly after Abraham Lincoln signed the Emancipation Proclamation). But the slaves were not released until after a 10-year transition period. The ex-slaves then generally abandoned the plantations, mainly migrating to the country's major city, Paramaribo. They form the base for the mixed African-European Creole population (whose commonly spoken patois mixes Dutch and English with Spanish and African speech).

The Dutch imported contract labor from India and later from Java (then part of the Dutch East Indies, now part of Indonesia). Their Asian-Indian descendants comprise the populous Indo-Suriname ethnic group, who speak a Caribbean Hindi. Javanese descendants continue to identify with fellow Javanese and speak Javanese. In addition, during the late 1800s and early 1900s, smaller numbers of laborers were recruited from the Middle

East and China, which later developed into the basis of a shop-owning commercial class. Small groups of European descendants—not only Dutch and British but also French, Scandinavian, German (many of a long-established Moravian sect), and Jewish (many of whom escaped Portugal when taken over by the Spanish and subjected to the Inquisition). Suriname is one of the most ethnically diverse countries in the world, not only because of the waves of imported labor but also because the colony attracted religious fugitives from multiple continents.

Suriname's 20th century fractious fragmented politics, Rosemarijn Hoefte has stressed, has roots that

> go back to slavery and colonialism, but after emancipation and independence, authoritarian traditions continued to permeate many institutions . . . [not only] the state but also the employers, churches, schools, political parties, and families.

Strong ethnic, religious, and cultural ties continue to maintain tight family and church community bonds that impact party affiliations. Leadership domination of the fragmented Suriname society has long supported tightened community bonds and fierce political contestation. The social history of Suriname shares many features with its Guiana neighbors.

In 1901, the Dutch government amended the 1865 constitution, giving even more power to The Hague's colonial bureaucracy, limiting the power of the colony-based legislature elected by a narrow franchise of 2 percent or less of the population. By the end of the 19th century, Dutch Guiana's plantation-based economy, like many other Caribbean colonies, was in a gradual and steady state of decline. With the abolition of slavery, fewer large plantations cultivated sugar; the early 20th century efforts of many planters to cultivate cacao, coffee, bananas, and cotton did not provide a commercially viable alternative to sugar. Large-scale agriculture lost its overwhelming importance; new economic sectors including gold, balata bleeders, and bauxite changed the colony by offering nonagricultural, and often better paying jobs, and by opening up the vast interior. In 1916, Alcoa, a United States-based industry that produces bauxite and aluminum products, formed a subsidiary to exploit the bauxite reserves and obtained exclusive control of the "land found and still to be found" (i.e., land with the potential for the extraction of bauxite), beginning what became Suriname's major economic support.

The World War II occupation of the Netherlands by Nazi Germany affected its immediate and longer-term relations with its colonies. On November 23, 1941, just two weeks before Pearl Harbor, in agreement with the Netherlands' government-in-exile, the United States stationed troops to protect Dutch Guiana's bauxite mines, a critical source of the aluminum vital to the war effort. In 1942, the Dutch government-in-exile expressed its intent to review its relations with its colonies following the war.

Immediately after World War II, Culturele Unie Suriname (CUS), a movement hoping to unite all population groups, attempted to capitalize on the new situation by pressing for constitutional reforms. Faced with this challenge, the socio-economic elite needed to legitimize themselves on the political scene, both on the domestic front and with regard to the Netherlands. The most important of the new parties were as follows: the Protestant-oriented and largely creole National Party Suriname (NPS), the Catholic-oriented Progressive Suriname People's Party (PSV), the Muslim Party, the Verenigde Hindoestaanse Partij (VHP), and the Indonesian Peasants Union (SPI). In 1954, as a step toward home rule, Dutch Guiana became one of the constituent parts of the Kingdom of the Netherlands. The Netherlands' home government retained control of defense and foreign affairs. The Netherlands, thus, followed a path parallel to that of Britain and France in the progress of self-governance.

In 1973, the NPS-led Dutch Guiana government began negotiations with the Dutch government. Despite opposition by the Hindustanis and other minorities, the country was granted independence in 1975 and adopted the name Suriname. The introduction of universal suffrage, replacing a franchise limited to less than 2 percent, increased the voting population from fewer than 3,000 to about 96,000. Dutch aid initially provided substantial assistance to the newly independent country.

As a newly independent country the former governor, Johan Ferrier, was elected the first president; the NPS leader, Henck Arron, became the first prime minister. Apprehension regarding independence under the Creole-led NPS caused nearly a third of the Surinamese to migrate to the Netherlands—including many of Javanese as well as Dutch descent. The extent of fraud in the 1977 elections, in which Arron was reelected, led many more to depart. This lax political leadership contributed to 18 percent unemployment and increasing social malaise.

In 1980, the "Sergeants' Coup," led by Desiré Bouterse, overthrew Arron's government and substituted military leadership. Two successive attempted counter-coups, in 1981 and 1982, failed to restore an elected government. Following the second counter coup, the military, led by Bouterse, rounded up 13 prominent citizens who had confronted the military dictatorship, imprisoned them, and, three days later, executed them. The Dutch government protested and withdrew financial aid. The government soon had to deal with an economic crisis—made worse because, in 1982, bauxite revenues declined sharply.

A brutal civil war began in 1986 when Maroons loyal to rebel leader Ronnie Brunswijk fought the army. In 1987, elections were held, and a new constitution was adopted, which provided for an elected, 51-member National Assembly, which elected the president. Bouterse remained in command of the army. The subsequent civilian-military tandem arrangement, through which Bouterse dictated policy, led to discontent. By telephone, Bouterse dismissed the government.

In 1999, a Dutch court convicted Bouterse in absentia for drug smuggling, a case accompanying strained Dutch-Suriname relations and interruption of Dutch support of Suriname development programs. In the 2010 elections, a Bouterse-led coalition won 23 seats (compared to the National Front's 20) in the National Assembly; with the aid of other parties, Bouterse was elected president. Despite his incumbency, in 2012, he, along with others, was tried for the 1982 murders, but the Suriname parliament granted him amnesty before the conclusion of the trial. Bouterse was re-elected in 2015. In 2020 he ran again, despite being sentenced only a few months earlier to a 20 year prison sentence for the crimes he committed in 1982. He was succeeded by Chandrikapansad (Chan) Santukhi in 2020.[65]

Of the country's population of about 550,000, almost 50 percent reside in Paramaribo, the capital. An additional 30 percent live in the 4 adjoining districts along the coast—the ones with the country's towns, including the largest ones, Lelydorp as well as Paramaribo. The other 6 local government

---

65. A military court has convicted Bouterse, the president from 2010 to 2020, of murdering political opponents in 1982.

districts contain about 95 percent of the land but few people. Each of the 10 administrative districts is headed by a district commissioner, appointed by the president and subject to his dismissal.

Suriname's successive steps to recruit labor has fashioned a country with a plethora of ethnic groups. Asian Indians constitute 37 percent of the population, Creoles (mixed African-European) 31 percent, Javanese 15 percent, Maroons (mixed African-Amerindian) 10 percent, Amerindians 2 percent, Chinese 2 percent, and Dutch, German, Jews, and other whites 1 percent. Languages, religions, and political orientation of each of these groups continue to reflect their heritage. The major spoken languages are Caribbean Hindustani, the creole Sranan Tongo, and Javanese. Dutch is the official language and is used in most households. English is widely used. The major faiths are Hinduism (20 percent of the population), Protestantism (25 percent of the population)—mainly Moravian—Catholicism (23 percent of the population), Islam (20 percent of the population), and indigenous beliefs. With so many religions, it's no wonder its politics are so fractious.

Cutting across these ethnic-religious divides are multiple tiers of social stratification. Kruger explains that Suriname's bourgeoisie

> consists of the heads of large firms, the highest government officials, the trading elite, several major local business men, and a number of professionals, and is thoroughly permeated by foreign elements . . . The "petite bourgeoisie" consists of employees, minor officials, retailers, medium-sized farmers, etc., who identify with the ruling class.

The underclass includes most farmers, industry workers, street and market sellers, and beggars. Beginning in the 1940s, many blacks and Asian Indians—and some Indonesians, but fewer Maroons and Amerindians—have risen above this lower class.

Suriname has missions in many countries, including the United States, China, and Brazil. Among the regional organizations of government it belongs to are the Caribbean Community and Common Market (CARICOM), which consists of 15 independent countries and dependents, in addition to the United Nations (and several affiliate organizations). The economic growth slowed in the late 1990s but has since gained strength as its economy diversified. Suriname's $6.9 billion GDP economy (per capita $12,600) is principally dependent upon the bauxite industry (15 percent of all industry), which accounts for 70 percent of its export earnings. Other exports include rice, bananas, and shrimp. Recently, the country has begun exploiting oil and gold resources. Most exports continue to go to the United States and Belgium. Its imports continue to be mainly from the United States, Netherlands, and China.

Rampant government expenditures, poor tax collection, a bloated civil service, and reduced foreign aid have contributed to continuing fiscal deficits. A 2010s government budget was $940 million, of which $41 million supported its 1,840-troop army.

In a recently published study, Rosemarijn Hoefte predicts that, despite the continuing hybrid nature of its society, Suriname's "increasing integration in the Caribbean and South America will undoubtedly lead to new opportunities, insights, and challenges . . . [which are] reasons for guarded optimism."

15. OCEANIA: THE SOUTHWEST PACIFIC

"To promote his grandiose westward expedition halfway around the world to the Indies, [Ferdinand] Magellan played his cards shrewdly. He . . . secured the enthusiastic approval of . . . the powerful bishop of Burgos, organizer of the Council of the Indies . . . and, for funds, he cultivated a representative of the international banking firm of Fuggers . . . On March 22, 1518, Charles V announced his support for Magellan's expedition. The familiar objective was to reach the Spice Islands by sailing westward. This time, the plan was more precise—to find a strait at the extreme tip of South America . . . Magellan's feat, by any measure—moral, intellectual, or physical—would excel even that of Gama or Columbus or Vespucci."

—Daniel J. Boorstin

Map 15. Oceania Micro-countries

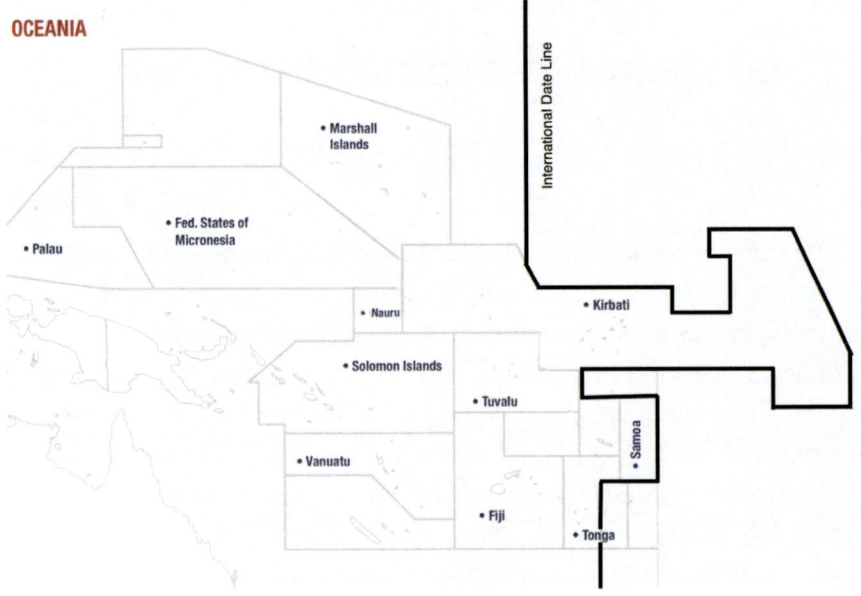

Source: world Map by *The Times Atlas of the World*

EXPLORING, COMMERCIALIZING, AND COLONIZING
THE SOUTH SEAS

The maze of islands (excluding Australia and New Zealand) clustering the South Pacific—which stretch about 6,000 miles from Palau and the Northern Mariana Islands in the northwest South Pacific to Kiribati, French Polynesia, and the Pitcairn Islands in the southeast—are populated by about 12 million people. The South Pacific includes not only 11 micro-countries but also Australia, New Zealand, Papua New Guinea, and many dependencies: in the United States, American Samoa, Guam, and the Northern Mariana Islands; in New Zealand, the Cook Islands and Tokelau; in France, French Polynesia, New Caledonia, and Niue; and, in Australia, Wallis and Futuna and Norfolk Island.

Polynesians, a physically, culturally, and linguistically related people, with their twin-hulled vessels and skill in navigating open seas, settled scattered islands in the Pacific beginning in about 1200 BCE. They spread and settled so quickly and cohesively that, "until the era of mass migration, Polynesians were both the most closely related and most widely dispersed people in the world." Hybridization of peoples with other ethnic strains moving out from Asia appears to have resulted in the emergence of three related groups of peoples: Polynesians settled on a triangle with many islands including Hawaii, Tahiti, and micro-countries Tonga, Samoa, and Tuvalu furthest from Asia in the east; Melanesians settled from New Guinea to Fiji, which includes micro-countries the Solomon Islands, Fiji, and Vanuatu in the southwest; and Micronesians settled on islands including micro-countries Kiribati, Nauru, Marshall Islands, Micronesia, and Palau in the north. Their languages are related but mutually unintelligible.

Numerous adventurers of various nationalities navigated the exploratory voyages that led to the colonization of Oceania. Ferdinand Magellan led a circumnavigation of the world, which landed in the Marianas and Caroline Islands in 1521. The Spaniard Álvaro de Mendaña de Neira, sailing from the Spanish west coast of South America, landed in the Solomon Islands, and may have sighted Tuvalu in 1568. The Dutch navigator Abel Tasman (after whom Tasmania is named) visited the Fiji archipelago in 1643, and Jacob Roggeveen probably visited the Samoan archipelago in

1722. The Englishman James Cook visited Tahiti, the New Hebrides, and New Caledonia between 1768 and 1776.

Three emerging powers—the United States, Germany, and Japan—joined Pacific empire-expansion colonial sweepstakes in the 20th century. The United States, after defeating Spain in the Spanish-American War, seized the entire Philippine archipelago (and Puerto Rico) from Spain, and annexed other territories including Guam, Wake, and the Hawaiian Islands. Andrew J. Bacevich points out, "By annexing the Philippines, the U.S. . . . enlisted in a high-stakes competition to determine the fate of the Western Pacific, with all the parties involved viewing China as the ultimate prize."

Spain's 1898 Spanish-American War loss of the Philippines, anchor of its Pacific presence, led Germany to buy the Caroline Islands from Spain and add them to its other already acquired Pacific possessions (part of Samoa, the Marshall Islands, Palau, and part of New Guinea). Japan purchased Nauru and Palau from Spain. Theodore Roosevelt's 1905-brokered treaty ending the Russo-Japanese War gave Japan political and economic hegemony in Korea and Manchuria. Within five decades, Spain's Spanish-American War defeat, Germany's World War I defeat, and Japan's World War II defeat lost them all their Pacific possessions.

Since 1962, beginning with Samoa, 11 micro-countries gained their independence. These new countries have maintained close connections with their former colonial overlords. Micronesia, Marshall Islands, and Palau gained independence under a compact of 'free association' with the United States, by which the U.S. handles their military and international affairs. In the mid-to-late 1980s, events unsettling the region included coups in Fiji, the blowing-up of a Greenpeace vessel by French agents, and a major revolt on Bougainville Island.[66] The Cold War increased colonial assertiveness. The United States considers itself the hegemon of the Pacific, Australia considers the South Pacific its backyard, and, increasingly, China asserts its strength. Its Belt and Road Initiative has expanded its investments in Oceania, and elsewhere, in recent years—not only in building roads, ports,

---

66. Bougainville, a region of 250,000 inhabitants of Papau New Guinea, in December 2019, voted almost unanimously (98 percent) to become independent. Since the 1970s, economic grievances have led mineral-rich Bougainville to strive for independence. If successful, it would become the 42nd micro-country.

schools, hospitals and other infrastructure but also in financing commercial ventures including bauxite mines in Fiji.

These investments have led South Pacific micro-countries to accrue substantial external debt, most of which is owed to China. President Trump's withdrawal of the United States from the Trans-Pacific Partnership, a trade pact of 12 countries—but not China—left the other countries little option but to join the Asian Infrastructure Investment Bank (AIIB), a new Chinese venture. Trump's withdrawal leaves a vacuum that China is eager to fill, extending its influence with its Belt and Road Initiative throughout Asia, Oceania, Africa, and beyond.

The Oceania microstates, as has been noted, divide generally into three groups: Melanesia (from the root *melas*, meaning black), Polynesia (from the root *poly*, meaning many), and Micronesia (from the root *micro,* meaning small). The Melanesian ones, which include micro-countries the Solomon Islands, Fiji, and Vanuatu, are the ones closest to Australia and north of New Zealand. The Polynesian Islands—including micro-countries Tonga, the Independent State of Samoa, and Tuvalu—straddle the International Date Line. The Micronesian ones, which include micro-countries Kiribati, the Federal States of Micronesia, Marshall Islands, Nauru, and Palau, lie northwest of Polynesia, and most are north of the Equator.[67]

Oceania is increasingly affected in the mounting rivalry between China, aggressively asserting its growing economic and political prowess and the United States trying to maintain its long-established primacy in the Pacific. Australia has stepped up its efforts to engage more closely with Oceania countries, especially with Papua New Guinea, Vanuatu, Fiji, Tonga, Kiribati, and the Solomon Islands, with aid and investments. In 2019, two Pacific countries, the Solomon Islands and Kiribati, lured by China's One Belt, One Road investment in international infrastructure and other construction, severed ties with Taiwan despite intense lobbying by the United States. A blatant example of China's aggressive actions is their activity in the South China Sea (of which six countries including micro-country Brunei claim overlapping parts). China has printed maps that propagandize their claim to the whole sea and is building fortifications on selected shoals.

---

67. New Caledonia and French Polynesia remain French dependencies. Niue and the Cook Islands are dependencies in "free association" with New Zealand.

The United States reacts to these and other Chinese aggressive actions so negatively that it is not unfortunate that six South Pacific countries have formal diplomatic relations with Taiwan, not China, despite U.S. recognition of China instead of Taiwan. The Solomon Islands, tempted by massive investment, is considering becoming the seventh.

The populations of the Oceania mini-states vary significantly from about 900,000 in Fiji and 500,000 in the Solomon Islands to 20,000 in micro-countries Tuvalu, Nauru, and Palau. All of these micro-countries have GDPs per person of less than $9,000. Those in Palau (GDP $8,100) and Tonga (GDP $7,400) do better than those in Micronesia (GDP $2,200), Marshall Islands (GDP $2,500), and the Solomon Islands (GDP $3,200), which are the least prosperous.

In Oceania, only Fiji has a national budget of over $1 billion. The others range from Nauru ($13.5 million) and Tuvalu ($23.1 million) to Vanuatu ($192 million) and Samoa ($302 million)—budgets hardly capable of supporting the development of infrastructure, economies, and global affairs capable of managing interests in the global community. The distances separating the islands from each other and from continental mainlands inhibits the development of manufacturing firms and tourism. Export income and tourist revenue are seldom sufficient to meet the costs—costs of imports and investments. Investments and assistance from countries contending for economic hegemony in the Pacific and continued growth in tourism (especially from Chinese tourists which increased ten-fold from 2001 to 2018) will fractionally mitigate this imbalance for some Oceania micro-countries

The political system of many Oceania micro-countries exhibits some degree of representative governance. But their working is strongly affected by tribal chiefs, church leaders, and ethnic factions. Hereditary chiefs have maintained widespread respect.[68] Evangelical churches with strongly held doctrinaire views continue to demand the allegiance of many voters. Fiji has experienced faction-led coups. A desire to maintain traditions drives the politics, the policies, and programs of many Oceanic micro-countries.

---

68. For instance, in Samoa, only chiefs can stand for election.

Inundation threatens many countries' very existence. Pacific storms have clearly been getting ever more menacing in recent decades. If the global community takes steps to reduce pollution inundation, or just severe damage to oceanfront areas—sufficient to discourage investment in relief and restoration—the countries may become less vulnerable. Can any amount of economic or political development change these conditions?

THE SOLOMON ISLANDS

"There is no effective cabinet process, [because] real power and decision-making occurs outside the formal political arena. A shadow state has emerged in the Solomon Islands—a patronage system centered on the ruling cabal's control over resources. The state has been gutted from the inside, and parliament largely serves as an avenue for access to dwindling resources by political players."

—Elsina Wainwright (2003)

The Solomon Islands consists of 10 large volcanic islands and 4 groups of smaller islands including the Santa Cruz Islands but not Buka and Bougainville in the northern Solomon Islands, which are part of Papua New Guinea. The country lies adjacent to Papua New Guinea to the east, Vanuatu and French New Caledonia to the southwest, and Australia further to the southwest. Its roughly 600,000 inhabitants, almost exclusively Melanesian, live on about 11,000 square miles of land. The Solomon Islands are severely threatened by rising seas, which have risen about 7 to 10 millimeters a year for over a decade; three islands have already sunk beneath the waves.

Migrants from other Pacific islands may have begun settling on what is now known as the Solomon Islands more than 30,000 years ago. The first recorded European visitor was the Spanish navigator Álvaro de Mendaña de Neira, sailing from Peru in 1568, but his effort to found a colony in 1595 failed. Not until the late 1700s did voyages led by the British captain Philip Carteret, the French explorer Count Louis-Antoine de Bougainville, and others visit these islands.

Missionaries began converting natives in the mid-1800s, but their efforts were impeded by blackbirding, British high-pressured recruitment and kidnapping of indentured laborers from the Solomon Islands for the sugar plantations in northeastern Australia and Fiji. The British established a protectorate over the southern islands in 1893, after a century of abuses. The British Solomon Islands Protectorate was placed under the jurisdiction of the British Western Pacific High Commission (WPHC) in Fiji.

Germany established a protectorate over the northern Solomon Islands in 1885. But, in 1900, all except Buka and Bougainville were ceded to Britain in exchange for British recognition of German claims to Western Samoa (now the Independent State of Samoa). While Germany's attention was distracted from the Pacific during World War I, Australia occupied Bougainville and Buka. Following the war, the League of Nations awarded the northern Solomon Islands, including Bougainville and Buka, to Australia to administer. The first self-rule movement, Magazine Guru (Marching Rule), arose in the late 1940s on Malaita Island and was suppressed by the British with mass arrests.

At the beginning of the Pacific phase of World War II, most planters and traders were evacuated to Australia. Japanese forces occupied the Solomon Islands. The fierce and bloody Battle of Guadalcanal (a.k.a. the Guadalcanal campaign), a key island in the Solomon chain, was a major part of the American-led Allied island-hopping campaign to regain the Pacific and defeat Japan. The devastation left by World War II was mainly confined to Guadalcanal. Following the war, Britain regained the islands it had administered before the war, including Guadalcanal, the southern Solomon Islands, and the Santa Cruz Islands.[69]

In the 1950s, the British began organizing local councils that interacted—and encroached upon—the traditional authority of the traditional leaders. In 1960, the British began a series of steps leading to self-governance, introducing legislative and executive councils. In 1970, another new constitution introduced an elected Governing Council. In 1974, yet another new constitution instituted an elected unicameral Legislative Assembly, which elected Solomon Mamaloni, leader of the center-left People's Progressive Party (PPP) as chief minister. In 1976, the islands gained full self-governance, elections were held for an enlarged Legislative Assembly, and Peter Kenilorea was elected as chief minister.

In 1978, the Solomon Islands became independent and a member of the British Commonwealth and the United Nations. Kenilorea became the first prime minister; Tuilaepa Sailele Malielegaoi became prime minister in 1998. The constitution provided for a constitutional monarchy headed

---

69. Australia once again administered the northernmost islands, Bougainville and Buka, as part of the New Guinea trust territory, which became independent Papua New Guinea in 1975.

by the British monarch with a governor-general as the local representative, a prime minister as head of government, and a unicameral parliament, which elects the prime minister and to which the prime minister is responsible. The judiciary is independent with the chief justice appointed by the governor-general upon the advice of the prime minister and leader of the opposition. The constitution enshrined, in 1978, a complex legal pluralism which combined two distinct sources of law—British and local custom (*kastom*)—which have not sat together comfortably.

The country is divided into 10 administrative districts, of which 9 are provinces with provincial assemblies and the 10th is the capital Honiara of the island of Guadalcanal. Solomon Islands maintains missions at the EU and at the UN where it shares a floor with several Commonwealth countries. The urban elite versus rural proletariat dichotomy challenges Solomon Islanders as well as many others who live in the South Pacific. The Solomon Islands belongs to the United Nations and several related global organizations, and the Pacific Islands Forum.

The elections in the initial decades of independence were contested with much crossing of the aisle, dismissals of ministers, votes of no-confidence, and unstable coalitions between the three major but weak parties that succeeded the pre-independence PPP: the People's Alliance Party (PAP), the Group for National Unity and Reconciliation (GNUR), and the National Coalition Partners (NCP). Kenilorea was elected as prime minister in 1978, lost in 1981 to Francis Billy Hilly of NCP, and won again in 1984; Mamaloni, successive leader of the PPP, the PAP, and NCP, was elected in 1994. Bartholomew Ulufa'alu was elected in 1997.

In 1998, ethnic-based riots broke out in Honiara in Guadalcanal, where the capital city was re-located after World War II. In 1999, Prime Minister Ulufa'alu declared a four-month state of emergency and requested assistance from the Australian-led Regional Assistance Mission to Solomon Islands (RAMSI) and New Zealand, but his request was denied. In 2000, Ulufa'alu was kidnapped; he resigned in exchange for his release. Manasseh Sogavare succeeded him as prime minister.

In 2003, rampant lawlessness, widespread extortion, insolvency, and ineffective police led to a formal request by the Solomon Islands' government for outside help. The Australian government responded, due largely to the fears of continuing deterioration in a neighboring country and the consequent threat to regional security. Under the auspices of RAMSI once again, more than 2,000 police and troops from Australia, New Zealand, and about 20 other Pacific countries arrived in the Solomon Islands—first to restore law and order, and then to assist in rehabilitating the country.

In 2006, Sogavare was again elected as prime minister; during his term, he ejected both the Australian high commissioner and an Australian police chief, straining relations. Sogavare was then toppled by a vote of no confidence, triggered by five ministers crossing the aisle shortly after Snyder Rini was elected prime minister in 2007. He was forced to resign for taking bribes. Derek Sikua was then elected prime minister. Danny Philip was elected in 2011, but, five days later, he resigned; Gordon Darcy Lilo took his place.

Sogavare became prime minister again in 2014 amid a heavy security guard of local and Australian police. Officers, talk of forging links with China, and wanting to work constructively with RAMSI. Sogavare, a black belt and a nationalist, was impetuous in the trying situation. In 2015, RAMSI was already a scaled-back operation proving police assistance, run by a junior foreign-service officer; in 2017, RAMSI departed. In 2017, after Sogavare lost a no-confidence vote, Rick Houenipwela (a.k.a. Rick Hou) became prime minister.

In 2019, in the wake of the Solomon Islands' 10th election since independence, riots once again broke out in the capital of Honiara. The new party, organized by Sogavare, which won 34 of the 50 legislature (MP) seats, selected him for his fourth term as prime minister.

About 95 percent of the population of Solomon Islands is Melanesian—only 3 percent is Polynesian and 1 percent is Micronesian clustered on a few specific islands. But the apparent demographic homogeneity masks a significant heterogeneity among the population scattered through the islands, as the number of languages reveals. While English is the only official language—spoken by less than 2 percent of the population—emphasizing the distance and distrust between the capital and others, especially those in the other islands. The lingua franca is Solomon Pidgin.

More than 70 distinct Melanesian languages are spoken in most of the islands, but Polynesian ones are spoken in the southern Solomon Islands. Immigrant Gilbertese from Kiribati speak a Micronesian tongue. The success of early missionary efforts is demonstrated by the fact that 74 percent of the population profess they are Protestant (33 percent from the Church of Melanesia, 20 percent South Seas Evangelical, 11 percent Seventh-day Adventist, and 10 percent United Church of Christ), and 19 percent are Roman Catholics.

The cultivation of coconuts, and export of its copra derivative, had begun in the mid-1800s. By the early 1900s, several British and Australian firms began large-scale coconut planting. Copra has remained the islands' chief source of export earnings. The Solomon Islands' GDP of $2 billion and consequent GDP per capita of only $3,400 ranks it among countries with the lowest GDPpc in the world. Subsistence farming (cocoa, coconuts, and palm oil) and fishing (tuna) account for 75 percent of its labor force. But repeated earthquakes, like the most recent one in 2016 that generated a tsunami, and cyclones do not offer favorable economic prospects.

Most manufactured goods and petroleum products must be imported, but the cost of imports ($360 in 2010), mainly from China and Singapore, is almost matched by its exports of $297 million, mainly to China. Until 1998, when world timber prices for tropical timber fell steeply, timber had long been the islands' principal export cash product. Its crash—plus the crisis that led to the RAMSI operation—drove the Solomon Islands' government into insolvency. The growth of tourism (mainly attracted by divers from Australia, New Zealand, and mainland China), the development of mining, and the improvement of farming and fishing technologies provide hope for the country's development.

In 2019 the Solomon Islands, like Vanatu, decided to recognize China instead of Taiwan and signed an agreement with a Chinese conglomerate to make a significant investment. A 75-year lease was signed for Tulagi, a small island that was the capital of the Solomon Islands before World War II; construction of an oil and gas terminal, a fishing harbor, and a "special economic zone" were pledged. But a *New York Times* article denouncing the lease led to the Solomon Islands' official who signed the agreement to disown it.

The Solomon Islands is a "failed state," Elsina Wainwright stated in 2003. That is to say it is "not yet [a] formed polity." Solomon Islanders, like many Pacific Islanders, confront the challenge of how to maintain their diverse and rural lifestyles while developing a government capable of meeting 21st-century challenges.

THE REPUBLIC OF FIJI

"Fijian Administration claimed to be based on the traditional system and yet to have the aim of development into a modern institution . . . It seems to me that one of the greatest obstacles facing the Fijians today is the failure to recognize that there is a contradiction: they must make the momentous choice between preserving and changing their 'way of life.' The belief that they can do both is a monstrous nonsense with which they have been saddled for so many years that its eradication may be very difficult to achieve."

—R. R. Nayacakalou

Fiji consists of more than 900 islands, but only about 100 are inhabited. The International Date Line has been bent eastward so that all the Fiji islands share the same time zone—and, thus, the same day. The two largest, most populous, and most important islands are Vanua Levu and Viti Levu, which includes Suva, the capital city, and three-quarters of the country's population. Fiji, with a population of over 900,000, is adjacent to Solomon Islands and Vanuatu on its west and Samoa and Tonga on its east. To reach many South Pacific islands, one changes planes in Fiji. Fiji has had four coups since 1987. During the fourth, in 2006, Commodore Frank Bainimarama assumed power. Prime Minister Bainimarama has significantly reoriented Fiji's politics.

Polynesians may have been the first to settle the Fiji islands more than 3,000 years ago; if so, most quickly moved on to Tonga and Samoa to the east. Melanesians have supplanted them.[70] The eastern Fiji islands have tended to retain traditional chieftaincy-based governing traditions, indicating the long-lasting impact of the pre-Melanesian-Polynesian culture. A distinctive Fijian culture and lifestyle has developed over centuries. Constant warfare, allegedly accompanied by cannibalism, was rampant.

In 1643, Dutch navigator Abel Jansen Tasman was the first recorded European to visit the islands. More than a century later, in 1774, the Brit-

---

70. Rotuma Island, more than 200 miles north of the Fijian islands, yet part of the Republic of Fiji, remains ethnically Polynesian.

ish explorer James Cook visited the Fijian islands. Early traders bought sandalwood from the islanders. Not until the early 1800s did Europeans found settlements. By the 1830s, settlers cultivated coconuts, traders sold coconut oil, and missionaries converted souls.

The initial importation of firearms into Fiji in the mid-1850s enabled the chief of Bau Island to unite many island tribes under his leadership. In the mid-1850s, Thakombau, the *Ratu* (chief) of Bau, accepted Christianity and, with the aid of the Christian king of Tonga, conquered the western Fijian islands and declared himself king of Fiji. In 1833, King George Tupou I of Tonga sent a prince to rule over the Fijian-Tongan community; by 1869, the prince ruled eastern Fiji. In 1858, the first British consul arrived in Fiji. In 1874, at the request of Thakombau and 12 other chiefs, a British warship suppressed slaving and other desecrations imposed upon by white adventurers and made Fiji a crown colony.

In 1879, the British governor of Fiji—not wanting to interfere with the traditional native governance, culture, and way of life—ceased the forced use of native labor and, instead, imported indentured Indian labor. In 1928, following the British suppression of strikes by Indian workers, Mahatma Gandhi sent A. D. Patel to Fiji to lead the Indo-Fijian cause. By the late 1930s, during World War II, about 120,000 of Fiji's approximately 210,000 inhabitants were native Fijian, about 90,000 were Indo-Fijian, 5,000 were European, and 2,000 were Chinese.

Before World War II, the British took several incremental steps developing quasi-representative institutions. In 1904, a Legislative Council, comprised of six elected Europeans and two native Fijians nominated by the Council of Chiefs, was established—an Indo-Fijian member was not appointed until 1916. Not until 1929 were Indo-Fijians elected to the Legislative Council. In 1937, Indo-Fijians, native Fijians, and Europeans were allotted an equal number of seats on the Legislative Council.

New Zealand and United States forces occupied Fiji in World War II. Nationalist-spurred anti-colonial momentum drove unrest in the late 1950s. In 1963, the Legislative Council was enlarged, the predominantly native Fijian Alliance Party (AP) and the mostly Indian National Federation Party (NFP) were formed—the latter led by A. D. Patel. In 1966, the British government introduced a new constitution providing for internal self-governance, led by a chief minister and a Legislative Council representing separate ethnic-based electoral rolls. In 1970, Fiji became independent; the new constitution provided for a governor-general representing the British monarch, a prime minister, and a bicameral legislature.

The cast of parties and prime ministers followed, enabled by successive constitutions and coups. AP leader Sir Kamisese Mara became Fiji's first prime minister. He was reelected in 1977 and 1982. In the 1987 elections, a NFP and (the 1985-formed) Fiji Labour Party (FLP) coalition won the majority of seats, 28—19 of which were Indo-Fijian—and formed the country's first Indo-Fijian dominated government, although its prime minister was Dr. Timoci Bavadra, FLP leader and a native (Melanesian) Fijian. The new coalition government lost no time in opposing the long-time dominance of the tradition-oriented native Fijian interests, including retention of traditional communal land tenure and the traditional role of chiefs. In opposition, Lieutenant Colonel Sitiveni Rabuka led a coup deposing the coalition, revoked the 1970 constitution, proclaimed the Republic of Fiji, and declared himself head of state. The British Commonwealth declared that Fiji's membership had lapsed. With the Queen no longer head of state, Commonwealth membership severed, and many foreign missions withdrawn, Fiji was diplomatically isolated. By the end of 1987, Rabuka resigned and was succeeded Prime Minister Mara.

A 1990 coup, led by native Fijians, enabled the unilateral adoption of a new constitution (condemned by Australia, India, New Zealand, and others). Of the 70 seats in the House of Representatives, 37 were reserved for native Fijians, securing ethnic Fijian domination of the political system. In the 1992 elections, Rabuka, as the leader of the newly formed (basically native Fijian) Fijian Political Party (SVP), became prime minister. During the 1992 elections, the first since the 1990 coup, the SVP won 30 seats. With the support of the FLP, Rabuka was again elected prime minister. In 1994, he was once again elected prime minister with the support of a minor third party and headed an all-native Fijian cabinet. International protests led Rabuka to agree to a Constitutional Review Commission. In 1997, the legislature passed a Constitution Amendment Bill, providing for a 71-member House of Representatives—46 seats were reserved on a communal basis (native Fijian 23, Indo-Fijian 19, and other 4) and reduced the power of the Council of Chiefs. The Commonwealth readmitted Fiji.

In the 1999 elections, the Indo-Fijian FLP won 37 seats, soundly defeating the VP, and elected Mahendra Chaudhry prime minister of a coalition government. In 2000, George Speight led a coup, took Chaudhry and
30 others hostage, and suspended the constitution, claiming that he had
reclaimed Fiji for the indigenous Fijians who were threatened by outsiders
taking over their government. Fiji was once again ejected from the Commonwealth. As the hostage crisis continued, an interim administration
composed totally of indigenous Fijians led by Laisenia Qarese was installed,
despite the opposition of Speight. In the 2001 elections, the Qurase-led,
newly created Fiji People's Party (FPP) won. In 2005, the FPP proposed a
resolution and Unity Commission (with power to recommend amnesty to
the leaders of the 2000 coup) that encountered strong opposition.

In 2006, the country's military commander, Commodore Bainimarama
led a successful coup, which he said in a speech announcing the coup was
justified by "necessity." In 2009, Fiji was again suspended from the Commonwealth for its failure to hold elections. It was also the first country suspended from participation in the Pacific Islands Forum (PIF). A new 2013
constitution, with a bewildering electoral system, and the advantageous incumbency secured victory for their new party. Ratuva has noted, "To some
extent, the coups revealed the paradox of Fiji's post-colonial multi-racial
experimentation . . . Cultural distinctiveness, fused with formal political
segregation as entrenched in the constitution, embedded in economic life,
and fueled with ethno-nationalism, created situation of potential ethnic
hostility." Native Fijians cling to their chiefs and communal land tenure;
Indo-Fijians do not.

In the 2014 elections, the first since the 2006 coup, Bainimarama's
Fiji First Party (FFP) secured about 60 percent of the vote, defeating the
pro-ethnic Fijian Social Democratic Liberal Party (SODELPA), headed by
one of Fiji's three paramount chiefs. While Bainimarama has lost some of
the pro-ethnic Fijian support by his efforts, his infrastructure development
programs and anti-racist messages have gained him popularity extending
into the Indo-Fijian community. He also abolished the Council of Chiefs
and declared all citizens Fijian, a term not previously used for Indo-Fijians.
The new start facilitated a return to the Commonwealth and PIF and foreign investments.

Bainimarama boycotted PIF meetings, protesting Australia and New Zealand membership, and launched the Pacific Island Development Forum, from which Australia and New Zealand were excluded—citing that these countries were insufficiently committed to reducing the pollution speeding up global warming as the reason. In 2018, he won a second election with 52 percent of the votes; his main opponent was former army commander and coup leader Rabuka. In the polling for 51-member parliament, who is elected from a single nationwide constituency, Fiji first won control. Bainimarama continues to practice nepotism and bully opponents.

Fiji is composed of 14 provinces and 1 dependency. Three traditional Fijian confederacies are headed by a paramount chief assisted by the other chiefs. The chieftaincies, which are inherited within the traditional chieftaincy families, continue to play an important role in the lives of indigenous Fijians. The population of Fiji consists of 57 percent indigenous Fijians and 37 percent Indo-Fijians.[71] The other 5 percent of the population is comprised of economically significant groups of Europeans, Chinese, and other Pacific Islanders. The islands' indigenous Fijian people, known as iTaukei, communally own most of the land. There is little social interaction of iTaukei with others. iTaukei resent the Indo-Fijians, suspect an Islamic conspiracy to take over the islands, and are bitterly opposed to Bainimarama's FFP's military-based dominance of the political process.

Indo-Fijians are descendants of the contract-indentured laborers, many of whom departed from Fiji after the coups. The native Fijians are overwhelmingly Melanesian, although many also have Polynesian ancestry (e.g., those living on Rotuma Island are Polynesian). The political tension between the two groups reflects their contrasting ethnic and cultural differences. Polynesia tends to have had a greater impact upon eastern native Fijians, in their genealogy and in their maintenance and respect for chiefly traditions. Indo-Fijians not only remain conscious of the Hindu-Muslim divide but also of the various strands of Hinduism. Fijians are not only divided among themselves but also from other Pacific Islanders. The elite are more likely to have traveled to a continental Asian, European, or American country than another Pacific island country.

---

71. Indo-Fijians once made up nearly half the population but many have migrated to New Zealand, Australia, and North America.

There are significant sociopolitical differences between eastern and western native Fijians as well as between them and the Fijian Indian community. They have long been divided along religious, linguistic, and socioeconomic lines. Religious affiliation breaks down similarly, with 54 percent professing Christianity—Protestantism (45 percent, including Methodism, Assembly of God, and Seventh-day Adventism) and Roman Catholicism (9 percent)—Hinduism (26 percent) and Islam (6 percent). More than 60 percent of the Fijian are urban. Both English and Fijian are official languages. Hindustani, though, is widely used among the Indo-Fijians.

Fragmented ethnicities and fractious politics have not prevented Fiji, with its forest, mineral, fish, offshore oil, sugarcane, coconut, and tourism resources from becoming one of the more developed Pacific island economies. Sugarcane was long the primary source of revenue. Of its labor force, 70 percent works in agriculture, 30 percent in services and manufacturing. About 8 percent is unemployed. The service sector, especially tourism, blessed with beaches, a surge in hotels and other tourist amenities, a favorable spot for trans-Pacific flight stops, and the completion of a new airport in 2017 is expected to increase tourism revenue.

Fiji had a 2016 GDPpc of $6,000. Its imports of $2.2 billion were mainly from Singapore, France, Australia, China, and New Zealand. Its exports (principally Fiji Water) of $1 billion are mainly to the United States and Australia. The former Fijian ambassador to the European Union and several European countries once pointed out that, if the United States did not levy import duties on sugarcane, Fiji would not need to rely on foreign aid. Fiji's early 2010s budget was $1.2 billion—$26 million was for defense.

In common with many other island economies, Fiji faces the continuing threat of weather and rising sea levels. In February 2016, winds from Cyclone Winston tore through the Pacific island chain, reached 177 miles per hour and left many people without power, fresh water, or communications. Predictions of more frequent and fiercer storms threaten the future of Fiji's tourism and its agriculture and other enterprises. Fiji has already relocated persons from several low-lying communities, and more are scheduled for relocation.

The steady stream of ships and planes docking at Fiji indicates its role as a Pacific hub and tourism center. But its future is threatened by climate change's higher temperatures raising seas, eroding coastal areas, destroying coral reefs, and salinizing farmland. The country's vigorous deforestation, imprudent overfishing, and uncontrollerd village pollution run-off threatens Fiji's future.

Fiji's ethnic divide, its economic, trade, and budgetary handicaps add to the challenges facing the country. Its threats, more sharply defined than those in many other island micro-countries, lead one to pray that hope triumphs over experience.

THE REPUBLIC OF VANUATU

"This fragility of this style of governance . . . [in which] tenuous party allegiances have led to frequent no-confidence votes in parliament, disrupting governance at the national level, and leaving the court to uphold the rule of law and protect the constitution . . . The persistence of this pattern since independence suggests that 'elite conflict' is the biggest threat to the security of the state."

—Graham Hassall

Vanuatu lies north of the Caledonia Islands (a French possession), south of the Solomon Islands, west of Fiji, and closer to Australia and New Zealand than the other Oceania micro-countries. Impoverished Vanuatu spreads over 277,000 square miles of ocean; its 83 volcanic islands contain 272,000 people in a land area of 4,706 square miles. Many villages lack electricity, are as far as an hour plane ride from the country's capital, and are reachable only by storm-vulnerable "banana boats," single-engine planes, or mountain footpaths impassable during frequent rains.

Melanesian people possibly settled the islands as early as 5,000 years ago. A Portuguese explorer, Pedro Fernandes de Queirós, led an expedition to the islands in 1606 and claimed the land as far south as the South Pole for Spain. Europeans did not return until 1768 when the French explorer Louis-Antoine de Bougainville rediscovered the islands. In 1774, the British explorer, James (later Captain) Cook charted the islands and named them the New Hebrides.

In 1825, the trader Peter Dillon found sandalwood on one of the islands, a discovery that led to a rush of immigrants; however, by the mid-1860s virtually all the sandalwood had been cut. In the 1860s, planters in Australia, Fiji, Samoa, and New Spain (the Americas), in need of labor, engaged in recruiting long-term indentured labor (blackbirding) from the islands, severely reducing the native population.

In 1882, a French company purchased large sections of land and attracted many French settlers. In 1887, as the number of British and even more French settlers increased, the United Kingdom and France created a

Joint Naval Commission to oversee the islands. In 1906, the increasing jumbling of interests led the British and French to jointly establish the islands as the Anglo-French Condominium of the New Hebrides, with each colonial power responsible for its own citizens and jointly for other non-islanders.

Rising nationalist pressures for self-governance led to the establishment of an Advisory Council being established in 1967 and its replacement by a Representative Assembly in 1974. In the early 1970s, Father Walter Lini formed the first political party, initially called the New Hebrides National Party and then the Vanua'aku Pati (VP), which pressed for independence. Preparatory to independence promised for 1980, elections were held in 1978, limited self-governance began, and a government of national unity was formed with Father Gérard Leymang as chief minister.

The constitution provided for a modified Westminster system with a (mostly ceremonial) president as head of the Republic elected for a five-year term by two-thirds majority of an electoral college consisting of the legislature supplemented by the council of provincial presidents/chiefs. The prime minister, who heads the government and the cabinet, is elected by three-quarters of the parliament and is responsible to it.

The unicameral parliament consists of 54 members elected for four-year terms. It is advised by the national council of chiefs, especially in a country in which many regard the chiefs as a "moral anchor." They wield considerable influence at the national as well as local level. The country is divided into six provinces. Each province has a popularly elected provincial council (advised by a council of chiefs), a council-elected president, and an executive officer appointed by the Vanuatu president on the advice of the prime minister. The provinces are divided into municipalities. The judiciary is independent.

Vanuatu belongs to the United Nations, the World Trade Organization, the Commonwealth, and the Pacific Islands Forum. It does not maintain a mission at the European Union, but does maintain a United Nations mission, sharing a floor of offices with several other Commonwealth countries in New York.

Vanuatu's governments have been remarkably unstable since independence. Votes of no-confidence have been the norm, facilitated by volatile coalitions with one or more of the numerous party groups: the more anti-colonial and socialist VP, the West Union of Moderate Parties (UMP) and the Melanesian Progressive Party (MPP) that are more supportive of closer relations with France, the People's Democratic Party (PDP), the Union of Republic Party (URP), Nagriamel, the Tan-Union, the Greens Confederation, French Melanesia, and the cult-turned-poitical party John Frum Movement.[72]

A rapid succession of prime ministers followed. Walter Lini won in 1980, 1983, and 1987 but President Sokomanu replaced him with an interim president, insurgent Barak Sopé, the former VP general secretary who had formed the MPP.[73] In 1991, Donald Kalpokas was elected president; four months later the UMP leader Maxime Carlot Korman was elected prime minister with the help of the NUP. In 1994, Jean-Marie Léyé became elected prime minister to be succeeded in 1995 by Serge Vohor of the UMP heading a four-party coalition. In 1996, Maxime Korman was again elected prime minister—succeeded within the year by the UMP leader Vohor. Bob Loughman became prime minister in 2020.

To force payment of unpaid allowances, Korman was abducted by the Vanuatu Military Force (for which they were arrested). The country has had six heads of government since independence. Sato Kilman took office in 2015, and Charlot Salwai succeded him 2016; in 2017, Tallis Obed Moses was elected president. The rapid changes in leadership have reflected the deep political schisms, which have ensured inconsistency, inefficiency, and ineffectiveness in governance.

The inhabitants of Vanuatu are called Ni-Vanuatu, a term of recent coinage. The Ni-Vanuatu are 98 percent Melanesian, but the rest are made up of a mix of Europeans, Asians, and other Pacific Islanders.[74] Over 70 percent of Ni-Vanuatu are Protestant, mainly Presbyterian, Anglican, and Seventh-day Adventist. Most of the rest are Roman Catholic.

English, French, and Bislama are the official languages. Bislama, a widely spoken tongue, combines Melanesian grammar with mostly English vocabulary. The embracive demographic term, Ni-Vanuatu, and the

---

72. Belief in a mythical figure named John Frum inspired an indigenous cult dates back 100 years and reached its zenith during and after World War II. Its followers say David Lea and Colette Milward have "advocated a return to traditional ways, rejecting European currency, religion, and Europeans . . . supporting traditions and customary [collective tribal] ownership of lands."
73. Sopé was later tried, convicted, and imprisoned for seditious conspiracy.
74. The three minute islands are historically populated by Polynesians.

widespread use of Bislama obscures the extent of ethnic and linguistic heterogeneity of the country. Over 100 different tongues are spoken locally. The people identify strongly with their local community and its customary leaders, tending to call themselves by the name of their island or island group. Tribalism resists efforts to reform Vanuatu's political institutions, reduce its chaotic politics, and develop its infrastructure and economy.

Vanuatu's economy is troubling. 65 percent of Vanuatu's labor force remains in rural subsistence agriculture (beef, coconuts, copra, cocoa, and coffee), 30 percent in services, and only 5 percent in industry (food and fish processing and canning). The country's GDP per capita was $2,600 in the mid-2010s. The tourism and finance sector constitute the largest sources of foreign exchange earnings. Located in Port Vila, the capital and commercial center has an expatriate population with significantly higher incomes than the rest of the country.

Vanuatu's imports in the early 2010s ($312 million) were principally from China and Singapore, less from the U.S., Japan, and Australia. Its exports are only $43 million—mainly to Thailand and the Ivory Coast. In the mid-2010s, Vanuatu's budget was about $201 million, and its GDP was $2.1 billion. Hassall reports the following:

> In 2004, the ADB [Asian Development Bank] assessed the country as having a range of good prospects. Vanuatu's proximity to Australia and New Zealand offers the tantalizing prospect of developing tourism, but in the context of an unstable and at times fragile policy and legal environment.

Little is changing. China supports the development of Vanuatu as part of its Belt and Road Initiative. Assistance includes loans from the government-owned China Exim Bank that can be called, in full, in the event of non-payment. What China gains from the Belt and Road development strategy and its constituent projects is not only gratitude but also long-term investment in access to local facilities and development of trade.

Climate change presents a constant threat. Cyclones regularly threaten the islands, overflowing rivers and destroying crops. Cyclone Pam, which ripped through Vanuatu in 2015, provides a forecast of what the future may bring. Climate change is not the only menace. In 2017, the threat of an earthquake in Ambae (one of the Vanuatu islands) forced the evacuation of its 11,000 inhabitants.[75] The future of Vanuatu is indeed bleak.

---

75. A threatened earthquake in Bali at the same time forced the evacuation of 140,000 Bali residents.

THE KINGDOM OF TONGA

Tonga has "been the only island country in the [Pacific] region to escape formal colonization during the period of European expansion in the 19th and 20th centuries and was therefore spared the turbulence which has typically accompanied most moves to decolonization."

—Stephanie Lawson

The Kingdom of Tonga is adjacent to Fiji to the west and the Independent State of Samoa and American Samoa to the east with a total land surface of 750 square miles scattered over 270,000 square miles of the South Pacific. The kingdom comprises 177 islands—52 of which are inhabited by its total 106,000 people. The archipelago is divided into three groups: Vava'u, Ha'apai, and Tongatapu (the largest, the most populous, and the site of the capital, Nuku'alofa).

Polynesian people settled the Tongan islands sometime between 1500 and 1000 BCE. "The first Tu'i [King] Tonga, 'Aho'eitu, was said to be the son of Tongoloa, a god in Tongan (and Polynesian) mythology . . . He began his rule in 950 CE. For many years, the Tu'i Tonga dynasty ruled absolutely and unchallenged by the chiefs." During the 1100s and 1200s, the Tongans and their paramount chief established quasi-hegemony from Samoa to Fiji, leading some historians to speak of a Tu'i Tongan Empire.

In the 1600s, Dutch explorers Willem Schouten and Jacob Le Maire visited a northern Tongan island, and Abel Tasman visited the southern island of Tongatapu. In the 1770s, the British explorer James Cook visited (calling the archipelago "The Friendly Islands"), and the Spaniard Alejandro Malaspina came in 1793. London-based missionaries came in 1797. After 70 plus years of intermittent civil war from the late 18th to the 19th century, Tāufa'āhau, a young chief of a leading dynasty, defeated the other chiefs and reunited the kingdom. At his baptism, by Wesleyan Methodist missionaries in 1831, Tāufa'āhau took the name George Tupou.

In 1862, the British encouraged Tupou to proclaim a legal code guaranteeing land tenure to natives. In 1875, the king promulgated a constitution securing freedom of speech, press, and religion. In 1876, a treaty was signed with Germany, recognizing Tongan independence. Similar treaties were signed with Britain in 1879 and the United States in 1888. King George Tupou II, who succeeded his grandfather in 1893, signed another Treaty of Friendship with Britain, one that established Tonga as a British Protectorate with control over Tonga's foreign affairs and defense. Tonga's centralized political system—with its long-established indigenous monarchy and landed nobility, entrenched by the 1875 adoption of a formal constitution designed to sustain its ruling class—had long enabled Tonga's rulers to delay colonial annexation and avoid the development of representative institutions. By the time of the treaty in 1900 (and its revision in 1958) Tonga became a British protectorate, self-governing but consulting with the British on external matters.

Queen Sālote Tupou III, in 1918, succeeded her father, King George Tupou II, and, in 1923, appointed her husband prime minister. When World War II began in 1939, New Zealand became responsible for the defense of Tonga. After World War II, the British superimposition of certain Westminster-style institutions, such as prime minister, cabinet, legislature, judiciary, and electoral system, gave the system some additional European legitimacy. Though, by no means did they add up to a system of electoral responsible government.

In 1965, the British protectorate was terminated. Tonga, Fiji, and (Western) Samoa established the South Pacific Forum—the forerunner of the Pacific Islands Forum.

That same year, King Tupou IV succeeded his mother and his brother became prime minister. To increase revenue, the king and his brother made some questionable financial decisions, including selling Tongan passports, which later forced Tonga to naturalize the purchasers, sparking ethnicity concerns. An airport and casino were built with support of an Interpol-accused criminal.

In 1993, the forerunner of the Human Rights and Democracy Movement (HRDM) party was founded, which won elections in 1999. Continuing criticism led to the passage, in 2003, of a constitutional amendment and legislation limiting freedom of the press. Members of the HRDM reacted immediately and canvassed for support against these measures.

King George Tupou V succeeded his father in 2006. Discontent with government programs led to riots that destroyed much of the capital's downtown.[76] The king supported the government's political reform keeping pace with its economic development. In 2010, a new legislature was elected under a revised constitution. In 2012, King George Tupou VI succeeded his father. Tonga is the only Pacific island country to remain a monarchy.

The prime minister and cabinet are selected by and responsible to the monarch. Not until the 21st century did a Tongan king suggest including any members of the legislature to the cabinet. Tonga traditional forces have long succeeded in preserving their prerogatives by opposing the development of representative institutions. Lacking the presence of a foreign colonial power upon which to vent anti-colonial, nationalistic rhetoric and rage, progressive forces have been handicapped in their efforts to secure representative institutions from an indigenous, myth-entrenched monarchy and nobility. Tonga has five administrative divisions. It is a member of the United Nations and several of its affiliates, the Commonwealth, and the Pacific Islands Forum.

The royal family and nobles continue to dominate Tonga's economy. With its mid-2010s GDP of $801 million and a GDPpc of $7,700, Tonga depends on agriculture for most of its employment and exports—mainly squash, coconuts and copra, bananas, cocoa, and coffee. Remittances from Tongans living overseas, mainly in Australia, New Zealand, and the United States, significantly help the economy. Manufacturing, consisting principally of handicrafts and small-scale industry, contributes only 3 percent to the economy. Agriculture accounts for 31 percent of the labor force, industry 30 percent, and service (mostly government and tourism) 18 percent. Imports (of $122 million), principally from Fiji and New Zealand, significantly exceed the exports (of $8 million) principally to South Korea, the United States, and New Zealand.

Several critical factors will affect Tonga's future. One is how effectively it manages the political transition away from royal dominance; how well it manages this transition will affect how it develops its economy. Another is how it deals with corruption. But the most important factor may be how

---

76. The British High Commissioner was withdrawn with budgetary restraints given as the reason.

the environment deals with Tonga! In 2018, the most powerful storm to hit this South Pacific island in at least 60 years destroyed many buildings, including the Parliament, and left thousands without power. Will such storms become more frequent? How will the threat of such storms and the continuance of the dominance of the royal family and nobles affect Tonga's future?

THE INDEPENDENT STATE OF SAMOA

"Radical nationalism and cultural conservatism often go hand in hand. Indeed, the latter is usually an important ingredient of most liberation movements. Davidson notes that the dominant tone of Samoan thinking on such matters as suffrage was characterized by 'a desire for cautious advance and for maintaining a firm link with the past.'"

—Stephanie Lawson

The Independent State of Samoa, lying east of Fiji (Melanesian) and Tuvalu (Polynesian), north of Tonga (also Polynesian), and about 1,000 miles northeast of New Zealand, occupies a 350-mile long archipelago with a population about 150,000. Southeast Asian Melanesians migrated to the islands more than 3,000 years ago. Between 950 and 1250 CE, Tongan paramount chiefs exercised hegemony over the Samoan Islands. The first European recorded visit to the Samoa Islands was Jacob Roggeveen, a Dutchman sailing on behalf of the Dutch East India Company who, in 1722, obtained water and herbs in Mantua. A Frenchman, Louis-Antoine de Bougainville, visited in 1768.

Contact with Europeans, write David Lea and Colette Milward, was limited to a few stranded beachcombers until 1830, when "Reverend John Williams of the London Missionary Society arrived in Samoa. The Islanders rapidly converted to Christianity, resulting in the suppression of many of their previous beliefs and a loss of much of their oral traditions of history." Extant native traditions were fused into their Christianity. Laws in the early Christian kingdom were promulgated under missionary guidance.

In the late 1800s, Germany, the United States, and Britain developed commercial interest in the Samoan Islands. In 1856, a German trading firm, Godeffroy & Co., established a post for trading coconut oil; the company bought land to be settled by Germans and to be worked by indentured Chinese.[77]

---

77. In 1872, the United States developed a naval base and established a protectorate of several islands to the east of the German-controlled islands. A constitution was drafted, Malietoa Laupepa, a descendant of tribal chiefs, became king but was dismissed in 1961, and an American official was appointed prime minister.

In 1884, the German consul in Samoa forced an agreement on the *de facto* paramount Samoan rulers, which virtually handed the government over to the Germans. The British and American consuls protested strenuously, but to no avail. In 1887, the Germans exiled Laupepa. An official of Deaths Handles (the successor of Godeffroy & Co.) became prime minister, sparking a Samoan revolt. Germany, the United States, and Britain sent warships. While these powers were dubious regarding the value of the Samoan Islands, they did not want another power possessing them.

Consequently, the 1889 Treaty of Berlin provided that Samoa would retain political independence, with its monarch advised by three consuls, German, American, and British and the two principal administrative officials appointed by the king of Sweden. Ten years later, unsatisfied with this arrangement, the Samoan Islands were divided between Germany and the United States. Germany took the western islands, which became Western Samoa, and, now, the Independent State of Samoa. The United States took the eastern islands, now American Samoa. The British vacated claims to Samoa upon Germany vacating claims to Tonga.

With the outbreak of World War I, New Zealand troops replaced the German presence in Western Samoa. At the conclusion of which, the newly created League of Nations granted New Zealand the mandate to govern the islands. Severe inflation, the catastrophic spread of the 1918 flu epidemic, and the perception of misrule led the Western Samoans to resent New Zealand's rule. By the 1920s, anti-colonial fervor had gathered widespread support.

Following World War II, the newly created UN made Western Samoa a UN trust territory with New Zealand continuing as the administering power. In the 1950s, measures of internal self-governance were introduced. In 1959, Mata'afa Faumuina Fiame Mulinu'u, one of the country's four highest-ranking chiefs became head of government. In 1962, after repeated efforts by the Samoan Independence Movement, Western Samoa became independent; the 1960-negotiated constitution formally came into force, and the two paramount chiefs became the co-monarchs. When one of them died a few years later, the other remained as sole head of state. In 1997, the country changed its name to the Independent State of Samoa. Upon the death of the sole head of state in 2007, Samoa changed from a monarchy to a republic with an executive president. Tuilaepa Aiono Sailele Malielegaoi, the first prime minister, was still serving in 2019.

Before 1990, two parties competed for power: principally the Human Rights Protection Party (HRPP) and the Samoan National Development Party (SNDP). In 1990, Tofilau Eti Alesana, the HRPP leader and prime minister, resigned and was succeeded by his deputy prime minister, Tuilaepa Sailele Malielegaoi. Global concepts of democracy combined with the elected leaders' lack of respect for the traditional *matai* system led to the HRPP (despite strong opposition from the conservative SNDP) securing passage of legislation adopting universal suffrage in 1990 while continuing to restrict public office-holding to chiefs (*matai)*. The political advantages the HRPP secured from this arrangement contributed to Malielgaoi and his party winning the subsequent elections, serving without interruptions as prime minister from 1990, continuing to serve for 25 years.

The Samoan parliamentary system, based on the British model but modified to take account of Samoan customs (and that it is a republic), provides for a unicameral legislature (*Fono*) of 49 members elected for 5-year terms from single-member territorial electoral districts; of these 47 are *matai* titleholders. The other two are elected by non-ethnic Samoans on separate electoral rolls. The judicial system combines British common law and traditional local *matai* custom; the Supreme Court of Samoa is the court of highest jurisdiction. Samoa has 11 local government districts, traditional polities established well before Europeans arrived. Senior *matai* titleholders in each district select the district's paramount titleholder who continues to carry out traditional civic duties. Besides the UN and affiliated agencies, Samoa belongs to the Commonwealth and the PIF. Among its missions is one in Brussels accredited to the EU and several European countries and one in New York City accredited to the United Nations, the United States, and Canada.

Lawson writes the following regarding the the continuing debates on no longer limiting office-holding to *matai*:

The fact that the proposed suffrage reforms would bring Western Samoa more into line with widely accepted principles of democratic practice was almost beside the point. Indeed, it seemed more a matter of pride among Western Samoans, including many of those supporting reform, that their system was not modeled on foreign political practices—democratic or otherwise. The often-repeated justification for retaining the *matai*-only eligibility for office is the importance of preserving Western Samoa's unique cultural heritage.

Matai influence will long impact the Samoan political scene.

Of the approximately 200,000 Samoan inhabitants, 93 percent are ethnic Samoan, 7 percent are of mixed European and Polynesian ancestry, and less than 0.5 percent are European. A Samoan dialect of Polynesian is the official language; English is widely spoken. Regarding religion, the influence of missionaries in Samoa is evident: 32 percent of the population are members of the Christian Congregational Church of Samoa, 14 percent is Methodist, 8 percent is Assembly of God, 4 percent is Seventh-day Adventist, 15 percent is Mormon, and 19 percent is Roman Catholic.

Samoa's mid-2010s GDP was $1.1 billion, with a per capita GDP of $6,800. Its annual budget was around $260 million. Of its imports of $319 million, over 70 percent were from neighbors New Zealand, Singapore, Fiji, and China, far exceeding its exports of $14 million, principally to American Samoa and Australia. Traditionally, the Samoan economy has depended mainly upon copra, cocoa, bananas, and fishing, which continue to represent a significant share of the exports. A Yazidi wire harness factory and food processing comprise the manufacturing sector. Tourism is expanding, assisted by significant capital investment, and improved airline service. In the last few decades, development aid and overseas family remittances have become increasingly important in the economy. Limiting office holding to *matai* doesn't help. The immediate concern is inundation—even partial inundation!

TUVALU

"At a referendum held in the Ellice Islands [in 1974], over 90 percent of voters favored . . . [becoming] a separate British dependency under the indigenous name of Tuvalu."

—David Lea and Colette Milward

The Tuvalu Islands, formerly called the Ellice Islands, located north of Vanuatu and Fiji and south of the Gilbert Islands of Kiribati, consists of a chain of 9 low-lying islands, 360 miles long, inhabited by fewer than 11,000 people living on its 10 square miles of land. The earliest settlers were Polynesians who probably began spreading from Samoa and Tonga into what is now Tuvalu more than 3,000 years ago.

The first recorded European sightings of Tuvalu were in 1568 and 1595 by a Spanish navigator, Álvaro de Mendaña de Neira. In 1751, a British captain, John Byron, visited the islands (the islands were later named for the ship owner Edward Ellice). Other navigators who visited included a Russian explorer in 1820, an American whaler in 1821, Louis Isidore Duperrey when circumnavigating the world in 1824, and Dutch explorers during an expedition in 1825. From 1850 to 1875, slave traders (especially from Peru) captured islanders for the South American markets. David Lea and Colette Milward assess that the slave trade, "together with European diseases, reduced the population from about 20,000 to 3,000."

Trading companies became active from the 1850s. Protestant missions began in the1860s. In 1877, the British established the Western Pacific High Commission (WPHC), headquartered in Fiji, with oversight of the Ellice Islands and other island groups. In 1892, the British established a protectorate over the Ellice Islands, linking them administratively to the Gilbert Islands as part of the British Western Pacific Territories (BWPT). In 1916, Britain formally annexed the islands (as distinguished from considering them as a protectorate) to the Gilbert Islands (now Kiribati) in the Gilbert and Ellice Islands Colony (GEIC) despite the long distance and ethnic differences that have long separated these two groups of islands.

During World War II, the Japanese occupied the Ellice Islands; the British reoccupied them after the war.

In a 1974 referendum, the Ellice Islanders voted decisively to separate from the Gilbert Islands; they became a separate British dependency under the indigenous name of Tuvalu. In 1978, the country gained independence; in 2000, it was admitted to the UN. The new constitution provided for a constitutional monarchy headed by the British monarch with a governor-general, a 15-member unicameral parliament (which replaced the earlier established House of Assembly), and a government headed by a prime minister responsible to the parliament.

There is an independent judiciary, whose decisions are seldom appealed to the Privy Council in London. The nine local government districts each have their own high chief chosen on basis of ancestry. The sub-chiefs (*alikis*) and traditional assembly of elders (*Falekaupule*) share power with the elected village presidents (*pule a kuapule*). Tuvalu is a member of the Commonwealth and the Pacific Islands Forum and maintains a United Nations mission, but not a European Union one.

Toaripi Lauti, who had been chief minister before independence, was appointed the first prime minister. Following non-partisan elections in 1981, Dr. Tomasi Puapua was elected prime minister and was reelected in 1986. In 1989, Bikenibeu Paeniu defeated Puapua and became prime minister. In 1993, Kamuta Latasi defeated Paeniu in 1993, but, in 1996, Paeniu beat Latasi and won again in 1996 and 1998. He was forced to resign in 1999, and his successor died within a few months. Faimalaga Luka followed in 2001. Prime Minister Willy Telavi, elected in 2011, received a vote of no confidence in 2013 and was succeeded by Enele Sopoaga. Kausea Nataal became prime minister in 2019. The non-partisan community elections, in which traditional allegiances and family ties remain predominant, have been surprisingly eventful.

Residents of the Ellice Islands are over 96 percent Polynesian compared to those of the Gilberts who are over 98 percent I-Kiribati, a Micronesian ethnic group. Tuvaluan, along with English, is the official language. It is a Polynesian tongue, not a Micronesian one. About 97 percent of Tuvaluans are members of the Protestant Church of Tuvalu, many of whose members combine ancestral traditions with the mission-introduced evangelical

Christianity. The 2012 population was almost 11,000. The population has almost doubled since then.

About 65 percent of those formerly employed in the private sector are now on the public sector payroll. About 15 percent of adult males are employed as seamen on foreign-flagged ships, many of who trained in the local maritime school. Other Tuvaluans work on coconut plantations or in traditional subsistence farming and fishing. Tuvalu's small population and remoteness limit the development of airport access and hotel facilities. Thus, tourism prospects are poor. As of 2013, the GDP was $38 million, and the per capita GDP was $3,400, which indicates the perilous state of the economy. Government revenues depend largely upon sales of stamps, sales of fishing licenses, commercialization of its domain internet name, ".tv," that generates more than $2 million annually (guaranteed until 2021), and income from the Tuvalu Trust Fund established in 1987 by the British and supported by Japan, South Korea, Australia, New Zealand, and the European Union.

Climate change threatens Tuvalu's future. A drought beginning in 2010 forced the government to declare a state of emergency. Rising sea levels continue to threaten to inundate the low-lying islands. But the government appears to be content to continue to muddle along. Its remoteness, minuteness, and tradition-mindedness provide its leaders little choice but to continue to rely upon offshore sources of income for sustaining their public services and development.

THE REPUBLIC OF KIRIBATI

"To plan for the day when you no longer have a country is indeed painful, but I think we have to do that."

—Anote Tong, President of Kiribati (2003–2015)

The Republic of Kiribati, which includes the widely separated Gilbert, Phoenix and Line, and Banaba groups of islands, straddles both the Equator and the International Date Line. Almost all of the Republic's 100,000-plus people live in the Gilbert Islands.[78] Less than 10,000 people reside on the Line Islands. These are the islands for which the International Date Line has been moved east to ensure they are included in the same time zone as the rest of Kiribati (see Map 15). Fewer than 50 persons live on the one inhabited island among the eight Phoenix Islands.

The Republic of Kiribati spreads over 1.35 million square miles of the Pacific, with some of its inhabitants 2,000 miles from the capital, Tarawa. All of the islands making up Kiribati (except Banaba) are wafer-thin sand and reef rock atolls rising less than six feet above sea level. Kiribati shares mid-ocean boundaries with the Marshall Islands and Nauru on the northeast and Tuvalu on the southwest.

Kiribati (pronounced "KEE-reel-bags," as the islanders articulate "Gilberts"), was long inhabited by migrants from Tonga, Samoa, and Fiji who introduced Polynesian and Melanesian language and customs and integrated them into a Micronesian language and culture. In the 1600s and 1700s, European navigators, as they circumnavigated the world or sought sailing routes from the south to north Pacific, paid chance visits to the islands. In 1788, Thomas Gilbert (after whom the Gilberts were named), when sailing from Australia to China, sighted these islands. By the 1830s, Europeans, Chinese, Samoans, and others came as beachcombers, castaways, merchants, and missionaries. Trade in copra had by begun by 1870.

In 1892, the British secured agreement from local chiefs to establish a protectorate over the Gilbert Islands and nine Ellice Islands (the latter are

---

78. In recent decades, almost 50 percent of them have lived on its principal island, South Tarawa.

now Tuvalu). The two groups were administered together by the Western Pacific High Commission (WPHC), which was based in Fiji. In 1900, the British annexed Ocean Island (later renamed Banaba Island), west of the Gilbert group, following the discovery of significant phosphate resources. In 1915–6, the British annexed the Gilbert and Ellice Islands, converting the protectorate into the Gilbert and Ellice Islands Colony (GEIC), which was extended to include Ocean Island and two Line Islands to the east. Christmas Island (now Kiritimati) in the Line group was added to GEIC in 1919, and eight then uninhabited Phoenix Islands were added in 1937.

During World War II, Japanese forces occupied Tarawa and other Gilbert Islands from 1941 to 1943. Tarawa was the site of one of World War II's fiercest battles in the Pacific. From 1957 to the 1960s, Americans and British used Christmas Island to test nuclear weapons, including hydrogen bombs.

In 1963, the British government took the first of successive steps to prepare the GEIC for self-governance, creating an Advisory Council and an Executive Council. In 1967, GEIC achieved self-governance and elected a chief minister. In 1975, the British separated the Ellice Islands, whose non-Micronesian people preferred to seek their own self-governance, from GEIC into their own separate territory, called Tuvalu. In 1979, the Gilbert Islands became independent as Kiribati. Kiribati belongs to the United Nations and several of its affiliates.

The people of Kiribati have long lived with the threat of environmental catastrophe. In the 1950s, prolonged drought led colonial authorities to relocate hundreds to the Solomon Islands. The adverse effects of phosphate mining and the coming exhaustion of the mines on Banaba Island led to hundreds more being settled in Fiji in 1971 on Rabi Island. There, a municipality has been organized which is represented in the 2,000 miles away Kiribati legislature. By 1979, the Banaba mines were exhausted and closed, and the population of the island had declined to less than 300 by the 1990s. Severe rains and flooding in Kiribati in 1997 provided a preview of more climate change-driven violent storms and rising sea levels, which will inundate their low-lying islands. Anticipating the time when its 117,000 inhabitants will need to abandon their communities, Kiribati has gained possession of 80 square miles of Fiji for its population to resettle. Anote Tong, leader of the opposing *Boutokaan Te Koaua* Party (Pillars of Truth), then became president. Tong was reelected in 2007 and 2011. During his 12 years as president, Tong became a celebrity speaker around the world, pleading the cause of reducing carbon emissions, global warming, and the threat of the consequent rising sea levels to so many island countries. He has developed contingency plans, using nearly $7 million of government funds to buy 6,000 additional acres of land in Fiji for possible resettlement of Kiribati islanders forced to evacuate.

The Kiribati constitution, adopted upon achieving independence in 1979, established a government with a unicameral legislature and a president who combines the roles of head of state, head of government, and head of party. The legislature (renamed the *Maneaba Ni Maungatabu*) elected Ieremia Tabai as *Beretitenti* (president); he was reelected in 1982. In 1983, the *Maneaba Ni Maungatabu* was dissolved and new elections were held. This happened again in 1987 when Tabai defeated the 1985 newly created Christian Democratic Party. Having served the constitutionally prescribed maximum term in office in 1991, his vice president was elected as president. In 1994, the *Maneaban Te Mauri* Party (Protect the Maneaba) defeated the incumbent National Progressive Party and elected Teburoro Tito president. He was reelected in 2003, but soon after the legislature removed Tito on grounds of graft.

Tong's popularity abroad was not matched at home. Many felt that he could have done more about the high rates of unemployment and infant mortality. In a fervently evangelical country whose churches dominate politics, faith in climate science was castigated by the opposition as a brazen challenge to divine authority. Having reached his three-term limit in 2016, Tong was replaced by the opposition's candidate, Taneti Maamau, a former minister of finance.

The I-Kiribati, as the Micronesian native people of Kiribati are called, comprise about 90 percent of its people. Nine percent of the population is I-Kiribati mixed with other Micronesian people, thus the islands are almost 100 percent populated by Micronesians. Both English and Taetae ni Kiribati (an Oceanic tongue also called Gilbertese) are official languages and widely spoken in Kiribati. The fact that 56 percent of the I-Kiribati people are Roman Catholic, 35 percent are Kempsville Presbyterian, and 5 percent are Mormon testifies to the diligence of the early missionaries.

Kiribati, whose mid-2016 GDP was less than $188 million (and had a per capita GDP of $1,700), is one of the least developed countries of the world and one of the poorest prospects. Kiribati's exports (only $85 million in 2013) depend principally on copra and fish. Other sources of revenue include fishing licenses, worker remittances, and minimal tourism—hardly enough to pay for the essential foodstuffs and manufacturing items it imports ($182 million in 2013). Kiribati depends upon development grants from agencies including the European Union, the World Health Organization, the World Bank, the Asian Development Fund, and Australia, Japan, New Zealand, and Taiwan, which Kiribati recognizes, to fund its annual budget ($180 million in 2013).

Economic prospects are dire. Few are anxious to invest in a country whose low-lying atolls remain extremely vulnerable to inundation or sea-level flooding. Scientists predict Kiribati could be underwater in as little as a century. Kiribati faces some difficult choices. Evacuation appears to be the only realistic prospect.

THE REPUBLIC OF NAURU

"The central drama of Nauru's modern history could be told as a story of phosphate mining, and of sudden wealth and rapid economic decline."

—Max Quanchi

The Republic of Nauru, with a population about 9,500 (of whom 2,500 are foreign contract workers), is, next to the Vatican, the world's least-populated independent country. Nauru straddles the Equator. Its nearest neighbor to the southwest is the Solomon Islands. To the northwest, it neighbor is the Federated States of Micronesia, and, to the northeast, the Republic of the Marshall Islands. Nauru was successively ruled by Germany from 1888 until World War I, administered as a mandate principally by Australia between World Wars I and II, and from 1947 as a United Nations trust with Australia as principal trustee. It gained its independence in 1968.

Micronesian and Polynesian peoples inhabited the island about 3,000 years ago. The first European to report sighting Nauru was a British whale hunter, John Fear, who named it Pleasant Island. By the 1830s, the island was a stopping point for European whaling ships and other traders replenishing supplies and a haven for deserters. Firearms introduced by Europeans in the in the 1878–88 Nauruan Tribal War led to Germany ending the war by annexing Nauru to its Marshall Islands Protectorate.

Phosphate was discovered on the island in 1899 by a British prospecting company. With a go-ahead from Germany, the Pacific Phosphate Company began phosphate mining in 1906, with profits shared between Germany and Britain. In 1914, Australian forces occupied Nauru. In 1920, the League of Nations assigned Nauru as a mandate to Australia (co-administered with New Zealand and Britain).

At the outset of World War II, the Japanese occupied Nauru and deported 1,200 Nauruan's as laborers to Truk (now Chuuk in the Federated States of Micronesia). Nauru was bypassed in the American-led World War II Pacific campaign. In 1947, Australia, New Zealand, and Britain became the UN co-trustees, with Australia as the administering power. In 1964, the UN Trusteeship Council, anticipating the exhaustion of the island's phosphate deposits, proposed that the population of Nauru be resettled off the northeast Australian coast. Nauru rejected the proposal. During the years of foreign rule, Nauru protested against the unpopular resident administrators. Agitation by the advisory council of chiefs in international forums led to the establishment of a legislative council in 1925, then an executive council in 1951, which was strengthened in 1965 and 1966 when Nauru gained limited self-governance.

In 1968, Nauru gained independence and joined the United Nations and its affiliates in 1999. Under the new constitution, the Nauruan executive president combines the roles of head of state and head of government (with a 5- or 6-member cabinet). The president is elected by the 19-member legislature. Nauru has an independent court system with a Supreme Court and lower courts. Cases may be appealed to a High Court in Australia, but this is rare. There are 14 local governments. The Nauru head chief, Hammer DeRoburt, was elected executive president and was reelected in 1971 and 1973.

Following independence, Nauruan politics was dominated by the 2,000 or so family groups in the 12 local districts. Only four parties have been active in Nauru politics, but alliances have formed frequently around extended family ties. From 1968 to 2013 (only 45 years) there have been 29 changes of heads of government. Twenty rulers lasted less than one year. Six held office two or more times—one for six times. Baron Waqa, who became president in 2013, lasted longer. He was replaced by Lionel Aingimea in 2019. Few countries have endured *18 turnovers* in less than half a century. The strength of individual and family identification with their districts and their chiefs has intensified the Nauruan political "musical chairs." This phenomenon has aggravated mismanagement, nepotism, bloated bureaucracy, misuse and extravagant use of funds, corruption, and poor investments. The phosphate funds were squandered. The present authoritarian government, led by Baron Waqa, has removed most opposition MPs from parliament.

In 1967, Nauru purchased the assets of the British Phosphate Commissioners. In 1970, the locally owned Nauru Phosphate Corporation took control. Income from the phosphate mines peaked in the late 1900s, giving Nauru a GDPpc in 1997 of over $29,000; by the mid-2010s, it had dropped to $5,000. In 1987, President DeRoburt created a commission that focused on the extensive "Topside" area of the island that had been left virtually uninhabitable as a result of the mining; in 1988, the commission proposed that Australia, New Zealand, and Britain each pay one-third of the cost to fund Nauru's rehabilitation. When Australia refused to comply, Nauru appealed to the International Court of Justice. Australia has agreed to comply with the eventual ruling.

By 2005, the GDP was only $65 million and its budget only $13.5 million (the last figures available). Coconut farming provides a subsistence living for only a few. Offshore banking has been introduced. Phosphate, once dominating Nauru's economy, left Nauru struggling.

Nauru (formerly Pleasant Island), highly dependent on foreign aid, receives significant financial assistance from Australia for allowing boat migrants refused entry by Australian authorities to be detained in a Nauru center operated by private contractors, making the prison business a major industry. Since July 2003, Australia has imprisoned more than 2,000 people in two island prisons: one in Manus, Papua New Guinea, and Nauru. These were would be migrants who were intercepted by Australian naval vessels and deposited in these islands. In effect, Nauru is paid by Australia to do their dirty work, but the prison arrangement is lucrative, for each detainee the Nauru government collects USD $1,400. The United Nations High Commissioner for Refugees (UNHCR) has been highly critical of the arrangement. The United States has, as of 2019, resettled over 500 of these refugees, leaving about 200. Australia has said that the last remaining children on Nauru would also be resettled in the United States.[79]

The long-time emphasis on phosphate mining significantly affected the ethnic composition of the population. Only 58 percent of the present population is descended from the original inhabitants; 28 percent is from other Pacific islands, 8 percent is Chinese, and 8 percent is European. Most non-natives are residents under contract; most of the others were once were under contract—or their parents were. Nauruan is the official language, but English is widely used for commercial and government business. 60 percent of Nauruans' population is Protestant (including Nauru Congre-

---

79. President Trump has not rescinded an earlier pledge by President Obama to resettle many of these prisoners in the United States.

gational, 36 percent, and Assembly of God, 13 percent); 33 percent is Catholic.

In the short-term, Nauru, with its authoritarianism, very few people, very limited resources, no readily visible, promising development options can only continue its financial management struggle. The scarcity of professionals, and the lack of education and post-secondary opportunities, has undermined its ability to manage its industry and its governance. In the long run, is there any hope for an island not only no longer possessing a valuable resource but also deprived of usable land? Resettlement may be the only viable option for Nauru.

THE REPUBLIC OF THE MARSHALL ISLANDS

"We see the danger occurring now. We're trying to beat back the sea".

—Tony deBrum, the Marshall Islands' Foreign Minister

The Republic of the Marshall Islands' 1,156 islands lie north of the Equator and west of the International Date Line. The Republic shares nautical boundaries with the Federated States of Micronesia on the southwest, Kiribati on the southeast, and Wake Island (a U.S. territory that the Marshall Islands claims) to the north. More than 30,000 of its some 72,000 inhabitants live in the major town and capital, Majuro.

Micronesians settled on what are now the Marshall Islands more than 3,000 years ago. The Spaniard, Alonso de Salazar, sighted one of its islands in 1526. In 1592, Spain formally claimed the islands. In 1788, British explorers, Captains John Charles Marshall and Thomas Gilbert visited the islands. Several German trading companies began operations in the Marshall Islands in the 1860s despite long-standing Spanish claims to what they called the Spanish East Indies. In 1884, Germany acquired the islands from Spain through papal mediation and, subsequently, signed treaties with the paramount chief and seven other chiefs. In 1887, Germany entrusted the islands' administration to a German trading company, *Jaluit Gesellschaft*. When Spain lost the Philippines and, thus, its imperial status in the Pacific following the Spanish-American War, Germany acquired the Caroline Islands, Palau, and the Marianas, which—with the Marshall Islands—became part of German New Guinea.

Decades before Germany had acquired the Marshall Islands, Japanese traders and fishermen had irregularly visited the islands. The 1868 Japanese Meiji Revolution accelerated the country's aspirations to become a major East Asian economic and political power. Throughout World War I, Japan occupied the Marshall Islands. In 1919, the Marshall Islands were merged with adjoining former German Pacific territories to form a League of Nations mandate administered by Japan and were retained by Japan, even after it withdrew from the League in 1934. During its years of colonial rule, Japan settled more than 1,000 Japanese to the Marshall Islands. The local administration Japan developed weakened the status and influence of the traditional local leaders. Japanese language and culture were taught in the schools, but European Catholic and Protestant missionaries remained.

During World War II, following the 1943 battle of Guadalcanal (which severely damaged the islands), the United States occupied the Marshall Islands. In 1947, the UN created the Trust Territory of the Pacific Islands (TTPI), which embraced Palau—what is now the Federated States of Micronesia—and the Mariana Islands (which continues to remain a United States dependency) as well as the Marshall Islands.

In 1946, the United States began nuclear testing on the Bikini Atoll, leading to its evacuation. The 1952 test destroyed Elugelab Island. Testing continued until 1958; the 67 tests caused the evacuation of residents and subjected many to radiation exposure. Marshall Islands' claims against the United States are ongoing. Already several hundred million dollars have been paid to Marshall Islanders for their exposure to nuclear testing.[80]

In 1979, after long negotiations between the United States and the four parts of the TTPI (the Marshall Islands, Palau, the Marianas, and what became the Federated States of Micronesia [FSM]) determined to go their own way. The United States then established the Marshall Islands as a self-governing dependency. In 1986, the Marshall Islands negotiated Compact of Free Association (COFA) with the United States, which provides continuing aid and assistance, defense of the islands, oversight of foreign affairs, and the right of entry of Marshall Islanders into the United States.[81] In exchange, the United States gained continued use of the missile testing range at Kwajalein Atoll where 2,000 American personnel are stationed.

COFA closely resembles the pre-World War II "protectorate" status accorded to many dependencies and the free association status of the Cook Islands with New Zealand. With support of the United States, the Marshall Islands became a member of the UN in 1971.[82]

---

80. The Marshall Islands filed a suit saying the U.S. violated international law by failing to respect obligations under the 1968 non-proliferation treaty. The case was dismissed in 2016 when the International Court of Justice ruled that it had no jurisdiction.
81. Communities of Marshall Islanders have already formed in Springdale, Arkansas, and Salem, Oregon.
82. The Marshall Islands generally votes with the United States in the United Nations General Assembly.

Climate change threatens the existence of the Marshall Islands. In 2008, extreme waves and high tides flooded Majuro and other urban areas, which lie generally less than three feet above sea level. In 2013, heavy waves again flooded the city. Also, in 2013 and 2016, the northern Marshall Islands experienced droughts that led to crop failure and the spread of diseases. These emergencies led the Marshall Islands' minister of foreign affairs, Tony deBrum, to promote international efforts to take steps to reduce the practices causing climate change. In 2013, the Marshall Islands hosted a summit where deBrum stressed the increasing threat of climate change; seeking support, he pointed out that "the ground water that supports [their] food crops is becoming inundated with salt . . . Many island nations are looking into buying farmland in other countries to grow food and eventually to relocate their populations."

The Marshall Islands has a presidential-parliamentary system; the executive president, serving as both head of state and government, is elected from and by the parliament (the *Nitijela*), whose 33 members are elected by universal suffrage from 24 constituencies. The *Nitijela* is advised by a council composed of the 12 paramount tribal chiefs. There are four political parties: Aelōn̄ Kein Ad (AKA), United Peoples' Party (UPP), Kien Eo Am (KEA), and United Democratic Party (UDP). A coalition of the AKA (14 seats) and the UDP (9 seats) presently controls the government. In 2016, Hilda Heine succeeded Christopher Loeak as president. David Kabua was elected in 2020. Four of the five former presidents have been paramount chiefs. Among the international memberships Marshall Islands belongs to are the United Nations and several of its affiliates, the Pacific Islands Forum, and the Commonwealth.

In five decades, the Marshall Islands' population has increased fivefold. More than two-thirds of the population lives in the two major urban centers, Majuro and Ebeye, located in Kwajalein Atoll. Ninety-two percent of Marshallese are descended from Micronesians who migrated to the islands thousands of years ago. The remainder of the population is of mixed ancestry, principally Japanese stemming from the decades of Japanese rule of the islands. Having lived for centuries on isolated coral atolls and islands, families and communities are part of strong traditional structures whose respected chiefs dominate the social and political life of the communities.

Marshallese and English are the official languages. Many Marshall Islanders speak a rudimentary English. Reflecting the dedicated work of missionaries, 55 percent of the islanders belong to the Congregational Church (UCC), 26 percent to the Assemblies of God, and 8 percent to the Roman Catholic Church. The islands' labor force is divided among agriculture (11 percent), industry (16 percent), and services (73 percent). The Marshall Islands have long depended upon fishing and small farm subsistence agriculture (coconuts, tomatoes, melons, and breadfruits). Small-scale commerce is limited to fish, food processing, handicrafts, and tourism ($4 million). Imports far exceed the exports at $118 million and $50 million, respectively. The mid-2010s GDP was $86 million—$8,600 GDP per capita.

Remoteness, poor infrastructure, lack of resources, and droughts impede the islands' economic development. The prospect of continued support from the United States upholds its economic survival. But the prospect of future inundation threatens its existence.

THE FEDERATED STATES OF MICRONESIA

"[T]he Congress of Micronesia . . . with representatives of all the TTPI . . . was created by the USA in preparation for the granting of greater self-governance to Micronesian territories . . . A Commission . . . established to examine the future status of the islands . . . declared Micronesians' rights to sovereignty over their own lands, self-determination, the right to devise their own constitution and to revoke any form of free association with the USA."

—David Lea and Colette Milward

The Federated States of Micronesia (FSM) lies north of the Equator, northeast of Papua New Guinea, south of Guam and the Marianas (U.S.), west of Nauru, and east of Palau. While the FSM's 607 islands occupy a land area of only 271 square miles, the country occupies about 1,000,000 square miles of the Pacific (1,600 miles from east to west).

Micronesians settled in the Caroline Islands over 4,000 years ago and, eventually, developed a culture with economic, religious, and chieftaincy features distinctive to the Yap Islands; it collapsed around 1500. European explorers, first the Portuguese then the Spanish, reached the Caroline Islands in the 1500s. In the 1800s, the Spanish established commercial outposts concerned principally with copra and mission stations seeking converts. Spain did not incorporate the islands into the Spanish East Indies until 1874, when copra and increasing international rivalry forced the issue. Spain's loss of the Philippines in the Spanish-American War (and thereby the base of their Pacific presence) led to the 1899 sale of the Caroline Islands to Germany, which annexed them to German New Guinea.

The German presence in the Pacific was brief, lasting only until World War I, when Japan, which had long coveted the islands, allied against Germany and, taking advantage of Germany's total involvement in Europe, occupied the German-held Pacific islands, including the Caroline Islands. At the Paris Peace Conference in 1919, Japan sought full sovereignty over the territory, but Australia and the United States opposed Japanese annexation, awarding them to Japan as a mandate under the League of Nations.

The Japanese government, nevertheless, colonized and exploited this mandate as though it was part of its planned empire. After Japan walked out of the League of Nations in 1933, the territory became a closed military area, strengthened with a series of fortified land bases.

In 1944 (during World War II), the United States attacked the Japanese fleet based in Truk (a.k.a. Chuuk Lagoon) in one of the most important naval battles of the war. Following the war, the UN granted the United States the administration of the islands as the TTPI. The TTPI embraced what is now Palau, the FSM, and the Mariana Islands (which remains a dependency of the United States), as well as the Marshall Islands.

In 1979, after long negotiations between the United States and the four parts of the TTPI determined to go their own way. The fourth part ratified a new constitution, which divided itself into four states—becoming the Federated States of Micronesia. The FSM, after critical discussions concerning the terms of "free association," signed a Compact of Free Association with the United States, which came into effect in 1986 and marked the FSM's transition from trusteeship to independence. Its independence was internationally recognized in 1991 when, upon the recommendation of the UN Security Council, the United Nations General Assembly voted the FSM into membership. The Compact, comparable to the one with the Marshall Islands and Palau, provided that the United States would be responsible for the FSM's defense and oversee its foreign affairs, features circumscribing the traditional definition of independence.

The treaty provides that Micronesian citizens can live and work in the United States without a visa, yet two-thirds remain in Micronesia. The FSM consistently supports the United States' position in the UN General Assembly. It maintains relations with 56 countries and is a member of the Pacific Islands Forum. The Federated States of Micronesia, according to its constitution of 1979, has a 14-member unicameral congress elected by universal suffrage. Four senators—one from each state—serve four-year terms; the other members are elected from single-member constituencies for two-year terms. Congress elects the president and vice president from among the four senators for four-year terms. There are no political parties.

Nearly all the 106,000 FSM people are Pacific Islanders and Asians of various ethnic groups, which have distinctive cultures and languages. The major ones are based on the four states: Chuukese (48 percent), Pohnpeian

(24 percent), Yapese (10 percent), and Kosraean (6 percent). Many have some Japanese ancestry stemming from intermarriages during the Japanese colonial years. English is the official and common language. Also spoken are several local ethnic ones. Christian mission conversions and conflicts have divided the islands primarily between Roman Catholicism (55 percent) and Protestantism (41 percent)—mainly Congregational (39 percent).

The Federated States of Micronesia's economy depends, primarily, on subsistence farming (including coconuts, bananas, betel nuts, and cassava), fishing, and fish processing. Except for phosphate, few minerals are worth exploiting in the region. The remoteness of the islands and lack of facilities limits the potential for tourism. Financial assistance from the United States is the major revenue source. The GDP is $754 million, and the per capita GDP is $7,300. Two-thirds of the labor force is employed by the government. The FSM's political dependence, lack of economic prospects, and environmental threats make its outlook bleak.

THE REPUBLIC OF PALAU

"Palau (along with the Marshall Islands) rejected participation in a federal Micronesian state."

—David Lea and Colette Milward

The 21,000 population of Palau, one of the four least-populated independent countries in the world, is spread across 250 islands of the western Caroline Islands—west of the Federated States of Micronesia. Migrants from the southwest Pacific initially inhabited the islands about 3,000 years ago. Micronesian people migrated there about 900 years ago. In 1710–12, Spanish ships visited Palau. In 1885–86, Spain included Palau in the Spanish East Indies colony. Following Spain's loss of the Philippines in the Spanish-American War, Spain sold the Caroline Islands (including Palau) and the Mariana Islands (except for Guam) to Germany. The governor of German New Guinea administered Palau.

During World War I, the Japanese occupied Palau. Following the war, the League of Nations made Palau part of a Japanese-administered South Pacific Mandate. During World War II, American forces occupied Palau. In 1947, the newly created United Nations established the TTPI, which included Palau, administered by the United States. In 1967, The United States formed a Congress of Micronesia with representatives from all parts of the TTPI with the intent of granting self-government to Micronesian territories. A commission created to consider the future status of the Micronesian islands recommended full sovereignty. In a 1978 referendum, Palau rejected joining the proposed Federated States of Micronesia. In a 1979 referendum, Palau approved a constitution, which came into effect in 1981.

In 1982, the United States proposed and signed a Compact of Free Association by which the United States managed Palau's defense and foreign affairs. Negotiations were delayed due to the United States' insistence and Palau's resistance to the stipulation that the Palau constitution allow the United States the right to transport and store nuclear materials and operate nuclear-propelled vessels within Palau. Since, initially, the compact required a 75 percent vote, only after 8 referendum elections and a change in the Palau constitution was the compact ratified in 1993—coming into effect in 1994, concluding Palau's transition to independence and becoming a member of the United Nations. The Compact, similar to the ones with the Federated States of Micronesia and the Republic of the Solomon Islands, focuses on issues of economics, security, and defense. Under the Compact, the United States has access to the islands for 50 years, and Palau, which has no military, relies on the United States for its defense and other external affairs.

Palau has a political system modeled after that of the United States, with a president, a bicameral legislature, and an independent judiciary. Elections are non-partisan. From 1984 to 2016, there have been eight presidents, but only one served two terms. Palau is divided into 16 states (called municipalities until 1984) ranging in population from 11,670 in Koror, the largest town and the capital, to 42 in Sensorial and 30 in Hatohobei. At the local level traditional chieftaincy rule remains strong, undercutting the national government.

A high chief, who had inherited the honorific title "king of Palau," though not part of the government, is highly respected in the southern states; in the northern states, a hereditary high chief is revered. Palau not only belongs to the UN and affiliated agencies but also maintains missions at Washington, D.C. and a few Pacific states. In 2009, Palau tentatively agreed to accept 17 Uyghurs imprisoned in Guantanamo, 6 arrived; an aid agreement, concluded in 2010, was, reportedly, not related to the deal.

Seventy-two percent of the approximately 21,000 people in Palau are Palauan (Micronesian with Malayan and Melanesian mix), 16 percent are Filipino, and 2 percent are other (Asian and European). More than 40 percent of the population is Roman Catholic, 25 percent is Protestant (including 5 percent Seventh-day Adventist), 9 percent is Modekngei (a combination of Christianity and the traditional Palauan faith), and 2 percent is Buddhist (a result of the Japanese occupation). Palauan and English are official languages on most islands, but, on two islands, a distinctive local language is also official.

Transport between the islands is mainly via private boats and chartered air service—a fact that fragments the republic economically as well as socially and politically. The early 2010s Palau GDP of $221 million, with a

GDP per person of $10,500, is twice as large as the neighboring Federated States of Micronesia. The economy depends primary on scuba- and snorkeling-based tourism, with help from subsistence agriculture and fishing and United States subsidies. In 2010, the national budget was $94 million, supported significantly by U.S. subsidies.

Palau's challenges are both internal and external. Traditional chieftaincies continue to attract more respect, and with respect to some issues wield more influence than the more distant and more recently established federation office holders. Palau, like the other small island countries, faces the environmental and existential threat of inundation.

Palau's economic challenge may be posed as such: How long can a country in the mid-Pacific dependent on United States subsidies and aspirations for expanding tourism maintain the fiction of sovereignty? Free association has come with a price. It may be a member of the United Nations and thus considered an independent country, but Palau does not have a monopoly of power over its own affairs—internally or externally. Its dependence upon the subsidies and defense provided by the United States may continue only as long as the U.S. finds it convenient to have multiple military bases off the coast of Asia, for instance.

16. LOOKING AHEAD

"The end of colonial empires was conflictual and contingent. European empires gave up a sovereignty that was becoming too costly, and the new Founding Fathers took over sovereignties that they thought they could entrench. We live with the consequences of these uneven and broken paths out of empire, with the fiction of sovereign equivalence and with the reality of inequality within and among states."

—Jane Burbank and Frederick Cooper

WILL MICRO-COUNTRIES SURVIVE?

Assessing the micro-countries' biographies—from their empire-building abilities and involvement in historical events that shaped their evolution through dependence to independence to confronting the challenges of independence—leads to an appreciation of their survival! Micro-countries less than 60 years old have survived multiple challenges during their initial decades. Will they confront the economic disadvantages, environmental disasters, and ethnic and other "identity" divides of the coming decades successfully? They may have insufficient local and international commitment, resources, and resolve to cope with the coming challenges! The confluence of these threats may create "perfect storms," ones that some micro-countries may not survive.

Isolated from other polities by water, mountains, distance, and circumstance made micro-countries unable to consider favorably merging/federating with other nations. In fact, they started to realize that they did not possess shared languages, cultures, tribes, and acquaintances with their neighbors, which was the basis for collaboration. The more conscious the local inhabitants were of what made them a distinctive "us," the more likely they perceived their neighbors as a "them." Isolation-fed separatism led to merger resistance, which meant remaining small in resources as well as population. Nevertheless, despite skepticism, they survived the hazards of independence-cum-smallness in a world of increasing interdependence. What are the major challenges confronting the African, Asian, American, and Oceanic micro-countries in the coming decades? How will future historians assess the impact of the mid-2000s?

The social, economic, environmental, and political threats of the late 1900s and early 2000s continue, but some have a sharpened poignancy, jeopardizing their very existence. The passage of decades, the emerging of global perspectives, the development of new technologies, and the evolution of new organizational arrangements for dealing with the challenges confronting micro-countries in the early 2000s require a renewed perspectives and determination. Small size and isolation continue to reduce the opportunities for developing the infrastructures, the industries, and the institutions needed to cope with the economic, environmental, and ethnic challenges.

Their lack of population, combined with their distance from major market and inferior infrastructure limits the ability of micro-countries to develop what Alfred D. Chandler has described as the advantages of scope, and scale, without which the development of industry, trade, urbanization, infrastructure, and, hence, prosperity are handicapped. These features are increasingly vital in an increasingly interdependent world.

Their isolation by expanses of water or mountains has adversely affected the convenience of contact, and, thereby, the costs of transport and trading opportunities. Inadequate public works, public safety, public education, and public health do not facilitate development. Infrequent contact and lack of sustained development have abetted the continuance of conservative/tribal traditions. The cross-fractured nature of micro-countries emerging from centuries of colonialism presents a stronger set of schisms to overcome. Particularly troubling are the urban-rural frictions and inequitable income distributions that identity-dividing schisms aggravate. This split is especially evident between rural areas and the more entrepreneurial, rapidly growing urban areas.

Subsistence agriculture, fishing, food processing, and mining of near or already exhausted mines no longer are industries that micro-countries may depend upon to sustain a growing economy. Even micro-countries with large-scale farming and fishing industries are finding it more difficult to compete with producers in countries with protective (and sometimes prohibitive) tariffs. The scrutiny and crackdown on the abuses of offshore banking practices leaves less room for growth.

Tourism is a promising industry for many micro-countries. The general increase in prosperity has enabled many more people to become tourists. To compete, micro-countries not only require surf and sand (or snow and slopes), historic and exotic sites, and scenic wonders but also modern airports, comfortable hotels, and convenient schedules—and proximity to major tourist-producing areas. While the number of potential tourists is rapidly increasing, the number of countries increasing efforts to attract them is increasing even more rapidly. The proximity of potential tourists to tourist sites will shape the tourism trade. The Pacific islands are catching some of the booming tourism in Southeast Asia (especially from China).

Renewable energy sources such as wind, biomass, geothermal, solar power, promise to generate more power less expensively than fossil fuels. Widespread use of renewable energy may be expected to lead to less need to import power from abroad, thus reducing the costs of imports and of deficits.

Ships, planes, trains, and trucks and their infrastructure of roads, rails, ports, and digitalization are continuing to be developed to carry more cargo and more passengers, be faster, operate more economically, and offer more optimized schedules. As they increase the size, speed, and scale of exports and imports, transport time shortens, and costs reduce. More cargo will be shipped, more people will fly, and micro-countries will become less isolated and more globally competitive.

CHALLENGES

Impudent fiscal management, marked by family and friends and padded payrolls and contracts, challenge many governments. Resolute resistance to taxation is disadvantaged in contesting aggressive demands for more services and corruption. The poorer the taxpayers, the less developed the infrastructure, and the skimpier the services the greater the tension. Raising taxes risks revolts (at the ballot box or worse). Deficits produce debt; large debt frustrates securing loans. Minimizing investments undermines economic and human development. The longer projects are delayed, the greater the probability that efforts to increase economic prosperity will be indefinitely postponed.

Faced with insufficient income, governments are forced to make awkward investment choices, which can be divisive and debilitating. Decision makers must decide priorities, such as between funding capital projects and ongoing programs, between developing public works and improving human development and financing glamorous projects such as major highways, dams, hospitals, universities, and palaces—or those that are more sensitive to local needs such as community roads, water supply and distribution, sewerage collection and treatment, power generation and distribution, police and fire stations, pre-tertiary and vocational education, and health care clinics. Too often, it appears that assistance with aesthetic projects is easier for countries to ask for and for global inter-government organizations to grant and banks and other firms to loan than help with human development projects.

The impact of pollution-abetted climate change, which is triggering the relentless rise of oceans, more frequent hurricanes and cyclones, and earthquakes and volcanic eruptions threaten the viability of many micro-countries, if not their every existence. As global warming causes Antartica and Greenland to lose nearly all their ice, sea levels will rise, threatening to salt the arable coastal land and inundate coastal development and the economic viability of many countries. The countries that signed the 2015 Paris Agreement to slow the rate of pollution emissions and limit warming to near 2°C above preindustrial levels appear to be in no hurry to abide by them. The extent of past pollution has not only already caused significant damage but also set the pace for continued warming, and, thus, escalating threats to residents' livelihoods, their lives, and their people, and the land of these micro-countries.

Few expect meaningful action that reduces the continuation of the disastrous effects of climate change. Words and promises are easier and cheaper than strategizing the politics of developing effective programs now—especially since multinational action, in many cases, may merely postpone the effects of the rising sea. It appears inevitable that, no matter how much is invested to prevent inundation, several island micro-countries will not survive complete or even partial inundation. Constructing stronger buildings (including houses) with steel and concrete will help. Building dikes (like sea-threatened Holland did centuries ago or like many flood-threatened cities around the world have done) appears to be a prudent investment only in select situations.

A mixture of motives, including "do-good-ism" and "keep-them-on-our-side-ism," has helped the recently emancipated micro-countries. The donor stakeholders' moods, however, appear to be shifting, to be less generous and more protectionist. It may prove difficult to maintain the present level of assistance—except when the giving country receives some direct benefits, such as a military installation.

Increasing nationalist, populist, and protectionist political pressures affecting demagogic-led electoral results raises questions whether support for developing countries will be sustained. This myopic nationalist mentality, which provokes trade wars and undermines the ability of emerging countries to engage in international trade, does not bode well for their future viability—or, in some cases, for their very existence.

China's Belt and Road Initiative is extending its interests, investments, and influence beyond Asia and the Pacific by constructing roads, rails, ports and other strategic projects throughout Asia, Oceania, and Africa (including a military base in Djibouti). This extension of a colonial-power-like presence may spur the Western donor countries to overcome their temptation to shrink their foreign aid budgets. The eagerness of China to extend its investments to South Pacific island countries especially concerns Australia, New Zealand, and Taiwan as well as the United States because a "frosty" war may emerge between China and the United States (and others) in "courting" countries seeking grants in Africa, Asia, and the Pacific.

The outlook for isolated micro-countries is especially hazardous. For most, the prospects for developing their economies, continuing to secure significant international assistance, and escaping the existential threats of climate change are less than promising. Securing international support on these issues has been imperative for supporting most post-1960 created micro-countries. Small countries are not the only ones challenged by the threats, undermining efforts to the prosperity and ensure their continued existence.

But some of the smaller ones are less likely to possess the resources, the economic scope and scale, and the maturity of representative political institutions to cope with the escalating challenges.

The impact of the global COVID-19 pandemic has delivered a devastating blow to the already precarious existence of the most vulnerable micro-countries.

MICRO-COUNTRIES AND CONFEDERATIONS

Post-World War II empire dismemberment enabled many micro-countries, along with many more populous former dependencies, to gain independence. To survive, they have depended on the critical, benevolent support of post-World War II global organizations, the increasing collaborative role of regional organizations, and the corollary worldwide growth of the global economy. These developments have morphed the role of empires—from a 19th century empire dominated by governors, residents, and district officers, a military presence, and a merchant establishment to a 21st century one energized by diplomats and a business community.

Many countries that won their independence after World War II have coped with major socially divisive, economically frustrating, and politically disruptive issues in order to survive. Several factors were especially critical for surviving in a competitive world still managed by mega-countries. Especially evident was the indispensable aid and assistance of former colonial shelter countries and global inter-governmental institutions such as the United Nations and its various affiliates.

The development of regional organizations of governments, such as the EU, the AU, ASEAN, PIF, and OECS, has facilitated receiving assistance, the development of trade pacts and joint development of projects, mediating disputes, pressing their concerns upon the world community, and facilitating the extent to which global governance is becoming more inclusive! The cultivation and strengthening of micro-country representative-governing institutions was basic in developing confidence locally and globally. Will future historians be able to record continuing advancement?

The European Union has been a model for the many regional organizations, which have developed throughout the world and may continue to serve as a model as they grow in the ability to promote joint action in common programs and advocating regional interests. For example, Anthony Teasdale points out the following:

> The European Union has been a model for the many regional organizations, which have developed throughout the world and may continue to serve as a model as they grow in the ability to promote joint action in common programs and advocating regional interests.

Will regional organizations of government—the more developed ones perhaps rechristened confederations, as K. C. Wheare has suggested. A few of them may. The European Union already exerts power, prowess, and prestige comparable to that exercised by major countries and overshadows that of most countries. A Caribbean community of English-speaking countries, which would combine the structure and vigor of the Organization of East Caribbean States and the membership of the Caribbean Community and Common Market—perhaps with other English-speaking Caribbean countries—may meet the challenge.

The African Union—or the incipient Tripartite Free Trade Area, which merges 3 regional organizations of 26 countries strung from Cairo to the Cape shows promise. The Pacific Islands Forum, which already includes Australia, New Zealand, Papua New Guinea as well as the 11 Oceania micro-countries, may gain potency in multiple spheres as the third force in the escalating contest for hegemony in the Pacific, as the United States resists the hegemonic ambitions of China. The Association of Southeast Asian Nations has aspirations for developing even more cohesive working relations.

What are the prospects for continuing to improve their economic viability, increase creditable representative governance, and reduce the corruption that has crippled the development of infrastructure, representative institutions, and the economy? The answer depends upon the entrepreneurs at local, national, and supranational levels of governance. The future of micro-countries depends upon the extent of the continued support of the United Nations and other intergovernmental global organizations such as the World Trade Organization, the World Health Organization, the International Monetary Fund, and the World Bank. This will majorly depend upon the extent to which countries continue to strengthen regional organizations of governments.

SOVEREIGNTY AND GLOBAL GOVERNANCE

Sovereignty, as the term has for four centuries been used and understood, has mutated. In a world of power blocs, instant communication, rapid transportation, and individuals with multiple community identities, the word no longer conveys the meaning it once did. Sovereignty, whose presence has been the *sine qua non* of independence, has long been subject to the making and interpretation of treaties and other international agreements. Collaboration, facilitated by more potent global organizations and growing regional ones, can more aggressively and more effectively address not only regional issues of small insular countries but also such global existential ones as global warming.

Most of the post-World War II micro-countries, like most new countries, have depended on the proactive support of postwar-created global organizations, former colonial powers and other wealthy countries, and banks and other finance firms to provide investment assistance. The motivation may be genuinely benevolent, seeking economic and political hegemony, profit, or a combination of considerations. The motivations include continued exploitation, investment opportunities, and a recognition that assistance to disadvantaged countries benefits the global economy.

These developments have facilitated the strengthening and may continue to develop more effective participation of smaller countries in world governance through more effective global and regional organizations. With this may come a better recognition that assisting the development of the economies of less developed countries and increasing the markets for all countries benefits all!

The continuing survival of micro-countries depends upon their ability to support economic and political viability. This depends on their ability to develop industry and other sources of income, adapt to the hazards of hurricanes, cyclones, and rising sea levels, and secure against international trade and internal security threats. Such challenges require international collaboration of the kind facilitated by global organizations and regional organizations of countries.

As regional organizations of governments mature into confederations, as they develop their range of functions, their members' confidence, and the respect of major countries and global organizations and mature into confederations, they will facilitate evolutionizing a more inclusive system of global governance. Such a system may afford the smaller countries a more effective voice, a more effective part in the decision-making process, a more effective share in the allocation of the global resources—to cope effectively with economic competitive disadvantages, climate change-driven environmental disasters, and ethnic, elitist, and other identity divides threatening their fiscal viability and even their existence.

In 1944, the English political scientist, socialist, and visionary Harold Laski foresaw that "the age of the nation-state is over . . . economically, it is the continent that counts . . . the true lesson . . . is that we shall federate the Continent or suffocate." His visionary prediction was the triumph of an academic activist's hope over experience. But Europe has made strides toward developing what may be labeled "confederation" sometime in the future.

In the Americas, Africa, South Asia, East Asia, and Oceania regional organizations have expanded their membership to embrace all or nearly all the countries within the vicinity. Many have taken on significant collective programs as well as initiated trade pacts and lobbied regional interests. Connections with shelter countries, global institutions, and regional organizations may be ignored, or just postponed, but, without them, everything falls apart.

The momentum of long-term substantive, yet stable, changes in the means, methods, and manners of governance more often are evolutionary than revolutionary, favoring compromises and piecemeal alterations.

One may predict with some confidence that major steps will be taken to strengthen regional organizations of governments in Africa, the Asian sub-continents, and Oceania in the next several decades. National identities will continue to block efforts to form federations.

How many micro-countries will survive the economic, political, and environmental challenges of the 2000s? A freer trade global economy, a less mega-country dominated world political arena, and a more proactive global effort to slow down the rate of climate change will improve their odds! Continuing strides in improving the organizational effectiveness of micro-countries in working with global organizations and developing regional ones would be essential steps. Energizing the components of an emerging world system enhances the economic, environmental, and political viability of all countries, including the least of them.

The critical questions, therefore, may be as follows: will the smaller countries organize themselves into regional organizations to better manage their challenges? Will the more affluent countries and the United Nations and related global organizations sufficiently support the efforts to meet the challenges of global trade and climate change, which are threatening every country?

ABOUT THE AUTHOR

Samuel Humes IV, born in 1930, graduated from The Hill School, Williams College (BA, Hons.), the Wharton Graduate School of the University of Pennsylvania (MGA), and the Faculty of Law of the University of Leiden (Drs. and PhD). He began his career as executive director of the Metropolitan Washington Council of Governments, County Administrator of Baltimore County MD. He continued his career serving as Professor at the University of Ife (Nigeria), the Center for African Research and Administrative Development (Morocco), and Queen's University (Canada). He later directed graduate management programs at, the Nova University Graduate Program of Public Administration (Florida), Rider University (New Jersey), and Boston University's graduate programs in Belgium.

Adjunct appointments include: the University of Pittsburgh, the Graduate School of Public Administration of The George Washington University, Williams College, Wharton Graduate School, Colorado State University, the University of Chiang Mai (Thailand), International Management School (Germany), Institut Catholique des Hautes Études Commerciales (Belgium), and Institute Modorniho Risen Zaut (Czech Republic). He prepared the plan for the reorganization of local government in Western Nigeria, was a consultant to the Commission of Inquiry of Kenya on the reorganization of its government. He designed a graduate management program for the Christian Medical College and Hospital in Vellore, India.

After living and working 33 years in Europe, Africa, and Asia he lives in Williamstown, Massachusetts with his wife, Dr. Lynne De Lay. They have two sons and six grandchildren.

NOTES

PROLOGUE

"In order to survive . . . breathe and flourish . . ." Davies, p.737

PART ONE

Successful statehood. . . . of the world since time immemorial" Davies (2011), p.738

"One peculiar feature . . . taking us by surprise." Toynbee and Caplan (1972), p.13

INTRODUCING THE 41 MICRO-COUNTRIES

"We live . . . universally desired." and "What then is. . . . governed differently." Burbank and Cooper, pp.2, 8

FROM DEVELOPING TO DISMEMBERING GLOBAL EMPIRES (LATE 1400S – LATE 1900S)

"The Blockage . . . with fact." Boorstin, p.146

"Thus circumstanced . . . regions of the East." Prescott, p.352

"Never certainly did any nation . . . assume anything like so much responsibility." J. R. Seeley, The Expansion of England, pp. 205, 213

"entirely convinced . . . his immortality7." Whitfield, pp.64-5

"The defeat . . . The legacy of maritime success" Knight, p.21

"100 political units (not including 600 Indian princely states)." Darwin (2013), p.189

"If the rise of working-class parties . . . nationality defined group" Hobsbawn (1987), p.3

"The history of Sea Power is . . . effort was made to exclude others . . ." Mahon, p.1

SOVEREIGNTY, SKEPTICISM, AND SURVIVAL

"The land we live . . . as they always have" Tim Marshall, pp.1-2

"Many of the jumbled ingredients . . . genes or blood." Tombs, p.83

"absolute and perpetual . . . mark of a commonwealth." Bodin, p.25

"a 'state' is a human . . . a given territory." Weber, Max, in Gerth and Mills, p.78

"clearly impossible . . . on their own." Great Britain 1947, quoted in Clarke, "Grenada" in Clarke and Payne, p.83

"(T)hey face . . . economic growth." Worrell (2012), p.15

ECONOMIC COMPETITIVE DISADVANTAGES

"(T)he size . . . Literacy, etc." Alesia and Spolaore, p.3

"What helps . . . less populous," The Economist "Where does the aid go?" June 11, 2016, p.62.

CLIMATE CHANGE DISASTERS

"In the last year . . . across the region" Christopher Leak, President of the Republic of the Marshall Islands

"put millions . . . (or luckier) opponents." Rosen, pp.2-3

ETHNIC, CLASS, AND OTHER IDENTITY DIVIDES

"The smaller the society . . . plan of oppression" James Madison, 'No. 10' The Federalist Papers, Hamilton, Madison, and Jay, p.83

"when party control of the prime minister . . . work in practice." Diamond, p.49

"additional opportunities . . . needs will be served." Ylvisacker in Maas, p.32

"naïve" When Spiro T. Agnew was County Executive of Baltimore County and I was County Administrator (before he resigned as Vice President over charges of this corruption), he called me 'naïve' for expressing concerns about the extent of Christmas gift-giving of county employees by contractors and consultants and laxness regarding gratuities. In retrospect, this is somewhat ironic.

"Everton is corrupt." Adebayo Adedeji Interview, June 1068v

"th' on'y thing . . . on each other." Dunne, p.281

REPRESENTATIVE GOVERNANCE, INSTITUTIONS, AND INTERDEPENDENCE

"The basis of a democratic state is liberty" Aristotle

"The greater the opportunities . . . no feasible alternative can." Dahl, Democracy, 1987, p.88

"Dating back to Aristotle . . . order and stability." Diamond, p.2

"of myriad compromises . . . certain mystique" The Economist, "Britain, a lonely domino," September 29, 2018, p.52

"at the apex . . . decision-making among the governments", Teasdale, p.319

"where 'league' and . . . the only alternative left." Wheare, 1953 p.31

RETROSPECT AND PROSPECTS

"Only the map . . . last few decades." Alesina and Spolaore, p.224

PART TWO

"The task of the historian . . . less fashionable memory sites," Davies, 2011, p.8

DEPENDENCE, INDEPENDENCE, AND INTERDEPENDENCE

"The Westernization of the world . . . expansion." Toynbee, 1972, p.336

EUROPE: THE FIRST MODERN MICRO-COUNTRIES

"The Treaty of Westphalia . . . three centuries." Jane Burbank and Frederick Cooper, pp.182-183

Parry, J. H., The Establishment of the European Hegemony, Trade and Exploration in the Age of the Renaissance 1415-1715, (Third Edition, Revised), Harper & Row, 1961

THE VATICAN

"[T]he [spiritual and administrative] center . . . billion followers." Stille, Alexander, "Holy Orders", The New Yorker, Sep.14, 2015, p.54

"after 962 . . . of the 93Pope" Duursma, p.395

"the latest financial scandal . . . for dubious transactions." The Economist, October 10, 2019, p.78

SAN MARINO

"The survival of San Marino . . . its autonomous institutions." Duursma, p.210

"One of the remnant . . . to the 'Great and General Council'." Duursma, p.213

MONACO

"Due to lacking means . . . interpretation of the treaty of 1918." Duursma, p.314

"On 8 January 1297, Francois . . . over Monaco" Duursma, p.262

"Judicial power is . . . Criminal tribunal" Duursma, pp.268-9

ANDORRA

"The objective characteristics . . . no specific Andorran language." Duursma, p.372

"The 'Pariatges' are the founding . . . disputed question" Duursma, pp.318-9

LIECHTENSTEIN

"The place which . . . inhabitants are aliens." Duursma, p.15.

LUXEMBOURG

"the small size . . . European Theater." Jeanne J. K. Hey, "Luxembourg: Where Small Works" in Hey, p.75

ICELAND

"Iceland – with its past in world trade." Gunnlaugsson Interview, September 24, 2014

"The reconciliation . . . colonization." Griesel, p.20

"several hundred monks . . . ailed to Iceland." Morrison, 1971, p.26

"All these attempts failed . . . revision of the constitution . . ." Karlsson, pp.267-8

CYPRUS

"Cyprus is a scene . . . but never to be ignored." Purcell, p.73

"The two last . . . ancient Near East" Cline, p.61

MALTA

"Il'mir bahar u min andu Malta" ("Who holds Malta rules the sea") Maltese Proverb," Castillo, p.129

MONTENEGRO

"The Montenegrins . . . unification of the Serbs." Djilas, 1977, p.149

"one clear lesson . . . and help." Roberts, p.475

AFRICA

"In the aftermath . . . saint or a visible feature or an event." Crowley, pp.4-5

SEARCHING FOR SPICES, TREASURE, AND PRESTER JOHN

"He explored a hundred leagues further every year." Morrison, 1974, p.7

"Corruption and institutional . . . easier to address." Nyanzi, interview

CABO VERDE

"a small nation . . . external aid." Deirde Meintel, "Cape Verde: Survival without Self-Sufficiency" in Cohen, p.145

"In the decades . . . Sao Tome and Principe." Deirde Meintel, "Cape Verde: Survival without Self-Sufficiency" in Cohen, p.147

"When the Portuguese . . . cabinet ministers" Jones interview

EQUATORIAL GUINEA

"Abundant oil resources . . . and authoritarian repression." Kitti, The New York Times, March 29, 2015

SÁO TOMÉ AND PRÍNCIPÉ

"During Portugal . . . business interests." L. M. Denny and Donald I. Ray, "Sao Tome and Principe" in Jens Erik Torp, L. M. Denny, and Donald I. Ray, "Mozambique and Sao Tome and Principe," p.130

"The continuation of slavery . . . in 1909." L. M. Denny and Donald I. Ray, "Sao Tome and Principe" in Jens Erik Torp, L. M. Denny, and Donald I. Ray, p.133

THE COMOROS

"The process of . . . former colonial power, France." Cohen, p. 217 (in introduction to the article by Claude Gaspart "Les Survivances Colonial aux Comoros"

"are not alone in being small . . . with other states of like size." Newitt, p.129

SEYCHELLES

"Unfortunately, the best . . . beautiful of places." Franda, p.126

"To a greater extent . . . racial types." Franda, p.18

"When English . . . Tricolour." Raphael Kaplinsky, "Prospering in the Periphery: a Special Case – The Seychelles" in Cohen, p.200

"Rene emphasized . . . British neglect." Franda, p.15

DJIBOUTI

"Apparently some . . . For many reasons." Quoted in Thompson, Virginia and Richard Adloff, p.91

"The presence . . . reinforced." Thompson, Virginia and Richard Adloff, p.30

ASIA: ALONG SOUTH ASIA'S FRINGES

"Manuel . . . to reach the Indies" Crowley, pp.31-2

THE MALDIVES

"The Government . . . of public services." Stoddard, Theodore L. and others, p.38

"The modern republican . . . personal recognizance." Stoddard, Theodore L. and others, p.21

"Preserve our country's . . . islands being inundated." Sareer, Interview, November 2, 2015

BHUTAN

"Bhutan undoubtedly . . . ordinary lives." Phuntsho, p.565

"Generations of Bhutanese . . . Commonwealth of Nations." Phuntsho, p.527

BRUNEI

"Sir Hassanal . . . world's richest." Leake, pp.66, 73, 75

"Brunei's foreign reserves . . . annual Budget." Leake, p.151

THE AMERICAS: THE EXTENDED CARIBBEAN

"The Greater Caribbean . . . unprecedented scale." Mulcahy, p.9

EXPLORING AND EXPLOITING THE SPANISH MAIN

"These Caribbean . . . dependence became a habit. " Naipaul. V. S., 1981, p.275"

"clearly impossible . . . on their own" Clarke, Colin, "Grenada," in Clarke and Payne (eds.), p.83 (The quote is from "Report on close association of the British West Indian Colonies" HMSO, London, 1947.)

"The fact that it is . . . and stability" and "without being tied" Michael Baptiste, Interview February 13, 2016

"Small countries . . . common currency" A, J. Mediate, Interview

"the lessons are of a more general applicability"DeLisle Worrell, e-mail, July 1, 2018

BARBADOS

"[A]dventurers came, full of greed . . . of a complex racial problem." Harlow, pp.1-2

"It was in Barbados that . . . during the 1670s . . ." Mulcahy, pp.3-4

"The close ties, both economic . . . were now quickly broken. " Harlow, p.37

"The year 1810 . . . peaked as sugar producers." Knight in Meditz and Hanratty (ed.), p.18

"Barbados has a . . . the Central Bank" DeLisle Worrell, email, June 3, 2018

"the electorate simply seemed to have tired of a mucky status quo." The Economist, "A clean sweep" June 2, 2018, p.30

GRENADA

"The grim events. . . . by colonialism." Lewis, p.1

"Grenada was technically . . . to claim her as their possession." Brizan, p.15

"The emancipation of . . . Colony government was introduced" Brizan, p.197

"in theory one . . . cost several hundred thousand dollars." Celia Edwards Q. C., Interview, February 17, 2016

SAINT VINCENT AND THE GRENADINES

"The 1975 Lome Convention . . . are priced out of the market ", Gibson, p.329

"Although the political system . . . a few years later." Cosover, Mary Jo, "St. Vincent and the Grenadines" in Meditz and Hanratty (ed.), p.337

SAINT LUCIA

"St. Lucia . . . nineteenth centuries." John F. Hornbeck, "St. Lucia" in Meditz and Handratty, p.293

DOMINICA

"[T]he early . . . and Jamaica." Atherton Martin, "Dominica" in Meditz and Hanratty (ed.), p.264

"In 1968 the Leblanc . . . leadership of Mary Eugenia Charles." Martin, p.283

"declared war . . . Power with communism" Gibson, p.301

"Even before . . . domestic industry." Localio, Liam, Interview, January 18, 2018

ANTIGUA AND BARBUDA

"As sugar took hold in the Leeward Islands . . . Tortola, Virgin Gorda, and St. Croix." Mulcahy, p.60

"One parish in Antigua provides . . . while 4 owned more than 100." Mulcahy, p.61

SAINT KITTS AND NEVIS

"[A] 'small-island' . . . mind-set to the larger islands" Meditz and Hanratty (ed.), p.430

THE BAHAMAS

"The Bahamas . . . the United States." and "until 1649 . . . of the 17th century." Mark P. Sullivan, "The Bahamas" in Meditz and Hanratty (ed.), p.521

"The structure of law and order . . . behavior himself." Craton and Saunders, p.104

BELIZE

"Two themes . . . past events." O. Nigel Bolland, "Belize: Historical Setting" in Merrill (ed.), p.157

"Britain formally . . . Caribbean." The Economist,"Half of Belize, please" April 21, 2018, p.29

"development of British . . . in 1871" Thomson, p.117

"The mixed . . . and the Commonwealth." Peyrefitte, Michael, Interview, December 26, 2013

GUYANA

"In order to confront . . . repression and misery." Smith, p.97

"Planter political . . . Dutch rule." Macdonald, "Guyana: Historical Setting" Merrill (ed.), p.11

SURINAME

"Consociational . . . democracy." Lijphart, p.216

"continued for almost . . . peace treaty" and "Eventually . . . accommodation" Mann (2011), pp.470, 472

"go back . . . families " Hoefte, p.1

"consists of the heads . . . ruing class." Kruger, pp.156ff

"the country's increasing . . . optimism." Hoefte, p.218

OCEANIA: THE SOUTHWEST PACIFIC

"Ferdinand Magellan (1480?–1521) . . . or Columbus or Vespucci" Boorstin, pp.268-9

EXPLORING, COMMERCIALIZING, AND COLONIZING THE SOUTH SEAS

"until the era . . . People in the world." Searles, Damion, "The Tide-Beating Heart of Earth" Harper's Magazine, May 2019, p.82 (Quoting from Thompson.)

"by annexing. . . . as the ultimate prize." Bacevich, Andrew J.," American Imperium", Harper's Magazine, May 2016, p.33

SOLOMON ISLANDS

"There is no effective cabinet . . . resources by political players" Wainwright, p.24

FIJI

"Fijian Administration claimed . . . difficult to achieve." Nayacakalou, 1975, p.134

"To some extent . . . potential ethnic hostility." Ratuva, Steve, "The Paradox of Multiculturalism: Ethnopolitical Conflict in Fiji" in Brown, p.198

VANUATU

"[T]he fragility . . . the state." Graham Hassalt, "Elite Conflict in Vanuatu" in Brown (ed.), p.223

"advocated a return to traditional . . . ownership of lands." Lea and Milward, p.213

"In 2004 . . . legal environment." Graham Hassalt, "Elite Conflict in Vanuatu" in Brown (ed.), p.228

"A new China wharf . . . Brisbane", The Economist, "Gateway to the Globe" July 25, 2018

TONGA

"been the only . . . most moves to decolonize" Lawson, p.79

"first Tu'i . . . their line." Senituli, Lopeti, "Unfinished Business: Democratic Transition in Tonga" in Brown (ed.), p.270

"superimposition . . . of responsible government." Lawson, pp.79-80

SAMOA

"Radical nationalism . . . with the past.' "Lawson, p. 136 (Quoted from Davidson, J.W.)

"Rev. John Williams . . . tradition of history." Lea and Milward, p.176

"the fact that the proposed . . . unique cultural heritage." Lawson, p.118

TUVALU

"At a referendum . . . name of Tuvalu." Lea and Milward, p.211

KIRIBATI

"plan for the day . . . have to do that." Anote Tong, as quoted by Suchi Nairin, "Tiny Atoll in Pacific cries out for help" in The Times of India, June 6, 2008.

NAURU

"The central drama . . . rapid economic decline." Max Quanchi, "Troubled Times: Development and Economic Crisis in Nauru" in Brown (ed.), p.249

MARSHALL ISLANDS

"We see the danger . . . beat back the sea" Davenport, Coral, The New York Times, "Pacific Island Nation Struggles Against the Sea," December 2, 2015, pp.A1, A8

"'The ground water . . . their populations.'" Davenport, Coral, The New York Times, "Optimism Faces Grave Reality at Climate Talks," December 1, 2014, pp.A1, A13

THE FEDERATED STATES OF MICRONESIA

"The Congress of Micronesia . . . association with the USA." Lea and Milward, p.112

PALAU

"Palau (along with the Marshall . . . Micronesian state." Lea and Milward, p.145

LOOKING AHEAD

"The end of the colonial empires . . . among states," Jane Burbank and Frederick Cooper, p.458

"The EU has been a model for . . . regional interests." Anthony Teasdale, Foreword

"the age of the nation-state is over . . . federate the continent or suffocate" Laski, vol. 2, p.311

SOURCES

GENERAL

Acemoglu, Daron and James A. Robinson. *Why Nations Fail, The Origins of Power, Prosperity, amd Poverty*. New York: Crown Business, 2012.

Alderfer, Harold F. *Local Government in Developing Countries*. New York: McGraw-Hill Book Company, 1964.

Alesina, Alberto and Enrico Spolaore. *The Size of Nations*. Cambridge, MA: MIT Press, 2003.

Allison, Graham. *Destined for War: Can America and China Escape Thucydides' Trap?* Boston: Houghton Mifflin Harcourt, 2017.

Almond, Gabriel and James S. Coleman (eds.). *The Politics of Developing Areas*. Princeton, NJ: Princeton University Press, 1960.

Appiah, Kwame Anthony. *The Lies That Bind: Rethinking Identity: Creed Country, Class, Culture*, Liveright, 2018.

Aristotle (edited by Stephen Emerson). *Politics*. Cambridge: Cambridge University Press, 1988.

Armstrong, J. *Nations before Nationalism*. Chapel Hill, NC: University of North Carolina Press, 1983.

Ashley, Maurice. *The Golden Century, Europe 1598-1715*. New York: Frederick A. Praeger, 1968.

Auslin, Michel R. *The End of the Asian Century: War, Stagnation, and the Risks to the World's Most Dynamic Region*. New Haven, CT: Yale University Press, 2017.

Bailyn, Bernard. *The Barbarous Years, The Peopling of British North America: The Conflict of Civilizations, 1600-1675*. New York: Alfred A. Knopf, 2013.

Baldwin, Richard. *The Great Convergence: Information Technology and the New Globalization*. Boston: Belknap, 2016.

Bartlett, Christopher A. and Sumantra Ghoshal. *Managing Across Borders*. Boston: Harvard Business School Press, 1989.

Benedict, Burton (editor). *Problems of Smaller Territories*. University of London: Athlone Press, 1967.

Blaustein, A. P., and P. M. Blaustein (editors). *Constitutions of Dependencies and Special Sovereignties*. Vatican City State, 1988.

Bodin, Jean (abridged and translated by M. J. Tooley). *Six Books of the Commonwealth*. Oxford: Basil Blackwell, 1967 (first published in 1576).

Boorstin, Daniel J. *The Discoverers, A History of Man's search to Know His Word and Himself*. New York: Vintage Books, Random House, 1985.

Brands, H. W. *American Colossus, The Triumph of Capitalism 1865-1900*. New York: Anchor Books, 2011.

Briguglo, Lino (editor). *Handbook of Small States*. New York: Routledge, 2018.

Brotton, Jerry. *A History of the World in 12 Maps*. New York: Penguin, 2014.

Brotton, Jerry. *Great Maps*. London: Dorling Kindersley, 2015.

Bryson, Bill. *At Home*. New York: Doubleday, 2010.

Bulmer, Simon and Christian Lequesne (editors). *The Member States of the European Union*. Oxford: Oxford University Press, 2005.

Burbank, Jane and Frederick Cooper. *Empires in World History*. Princeton, NJ: Princeton University Press, 2010.

Canadice, David. *Victorious Century, The United Kingdom, 1800-1906*. New York: Viking, 2018.

Cantor, Norman F. *Medieval History, The Life and Death of a Civilization (Second Edition)*. London: The Macmillan Company, 1969.

Carlness, Walter, Thomas Risse, and Beth A. Simmons. *Handbook of International Relations*. London: Sage Publications, 2002.

Chandler, Alfred D. Jr. *Scale and Scope*. Cambridge, MA: Belknap Press, 1990.

Chapman, Brian, *Introduction to French Local Government*, George Allen & Unwin, London, 1953

Chapman, Brian, *The Prefects and Provincial France*, George Allen & Unwin, London, 1955

Clarke, Colin and Tony Payne (editors). *Politics, Security and Development in Small States*, London: Allen & Unwin, 1987.

Cobban, Alfred. *The Nation State and National Self-Determination*. New York: Thomas Y. Crowell, 1969.

Cooper, N. *The Breaking of Nations*. New York: Atlantic Books, 2003.

Crawford, J. *The Creation of States in International Law*. Oxford: Oxford, 1994.

Crowley, Roger (ed.). *What If?* New York: Berkeley Books, 2000.

Crowley, Roger. *Conquerors: How Portugal Forged the First Global Empire*. New York: Random House, 2015.

Crowley, Roger. *Empires of the Sea*. New York: Random House, 2008.

Curtin, Robert. *Polyarchy, Participation and Opposition*. New Haven, CT: Yale University Press, 1971.

Dahl, Robert. *Democracy and Its Critics*. New Haven, CT: Yale University Press, 1989.

Dalburg-Acton, John (Lord Acton). *Lectures in Modern History, Fontana Library – Collins (5th impression)*. London, 1964.

Dalrymple, William. *The Anarchy: The Relentless Rise of the East India Company*. London: Blooms, 2019.

Darby, H.C. and Fullard, Harold (eds.). *The New Cambridge Modern History Atlas*. Cambridge: Cambridge University, 1970.

Darwin, John. *The Empire Project, The Rise and Fall of the British World System, 1830-1970*. Cambridge: Cambridge University Press, 2009.

Darwin, John. *Unfinished Empire: The Global Expansion of Britain*. London: Penguin Books, 2012.

Davies, Norman. *Europe: A History*. New York: Oxford University Press, 1996.

Davies, Norman. *Vanishing Kingdoms, The Rise and Fall of States and Nations*. London: Penguin Books, 2011.

Davis, James C. *The Human Story, Our History, From the Stone Age to Today*. New York: Harper Collins, 2004.

Davis, Kenneth C. *The Hidden History of America at War*. New York: Hachette Books, 2015.

Demas, W. G. *The Economics of Development in Small Countries with Special Reference to the Caribbean*. Montreal: McGill University Press, 1965.

Diamond, Larry. *Developing Democracy – Toward Consolidation*. Baltimore: Johns Hopkins University Press, 1999.

Dieckhoff, Alain. *Nationalism and the Multinational State*. London: Hurst, 2016.

Diehl, Paul F. *The Politics of Global Governance*. Boulder, CO: Lynne Rienne Publishers, 2005.

Duchacek, Ivo D. *Comparative Federalism: The Territorial Dimension of Politics*. New York: Holt, Rinehart and Winston, 1970.

Dugard, J. *Recognition and the United Nations*. Cambridge: Cambridge University Press, 1987.

Dunne, Peter Finley. *Dissertations by Mr. Dooley*. Boston: Maynard and Co., 1898.

Dunne, Peter Finley. *In Peace and War*. Small, Maynard & Company, 1898.

Dunne, Peter Finley. *Mr. Dooley In Peace and War*. Boston: Small, Maynard & Company, 1898.

Duursma, Jorri C. *Fragmentation and the International Relations of Micro-States, Self-determination and Statehood*. Cambridge: Cambridge University Press, 1996.

Eide, Asbjorneand and Bernt Hagtvet (editors). *Conditions for Civilized Politics and Compliance with Human Rights*. Oslo: Scandinavian Universities Press, 1996.

Elliott, J. H. *Imperial Spain 1469-1716*. London: Penguin, 2002.

Ellis, Joseph J. *The Quartet, The Orchestration of the Second American Revolution, 1783-1789*. New York: Alfred A. Knopf, 2015.

Evans, J. D. *Malta: Ancient Peoples and Places (Volume 11)*. New York: Frederick A. Praeger, 1959.

Evans, Richard J. *The Pursuit of Power, Europe 1815-1914*. New York: Viking, 2016.

Fairbank, John King and Merle Goldman. *China: A New History (Second Enlarge Edition)*. Cambridge, MA: The Belknap Press of Harvard University Press, 2006.

Felsenstein, Daniel and Beris A. Portner. *Regional Disparities in Small Countries*. Berlin and Heidelberg, Germany: Springer Berlin Heidelberg, 2005.

Ferguson, Robert, *The Vikings, A History*, Viking, London, 2009

Finer, S. E. *Comparative Government*. New York: Basic Books, Inc., 1971.

Fisher, H. A. L. *A History of Europe*. London: Edward Arnold, 1936.

Franck, Thomas M. *The Power of Legitimacy Among Nations*. Oxford: Oxford University Press, 1990.

Freeman, Christopher and Bengt-Ake Lundvall (editors). *Small Countries Facing the Technological Revolution*. London and New York: Pinter, 1988.

Gal, Michal. *Competition Policy for Small Market Economics*. Cambridge, MA: Harvard, 2003.

Gellner, E. *Nationalism*. New York: New York University Press, 1997.

Gellner, E. *Nations and Nationalism*. Ithaca, NY: Cornell University Press, 1983.

George, Stephen and Ian Bache. *Politics in the European Union*. Oxford: Oxford University Press, 2001.

Ghemawat, Pankay. *World 3.0.: Global Prosperity and How to Achieve It*. Boston: Harvard Business Review Press, 2011.

Gilbert, William, [1600], (translated by P. Fleury Mottelay). London: Bernard Quaritch, 1893.

Goodell, Jeff. *The Water Will Come: Rising Seas, Sinking Cities, and the Remaking of the Civilized World*. New York: Little, Brown, 2018.

Goodman, Louis. *Small Nations, Giant Firms*. New York and London: Holmes & Meier, 1987.

Goodwin, Robert. *Spain: The Centre of the World 1519-1682*. New York: Bloomsbury Press, 2015.

Grynberg, Roman (editor). *WTO at the Margins, Small States and the Multilateral Trading System*. Cambridge: Cambridge University Press, 2006.

Hahn, Steven. *A Nation Without Borders, The United States and Its World in an Age of Civil Wars, 1830-1910*. New York: Viking, 2016.

Hamilton, Alexander, James Madison, and John Jay, The Federalist Papers, New American Library, New York, 1961, p.83.

Hanna, David. *Knights of the Sea*. New American Library, Penguin Group, 2012, p.21.

Harari, Yuval Noah. *Sapiens*. New York: Harper Collins Publishers, 2015.

Harden, Sheila (editor). *Small is Dangerous: Micro States in a Macro World*. London: Frances Pinter, 1985.

Heady, Ferrel. *Public Administration: A Comparative Perspective*. New York: Marcel Dekker, 1979.

Heidenhammer, Arnold J. *Political Corruption: Readings in Comparative Analysis*. New York: Holt, Reinhardt and Winston, 1970.

Hey, Jeanne A.K. (editor). *Small States in World Politics: Explaining Foreign Policy Behavior*. Boulder, CO: Lynne Rienner, 2003.

Hix, Simon and Bjorn Hoylsand. *The Political System of the European Union (Third Edition)*. New York: Palgrave MacMillan, 2011.

Hobsbawn, E. J. *Nations and Nationalism since 1780, programme, myth, reality*. Cambridge: Cambridge University, 1990.

Hobsbawn, Eric. *The Age of Capital: 1848-1875*. London: Abacus, 1975.

Hobsbawn, Eric. *The Age of Empire: 1875-1914*. London: Abacus, 1991.

Hobsbawn, Eric. *The Age of Extremes:The Short Twentieth Century, 1914-1991*. London: Abacus, 1994.

Hobsbawn, Eric. *The Age of Revolution: 1789-1848*. London: Abacus, 1962.

Hochschild, Adam. *To End All Wars: A Story of Loyalty and Rebellion*. Boston: Houghton Mifflin Harcourt, 2011.

Hume, David. "Of the Rise and Progress of the Arts and Sciences" in Miller (ed.) David Hume, *Essays: Moral, Political and Literary*. Indianapolis: Liberty Fund.

Humes, Samuel and Eileen Martin. *The Structure of Local Government: A Comparative Survey of 81 Countries*. The Hague: International Union of Local Authorities, 1969.

Humes, Samuel IV. *Local Governance and National Power*. London: Harvester Wheatsheaf, 1991.

Humes, Samuel and Eileen Martin. *The Structure of Local Governments Throughout the World*. Martinus Nijhoff, 1961.

Huntington, Samuel P. *The Third Wave: Democratization in the Twilight of the Late Twentieth Century*. Norman, OK: University of Oklahoma Press, 1991.

Huntington, Samuel P. *Who are We?* New York: Simon & Schuster, 2004.

Ingebritsen, Christine, Iver Neuman, Sieglinde Gstohl, and Jessica Beyer (editors). *Small States in International Relations*. Seattle: University of Washington, 2006.

Jainarain, Jordan. *Trade and Underdevelopment: A Study of the Small Caribbean Countries and Large Multinational Corporations*. Georgetown: University of Guyana, 1976.

Jones, Eric L. *Cultures Merging: a Historical and Economic Critique of Culture*. Princeton, NJ: Princeton University Press, 2006.

Jones, Eric L. *The European Miracle*. Cambridge: Cambridge University Press, 1981.

Kamen, Henry. *How Spain Became a World Power 1492-1763*. New York: Harper Collins, 2004.

Kamen, Henry. *Philip of Spain*. New Haven, CT: Yale University Press, 1997.

Kennes, Walter. *Small Developing Countries and Global Markets: Competing in the Big League*. New York: St. Martin's, 2000.

Key, V. O. Jr. *The Techniques of Political Graft in the United States*. Chicago: University of Chicago Libraries, 1936.

Kinzer, Stephen. *The True Flag: Theodore Roosevelt, Mark Twain, and the Birth of the American Empire*. New York: Henry Holt, 2016.

Knight, Roger J. B. *Britain Against Napoleon: The Organization of Victory 1795-1815*. New York: Barnes & Noble, 2013.

Landes, David. *The Wealth and Poverty of Nations: Why Some Are So Rich and Some So Poor*. New York: W. W. Norton, 1998.

Laycock, Stuart. *All the Countries We've Ever Invaded*. Chichester, West Sussex: The History Press, 2012.

Levin, Yuval. *The Fractured Republic: Reviving America's Social Contract in the Age of Individualism*. New York: Basic Books, 2016.

Levitsky, Steven. *Why Democracies Die?* Carmarthen, U.K.: Crown House, 2018.

Lijphart, Arend. *Democracy in Plural Societies: A Comparative Exploration*. New Haven, CT: Yale University Press, 1977.

Linz, Juan J. and Alfred Stephan. *Problems of Democratic Transition and Democratic Consolidation*. Baltimore: Johns Hopkins University Press, 1996.

Long, Edward Leroy, Jr. *Conscience and Compromise*. Philadelphia: The Westminster Press, 1954.

Maas, Arthur (editor). *Area and Power: A Theory of Local Government*. Glencoe, IL: The Free Press, 1959.

MacCormick, N. *Questioning Sovereignty*. London: Oxford University Press, 1999.

Maddick, Henry. *Democracy, Decentralization and Development*. Bombay: Asia Publishing House, 1963.

Mahan, A. T. *The Influence of Sea Power upon History 1660-1783*. Boston: Little, Brown & Co., 1944.

Man, John. *Kublai Kahn*. London: Bantam Press, 2006.

Mann, Charles C. *1491: New Revelations of the Americas Before Columbus (Second Edition)*. New York: Vintage, 2005.

Mann, Charles C. *1493: Uncovering the New World Columbus Created*. New York: Vintage, 2011.

Marshall, Tim. *Prisoners of Geography: Ten Maps That Explain Everything about the World*. New York: Scribner, 2015.

McKibben, Bill. *Falter: Has the Human Race Begun to Play Itself Out?* New York: Henry Holt, 2019.

McNeill, William H. *The Rise of the West*. Chicago: University of Chicago, 1963.

Miller, James. *Can Democracy Work? A Short History of a Radical Idea, from Ancient Athens to Our World*. New York: Farrar, Straus and Giroux, 2018.

Mokyr, Joel. *A Culture of Growth: The Origins of the Modern Economy*. Princeton, NJ: Princeton University Press, 2016.

Moore, Barrington, Jr. *Social Origins of Dictatorship and Democracy*. Boston: Beacon Press, 1966.

Muller, Jan-Werner. *What is Populism?* Philadelphia: University of Pennsylvania Press, 2016.

Mulroy, Pat. *The Water Problem: Climate Change and Water Policy*. Washington, D.C.: Brookings, 2017.

Nichols, Roy F., *Advance Agents of American Destiny*, University of Pennsylvania, Philadelphia, 1956

Ostler, Nicholas. *Empires of the Word: A Language History of the World*. London: Harper Perennial, 2006.

Ott, Dana. *Small is Democratic: An Examination of State Size, and Democratic Development*. New York and London: Garland Publishing, 2000.

Parrkinson, C. Northcote. *East and West*. London: John Murray, 1963.

Parry, J. H. *The Establishment of the European Hegemony: 1415-1715, Trade and Exploration in the Age of the Renaissance (Third Edition, revised)*. New York: Harper & Row, 1961.

Peyrefitte, Alan. *C'etait de Gaulle (2 vols)*. Paris: Fayard, 1994.

Polo, Marco (Translated and edited by William Marsden; re-edited by Thomas Wright). *The Travels of Marco Polo The Venetian*. New York: Doubleday, 1948.

Porter, Michael E. *Competitive Strategy*. New York: The Free Press, 1980.

Porter, Michael E. *The Competitive Advantage of Nations*. New York: The Free Press, 1990.

Potter, G. R. (editor). *The New Cambridge Modern History, Volume XII*. London: Cambridge University Press, 1961.

Prescott, William H. *History of the Reign of Ferdinand and Isabella, The Catholic, of Spain (Volume 1)*. London: George Routledge & Co., 1854.

Previtte-Orton (editor). *The Shorter Cambridge Medieval History*. Cambridge: Cambridge University Press, 1962.

Rapaport, Jacques, Ernest Muteba, and Joseph J. Therattil. *Small States & Territories: Status and Problems*. New York: Arno Press, UNITAR, 1971.

Rich, Nathaniel. *Losing Earth: A Climate History*. New York: MCD/Farrar, Straus & Giroux, 2019.

Ridley, Jasper. *A Brief History of the Tudor Age*. London: Constable & Robinson, 2002.

Riesenberg, Felix. *The Pacific Ocean*. London: Whittlesey House, 1940.

Rifkin, Jeremy. *The European Dream*. Cambridge, MA: Polity, Cambridge, 2005.

Riggs, Fred W. *Administration in Developing Countries: The Theory of Prismatic Society*. Boston: Houghton Mifflin Company, 1964.

Risen, Clay. *The Crowed Hour: Theodore Roosevelt, The Rough Riders, and the Dawn of the American Century*. New York: Scribers, 2019.

Rodden, Jonathon. *Why Cities Lose: The Deep Root of the Urban-rural Political Divide*. New York: Basic Books, 2019.

Rogow, Arnold and Harold D. Lasswell. *Power, Corruption, and Rectitude*. Englewood Cliffs, NJ: Prentice-Hall, 1963.

Ronen, Dev. *The Quest for Self-Determination*. New Haven and London: Yale University Press, 1979.

Rosen, William. *The Third Horseman: Climate Change and the Great Famine of the 14th Century*. New York: Viking, 2014.

Rougemont, Denis de. *The Meaning of Europe*. New York: Stein and Day, 1965.

Sabine, George H. (Revised by Thomas Landon Thorson). *A History of Political Theory (Fourth Edition)*. Hinsdale, IL: Darden Press, 1973.

Schumpeter, Joseph. *Capitalism, Socialism, and Democracy (Second Edition)*. New York: Harper, 1969.

Scott, James C. *Comparative Political Corruption*. Englewood, NJ: Prentice-Hall, 1972.

Shurman, J.C. *Empire of the Weak: The Real Story of European Expansion and the Creation of a New World Order*. Princeton, NJ: Princeton University Press, 2019.

Singer, Charles, et la. *A History of Technology*. Oxford: Oxford University Press, 1958.

Simon, Herbert A., Donald W. Smithburg, and Victor Thompson. *Public Administration*. New York: Alfred A. Knopf, 1950.

Sinclair, Upton. *The Jungle*. New York: Doubleday, 1905.

Smith, Anthony D. *Myths and Memories of the Nation*. London: Oxford University Press, 2000.

Smith, Anthony D. *Nationalism and Modernism: A critical survey of recent theories of nations and nationalism*. London and New York: Routledge, 1998.

Smith, Anthony D. *The Ethnic Origins of Nations*. Oxford: Blackwee, 1981.

Somin, Ilya. *Democracy and Political Ignorance: Why Smaller Government is Smarter*. Stanford: Stanford University, 2016.

Stuenkel, Oliver. *Post-Western World: How Emerging Powers Are Remaking World Order*. Cambridge, MA: Polity, 2017.

Taylor, Alan. *American Colonies*. New York: Viking Penguin, 2001.

Teasdale, Anthony and Timothy Bainbridge. *The Penguin Companion to the European Union (Fourth Edition)*. London: Penguin Books, 2012.

Thomas, Hugh. *The Slave Trade*. New York: Simon & Schuster, 1997.

Thomas, Hugh. *World Without End: Spain, Philip II, and the First Global Empire*. New York: Random House, 2015.

Thompson, E. A. *A History of Attila and the Huns*. Oxford: Oxford University Press, 1948.

Thompson, Virginia and Richard Adloff. *Djibouti and the Horn of Africa*. Stanford University Press, 1968.

Tocqueville, Alexis de (translated by Mansfield and Winthrop). *Democracy in America*. Chicago: University of Chicago Press, 2000 (originally published in 1835).

Tombs, Robert. *The English and their History*. New York: Vintage, 2014.

Toynbee, Arnold. *A Study of History* (a new edition revised and abridged by the author and Jane Caplan). New York: Weathervane Books, Oxford University Press, 1972.

Toynbee, Arnold. *The World and the West*. London: Oxford University Press, 1953.

Tsushima, Yuko (translated from Japanese by Geraldine Harcourt). *Territory of Light: A Novel*. New York: Farrar, Straus & Giroux, 2019.

Tuchman, Barbara. *A Distant Mirror: The Calamitous Fourteenth Century*. New York: Alfred A. Knopf, 1978.

Vital, David. *The Inequality of States*. Oxford: Clarendon Press, 1967.

Vollman, William T. *Carbon Ideologies: No Good Alternatives (Volume 2)*. New York: Viking, 2018.

Vollman, William T. *Carbon Ideologies: No Immediate Danger (Volume 1)*. New York: Viking, 2018

Wallace Peter G. *The Long European Reformation: Religion, Political Conflict, and the Search for Conformity, 1350-1750*. New York: Palgrave-Macmillan, 2004.

Wallace-Wells, David. *Life After Warming*. New York: Tim Duggan Books, 2019.

Weber, Max (translated, edited, and with an introduction by H. H. Gerth and C. Wright Mills). *From Max Weber: essays in sociology*. New York: Oxford University Press, 1958.

Weber, Max, (edited with an introduction by Talcott Parsons). *The Theory of Social and Economic Organizations*. New York: The Free Press, 1947.

Wheare, K. C. *Modern Constitutions*. Oxford: Oxford University Press, 1951.

Wheare, K. C. *Federal Government (Third Edition)*. London: Oxford University Press, 1953.

Wheare, K. C. *Government by Committee*. London: Oxford University Press, 1955.

Wheare, K. C. *Modern Legislatures*. New York: Oxford University Press, 1963.

White, George W. *Nation, State, and Territory, Origins, Evolutions, and Relationships (Volume 1)*. Oxford: Rowman & Littlefield, 2004.

Whitfield, Peter. *New Found Lands Maps in the History of Exploration*. New York: Routledge, 1998.

Williamson, Edwin. *The Penguin History of Latin America*. London: Penguin Books, 2009 (revised).

Wilson, Peter N. *The Thirty Years War: Europe's Tragedy*. Cambridge, MA: Belknap Press of Harvard University Press, 2009.

Wraith, Ronald and Edgar. *Corruption in Developing Countries*. London: George Allen & Unwin, 1963.

Ylvisaker, Paul. "Some criteria for a 'proper' areal (distribution) of (governmental) power, in A. Maas et la. (ed.), *Area and Power*. Glencoe, IL: Free Press.

Ziegler, Philip. *The Black Death*. Dover, NH: Alan Sutton Publishing, 2011.

Zinn, Howard. *A Peoples' History of the United States*. New York: Barnes & Noble, 1980.

EUROPEAN MICRO-COUNTRIES

Adois, Ferdinand. *The Eternal City: A History of Rome*. New York: Pegasus Books, 2018.

Arblaster, Paul. *A History of the Low Countries*. Hampshire and New York: Palgrave Macmillan, 2006.

Banac, Ivo. *The National Question in Yugoslavia: Origins, History, Politics*. Ithaca, NY: Cornell University Press, 1984.

Bradford, Ernie. *The Great Siege of Malta 1565*. Hertfordshire: Wordsworth, 1979.

Brook-Shepherd, Gordon. *The Austrians: A Thousand-Year Odyssey*. London: Harper Collins, 1996.

Brutails, J.A. *La Coutume d'Andorre, (Second Edition)*. Andorra: Monumenta Andoranna, 1965.

Bulmer, Simon and Christian Liquesce. *The Member States of the European Union (Second Edition)*. Oxford: Oxford University, 2013.

Castillo, Dennis. *The Maltese Cross: A Strategic History of Malta*. Westport, CT: Prager Security International, 2006.

Cline, Eric H. *1177 The Year Civilization Collapsed*. Princeton, NJ: Princeton University Press, 2014.

Crankshaw, Edward. *The Fall of the House of Hapsburg*. New York: The Viking Press, 1963.

Davies, Norman. *Vanished Kingdoms: The Rise and Fall of States and Nations*. New York: Penguin Press, 2012.

Denman, Roy. *Missed Chances: Britain and Europe in the Twentieth Century*. London: Cassell, 1996.

Djilas, Mawlovan (translated by Michael B. Petrovich). *Wartime*. New York: Harcourt Brace Javanovich, 1977.

Duursma, Jorri C. *Fragmentation and the International Relations of Micro-States*. Cambridge: Cambridge University, 1996.

Duverger, M. *La Reforma de les Institutions d'Andorra*. Andorra: Andorra de Vella, 1981.

Ferguson, Robert. *The Vikings A History*. London: Penguin Books, 2009.

Fine, John V. A. *The Early Medieval Balkans: A Critical Survey from the Sixth Century to the Late Twelfth Century*. Ann Arbor, MI: University of Michigan Press, 1991.

Fine, John V. A. *The Late Medieval Balkans: A Critical Survey from the Late Twelfth Century to the Ottoman Conquest*. Ann Arbor, MI: University of Michigan Press, 1994.

Fischer, David Hackett. *Champlain's Dream.* New York: Simon & Schuster, 2008.

Fisher, H.A.L. *A History of Europe.* London: Edward Arnold, 1936.

Galbraith, James K. *Welcome to the Poisoned Chalice: The Destruction of Greece and the Future of Europe.* New Haven, CT: Yale University Press, 2016.

Gyerset, Knut. *History of Iceland.* New York: Macmillan, 1924.

Holland, James. *Fortress Malta: An Island Under Siege.* Cassell Military Paperbacks, 2006.

Humes, Samuel. *Belgium: Long United, Long Divided.* London: Hurst & Co., 2014.

Jones, Dan. *The Templars: The Rise and Spectacular Fall of God's Holy Warriors.* New York: Penguin Books, 2017.

Karlsson, Gunnar. *The History of Iceland.* Minneapolis: University of Minnesota Press, 2000.

Kossman, E.H. *The Low Countries 1780-1940.* Oxford: Oxford University Press, 1978.

Laspina, Msgr. S. *Outlines of Maltese History.* Valetta, Malta: A. C. Aquilina, 1971.

Lebande, L.-H. *L'histoire de la Principaute de Monaco (Second Edition).* Monaco: 1934.

Macintyre, Donald. *The Battle for the Mediterranean.* New York: W.W. Norton, 1964.

Merrritt, Giles. *Slippery Slope.* Oxford: Oxford University Press, 2016.

Middelaar, Luuk van (translated by Liz Waters). *The Passage to Europe, How a Continent Became a Union.* New Haven and London: Yale University Press, 2013.

Moravcsik, Andrew. *The Choice for Europe, Social Purpose & State Power From Messina to Maastricht.* Ithaca, NY: Cornell University Press, 1998.

Nugent, Neill. *The Government of the European Union (Seventh Edition).* New York: Palgrave Macmillan, 2010.

Pinder, John. *The European Union: A Very Short Introduction (Third Edition).* Oxford: Oxford University Press, 2013

Purcell, H. D. *Cyprus.* New York: Frederick A. Praeger, 1969.

Roberts, Elizabeth. *The Realm of the Black Mountain: A History of Montenegro.* Ithaca, NY: Cornell University Press, 2007.

Rougemont, Denis de (translated by Alan Braley). *The Meaning of Europe.* New York: Stein and Day, 1965.

Sawyer, Peter (ed.). *The Oxford Illustrated History of the Vikings.* Oxford and New York: Oxford University Press, 1987.

Siddall, Henry. *Malta: Past and Present.* London: Chapman and Hall, 1970.

Sire, H. J. *The Knights of Malta.* New Haven, CT: Yale University Press, 1994.

Teasdale, Anthony and Timothy Bainbridge. *The Penguin Companion to European Union (Fourth Edition).* London: Penguin, 2012.

Thomson, P. A. B. *Belize: A Concise History.* Oxford: Macmillan Caribbean, 2004.

Thorhallsson, Baldur. *The Role of Small States in the European Union.* Aldershot: Ashgate, 2000.

Thurman, Judith. "Maltese for Beginners." *The New Yorker,* September 3, 2018, pp.48-55.

White, Sam. *A Cold Welcome: the Little Ice Age and Europe's Encounter with North America.* Cambridge, MA: Harvard University Press, 2017.

Winroth, Anders. *The Age of the Vikings.* Princeton, NJ: Princeton University Press, 2014.

AFRICAN MICRO-COUNTRIES

Bennett, Herman L. *African Kings and Black Slaves: Sovereignty and Dispossession in the Early Modern Atlantic.* Philadelphia: University of Pennsylvania Press, 2019.

Boahen, A. Adu. *The Horizon History of Africa, Volumes I and II.* New York: American Heritage, 1971.

Chabal, Patrick and others. *A History of Postcolonial Africa.* Bloomington and Indianapolis, IN: Indiana University Press, 2002.

Cohen, Robert (editor). *African Islands and Enclaves.* Beverly Hills, CA: Sage, 1983.

Coleman, J.S., and C. G. Rosberg. *Political Parties and National Integration in Tropical Africa.* Los Angeles and Berkeley: California University Press, 1966.

Crowder, Michael. *West Africa Under Colonial Rule.* London: Hutchinson, 1968.

Duffy, James. *Portugal in Africa.* Hammondsworth: Penguin Books, 1962.

Fage, J. N. and Roland Oliver. *Cambridge History of Africa, Volumes I-VIII.* Cambridge: Cambridge University Press, 1975-1986.

Fauvelle, Francois-Xavier (translated from French by Troy Tice). *The Golden Rhinoceros: Histories of the African Middle Ages.* Princeton, NJ and Oxfordshire: Princeton University Press, 2018.

Franda, Marcus. *The Seychelles: The Unquiet Islands.* Boulder, CO: Westview Press, 1982.

French, Howard W. *China's Second Continent: How a Million Migrants are Building A New Empire in Africa.* New York: Alfred A. Knopf, 2015.

Gann, Lily and Peter Duagan. *Colonialism in Africa 1870-1960, Volumes I-IV.* Cambridge: Cambridge University Press, 1971.

Gibson, Richard. *African Liberation Movements: Contemporary Struggles against White Minority Rule.* Oxford: Oxford University Press, Institute of Race Relations,1972.

Gomez, Michael A. *African Dominion: A History of Empire in Early and Medieval West Africa.* Princeton, NJ: Princeton University Press, 2019.

Green, Toby. *A Fistful of Shells: West Africa from the Rise of the Slave Trade to the Age of Revolution.* Chicago: University of Chicago Press, 2019.

Grovogui, Siba N'Zatioula. *Sovereigns, Quasi Sovereigns, and Africans: Race and Self-Determination in International Law.* Minneapolis: University of Minnesota, 1996.

Hailey, Lord Lothian. *An African Survey.* London: Oxford University Press, 1957.

Hallett, Robin. *Africa since 1875.* Ann Arbor, MI: University of Michigan Press, 1974.

Harlow, Vincent and E. M. Chilver assisted by Alison Smith (eds.). *History of East Africa, Volume II.* Oxford: Clarendon Press, 1965.

Hodges, Tony and Malyn Newitt. *Sao Tome and Principe, From Plantation Colony to Micro-State.* Boulder, CO: Westview, 1988.

July, Robert. *A History of the African Peoples (Fourth Edition).* Prospect Heights, IL: Waveland Press, 1992.

Meredith, Martin. *The Fortunes of Africa: 5000 Years of Wealth, Greed, & Endeavor, Public Affairs.* New York: Simon & Schuster, 2011.

Meredith, Martin. *The Fate of Africa: A History of a 50 Years of Independence, Public Affairs*. New York: Simon & Schuster, 2014.

Newitt, Malyn. *Portugal in Africa: The Last Hundred Years*. London: C. Hurst & Co., 1981.

Newitt, Malyn. *The Comoro Islands, Struggle Against Dependency in the Indian Ocean*. Boulder, CO and London: Westview Press, 1984.

Oliver, Roland and J. D. Fage. *A Short History of Africa*. London: Penguin, 1960.

Pakenham, Thomas. *The Scramble for Africa, The White Man's Conquest of the Dark Continent 1870-1912*. New York: Random House, 1940.

Reader, John. *Africa: A Biography of the Continent*. New York: Alfred A Knopf, 1998.

Torp, Jens, L.M. Denny, and Donald I. Ray. *Mozambique; Sao Tome and Principe*. London: Pinter, 1989.

ASIAN MICRO-COUNTRIES

Armitage, David and Alison Bashford (editors). *Pacific Histories: Ocean, Land, People*. New York: Palgrave, 2014.

Barwise, J.M. and N. J. White. *A Traveler's Guide to Southeast Asia*. New York: Interlink Books, 2001.

Cohen, Stephen Philip. *India: Emerging Power*. Washington, D.C.: The Brookings Institution, 2002.

Gunther, John. *Inside Asia*. New York: Harper & Brothers, 1939.

LaTourette, Kenneth Scott. *A Short History of the Far East (Third Edition)*. New York: The Macmillan Company, 1957.

Leake, David, Jr. *Brunei, The Modern Southeast Asian Islamic Sultanate*. Jefferson, NC: McFarland, 1989.

Orru, Maree, Nicvole Woolsey Biggart, and Gary Hamilton. *The Economic Organization Of East Asian Capitalism*. London: Sage, 1996.

Phuntsho, Lopen Karma. *The History of Bhutan*. London: Haus Publishing, 2013.

Ricard, Matthieu. *Bhutan, the Land of Serenity*. London: Thames & Hudson, 2008.

Robinson, J. J. *The Maldives: Islamic Republic, Tropical Autocracy*. London: Hurst, 2015.

Rose, Leo. *The Politics of Bhutan*. Ithaca, NY: Cornell University Press, 1977.

Saunders, Graham. *A History of Brunei*. Oxford, Singapore and New York: Oxford University Press, 1994.

Stoddard, Theodore L. and others. *Area Handbook for Indian Ocean Territories*. Washington, D.C.: American University, 1971.

Thomas, Nicholas. *Islanders: The Pacific in the Age of Empire*. New Haven, CT: Yale University Press, 2010.

AMERICAN MICRO-COUNTRIES

Beckles, Hilary McD. *A History of Barbados, From Amerindian settlement to nation-state*. Cambridge: Cambridge University Press, 1990.

Bergreen, Laurence. *Columbus: The Four Voyages 1492-1509*. New York: Viking Penguin, 2011.

Brizan, George. *Grenada: Island of Conflict, From Amerindians to People's Revolution 1498-1979.* London: Zed Books, 1984.

Burrowes, Reynold A. *The Wild Coast: An Account of Politics in Guyana.* Cambridge, MA: Schenkman, 1984.

Cambridge, 1921.

Chernow, Ron. *Alexander Hamilton.* New York: Penguin Press, 2004.

Chin, Hank E. *Surinam: Politics, Economics and Society.* London and New York: Francis Pinter, 1987.

Craton, Michael and Gail Saunders. *Islanders in the Stream, A History of the Bahamian People (Volume One; From Aboriginal Times to the End of Slavery).* Athens, GA and London: University of Georgia Press, 1992.

Craton, Michael. *A History of the Bahamas (Thirds Edition).* Waterloo, Canada: San Salvador Press, 1986.

Dew, E. *The Difficult Flowering of Surinam.* The Hague: Martinus Nijhoff, 1978.

Gibson, Carrie. *Empire's Crossroads: A History of the Caribbean from Columbus to the Present Day.* London: Macmillan, 2014.

Greene, Jack. *Imperatives, Behaviors, and Identities: Essays in Early Colonial History.* Charlottesville, VA: University of Virginia Press, 1992.

Harlow, Vincent T. *A History of the Barbados 1625-1685* (Initially published in 1926 by Clarendon Press). New York: Negro Universities Press, 1969.

Harris, Jane. *Sugar Money.* New York: Arcade Publishing, 2018.

Higham, C. S. S. *The Development of the Leeward Islands Under the Restoration 1660-1688, A study of the Foundations of the Old Colonial System.* Cambridge: Cambridge University Press, 2015.

Hoefte, Rosemarijn. *Surinam in the Long Twentieth Century: Domination, Contestation, Globalization.* New York: Palgrave Macmillan, 2014.

Jelly-Schapiro, Joshua. *Island People: The Caribbean and the World.* New York: Alfred A. Knopf, 2017.

Kruijer, G. J. *Suriname Neokolonie in Rijksverbond.* Meppel: J.A. Boom, 1973.

Lewis, Gordon K. *Grenada: The Jewel Despoiled.* Baltimore: Johns Hopkins University Press, 1987.

Lier, R. J. van. *Frontier Society: A Social Analysis of the History of Surinam.* The Hague: Martinus Nijhoff, 1971.

Lier, R. J. van. *Samenleving in een Grensgeheid.* The Hague: Martinus Nijhoff, 1949.

Meditz, Sandra and Dennis M. Hanratty (eds.). *Islands of the Commonwealth Caribbean.* Washington, D.C.: Federal Research Division, Library of Congress, 1989.

Merrill, Tim L. (editor). *Guyana and Belize, country studies (Second Edition).* Washington, D.C.: Federal Research Division, Library of Congress, 1993.

Morison, Samuel Eliot. *The European Discovery of America: The Southern Voyages A. D. 1492-1616.* New York: Oxford University Press, 1974.

Morrison, Dane. *True Yankees: The South Seas & The Discovery of American Identity.* Baltimore: Johns Hopkins University Press, 2013.

Mulcahy, Matthew. *Hubs of the Empire, The Southeastern Lowcountry and British Caribbean*. Baltimore: Johns Hopkins, 2014.

Naipaul, V. S. *The Overcrowded Barracoon and Other Articles*. London: Penguin, 1981.

Newby, Eric. *World Atlas of Exploration*. New York: Rand McNally & Company, 1975.

Newman, Peter. *British Guiana, Problems of Cohesion in an Immigrant Society*. London: Oxford University Press, 1964.

O'Shaughnessy, Hugh. *Grenada: An Eyewitness Account of the U.S. Invasion and the Caribbean History that Provoked It*. New York: Dodd, Mead & Company, 1884.

Scarr, Deryck. *Fiji: A Short History*. Sydney: Allen & Unwin, 1984.

Schomburgk, Robert A. *The History of the Barbados*. Portland, OR: Frank Cass, 1848 (reprint 1998).

Sheridan, Richard B. *Sugar and Slaves: An Economic History of the British West Indies 1673-1775*. Baltimore: Johns Hopkins University Press, 1973.

Singh, Chaitram. *Guyana: Politics in a Plantation Society*. New York: Praeger, 1988.

Smith, Raymond T. *British Guiana*. London: Oxford University Press, 1962.

Sullivan, Mark. *Our Times, Volume III Pre-War America*. New York and London: Charles Scribner's Sons, 1930.

Swan, Michael. *British Guiana, The Land of Six Peoples*. London: Her Majesty's Stationary Office, 1957.

Thomson, P.A.B. *Belize: A Concise History*. Caribbean and Oxford: Macmillan, 2004.

Williams, Eric. *From Columbus to Castro: History of the Caribbean 1492-1969*. New York: Vintage, 1984.

Wint, Alvin. *Competitiveness in Small Developing Economies, Insights from the Caribbean*. Barbados: University of West Indies, 2003.

OCEANIA MICRO-COUNTRIES

Asterman, Sylvia. *The Origins of International Rivalry in Samoa 1845-1884*. Stanford: Stanford University Press, 1934.

Beaglehole, J.C. *The Exploration of the Pacific*. London: A. & C. Black, 1934.

Boochani, Behronz (translated by Omid Tofighian). *No Friend But the Mountains: Writing from Manus Prison*. Toronto: Anansi Publishing, 2019.

Brown, M. Anne (editor). *Security and Development in the Pacific Islands*. Boulder, CO: Lynne Rienner Publishers, 2007.

Bunge, Frederica and Melinda W. Cooke (editors). *Oceania: A Regional Study*. Washington, D.C.: The American University, 1985.

Christian, F. W. *The Carolina Islands: Travel in the Sea of the Little Islands*. London: Methen & Co., 1899.

Coates, Austin. *Western Pacific Islands*. London: Her Majesty's Stationery Office, 1970.

Cooke, Stephanie. *In Mortal Hands: A Cautionary History of the Nuclear Age*. Collingwood, Australia: Black Inc., 2009.

Crocumbe, Ron (editor). *The Politics of Micronesia.* Suva, Fiji: Institute of Pacific Studies, 1989.

Davidson, J. W. *Samoa Mo Samoa: The Emergence of the Independent State of Western Samoa.* Melbourne, Australia: Oxford University Press, 1967.

Dunmore, John. *French Explorers in the Pacific.* Oxford: Clarendon Press, 1965.

Ells, Albert. *Ocean Island and Nauru: Their Story.* Sydney: Angus and Robertson, 1936.

Fairbairn, T., and D. Worrell. *South Pacific and Caribbean Island Economies: a Comparative Study.* Adelaide, Australia: Foundation for Development Cooperation, 1996.

Gailey, Christine Ward. *Kinship to Kingship: Gender Hierarchy and State Formation in the Tongan Islands.* Austin: University of Texas Press, 1987.

Ghai, Yash (editor). *Law, Politics and Government in the Pacific Islands States.* Suva, Fiji: University of the South Pacific, Institute of Pacific Studies, 1988.

Gillepsie, Alexander and William C. G. (editors). *Climate Change in the South Pacific, Impact and Response.* The Netherlands: Kluwer Academic Press, 1999.

Gowdy, John M. and Carl N. McDaniel. *Paradise for Sale.* Berkeley, CA: University of California Press, 2000.

Hawley, Pat. *Breaking Spears and Mending Hearts: Peacemakers and Restorative Justice in Bougainville.* Annandale, Australia: Federation Press, 2002.

Heine, Carl. *Micronesia at the Crossroads.* Honolulu: University Press of Hawaii, 1974.

Henningham, Stephen. *The Pacific Island States: Security and Sovereignty in the Post-Cold War Era.* New York: St. Martin's Press, 1995.

Lal, Brij V. *Broken Waves: A History of the Fiji Islands in the Twentieth Century.* Honolulu: University of Hawaii Press, 1986.

Lawson, S. *The Failure of Democratic Politics in Fiji.* Oxford: Clarendon Press, 1991.

Lawson, S. *Tradition versus Democracy in the South Pacific.* Cambridge: Cambridge University Press, 1996.

Lea, David and Colette Milward (editors). *A Political Chronology of South-East Asia and Oceania.* London: Europa, 2001.

Nayacakalou, R. R. *Leadership in Fiji.* Melbourne, Australia: Oxford University Press, 1975.

Niditauae, T. *Pastors in Politics: The Question of Political Involvement and Church Leadership in Vanuatu.* Suva Fiji: Pacific Theological College, 1985.

Norton, R. *Race and Politics in Fiji.* St Lucia: University of Queensland Press, 1994.

Oliver, Douglas L. *The Pacific Islands (Third Edition).* Honolulu: University of Hawaii, 1989.

Pomfret, John. *The Beautiful Country and the Middle Kingdom, America and China, 1776 to the Present.* New York: Henry Holt and Co., 2016.

Riesenberg, Felix. *The Pacific Ocean.* London: Whittlesey House, 1940.

Rutherford, Noel. *Tonga and Tongans: heritage and identity.* Melbourne, Australia: Oxford University Press, 1977.

Scarr, Deryck. *Fiji: A Short History.* Sydney: Allen & Unwin, 1984.

Sharp, Andrew. *The Discovery of the Pacific Islands.* Oxford: Clarendon Press, 1960.

Tegiogni, Eric Dungmo. Ministry of Finance, The Cameroons, March 10, 2016.

Thakor, Ramesh (editor). *The South Pacific Problems, Issues and Prospects.* Houndmills, Macmillan, 1991.

Thompson, Christina. *Sea People: The People of Polynesia.* New York: Harper, 2019.

Wainwright, Elsina. *Our Failing Neighbour: Australia and the Future of Solomon Islands.* Canberra, Australia: Australian Strategic Policy Institute, 2003.

Ward, R. Gerald and Elizabeth Kingdon (editors). *Land, Custom and Practice in the South Pacific.* Cambridge: Cambridge University Press, 1995.

Winchester, Simon. *Pacific.* New York: Harper, 2015.

JOURNAL ARTICLES AND PAPERS

Abi-Habib, Maria and Hassan Moosa. "Maldives Opposition Declares Upset Victory in Presidential Election." The New York Times, Monday, Sep. 24, 2018, p.A6.

Ahmed, Azam and Kirk Semple. "Hurricane Gains Force and Targets Caribbean." The New York Times, Tuesday, September 19, 2017, p.A8.

Aldermam, Liz. "A World without Ice? Iceland Is Preparing." The New York Times, Monday, August 9, 2019, pp.1, B4-5.

Allen, Matthew G. "Resisting SAMSI, Intervention, Identity and Symbolism in the Solomon Islands." Oceania-Volume 79, No. 1 March 2009, pp.1-17.

Amin, Mridula and Isabella Kwai. "Scars Surface for Asylum Seekers in Australia." The New York Times, Thursday, November 8, 2018, p.A4.

Appelbaum, Binyamin. "Little Noticed Trade Fact: It's No Longer Rising." The New York Times, Monday, October 31, 2016, pp.A1, A3.

Bacevich, Andrew J. "American Imperium." Harper's Magazine, May 2016, p.32.

Barbados Advocate. "Mitchell Leads NNP in clean sweep in Grenada." Thursday, March 15, 2018, p.13.

Bell, David A. "The Many Lives of Liberalism." The New York Review of Books, January 17, 2019, pp. 25-27.

Belson, Ken. "Paradise Threatened." The New York Times, Sunday, October 28, 2018, pp.TR1, 4-5.

Blanding, Richard. "A Rare Map or a Crafty Fake." The New York Times, Monday, December 11, 2017, pp.C1, C8.

Bosman, Julia. "The Storms Moved On. The Caribbean Islands Fear the Tourists Might, Too." The New York Times National, Sunday, September 24, 2017, p.19.

Brooks, David. "The Fragmented Society." The New York Times, Friday, May 20, 2016, p.A23.

Brown, Marley. "Marshal Islands Stick Chart" in "Tapping the Past." Archaeology Magazine, May/June, 2019, p.33.

Cave, Damien. "Island-Hopping In the South Pacific, Before the Seas Flood Them Out." The New York Times, Friday, July 27, 2018, p.A6.

Cave, Damien. "Climate Choices, and Costs Fall to Cities and Towns." The New York Times, Sunday, October 13, 2019, p.6.

Cave, Damien. "China Is Leasing a Pacific Island, Its Residents Are Shocked." The New York Times, October 17, 2019, p.A2.

Coetzee, J. M. "Australia's Shame" The New York Review of Books, September 26, 2019, pp.85-89.

Cohen, Roger. "Death by Numbers." The New York Times, December 31, 2016, p.A21.

Cohn, Julie. "Can a Blur Lead to Earhart?" The New York Times, Tuesday, August 13, 2019, pp.D1, D3.

Commission on British West Indies Colonies, Report on closer association of the British West Indian Colonies. HMSO, London, 1947.

Comparative Law Quarterly, Cambridge University Press on behalf of the British Institute of British and Comparative Law, July 1993, pp.710-718.

Cummings-Bruce, Nick and Andreas Rires. "U.N. Chief Says Deal to End Cyprus Divide Could Be Near." The New York Times, January 13, 2017, p.A3.

Davenport, Carol. "The Marshall Islands Can't Sue Nuclear Power." The New York Times, October 6, 2010, p.A5.

Davenport, Coral. "A Pacific Island Nation Struggles Against Relentless Rising Sea." The New York Times, December 2, 2015, pp.A1, A8.

Davenport, Coral. "Optimism Faces Grave Reality at Climate Talks." The New York Times, December 1, 2014, pp.A1, A13.

Dunn, Richard. "The Barbados Census of 1680: Profile of the Richest Colony in English America." William and Mary Quarterly 26, 1969, pp.3-30.

The Economist, "Africa's traditional leaders," December 23, 2017, p.73.

The Economist, "An old tongue' new tricks," December 23, 2017, pp.63-5.

The Economist, "Church of poor judgement," November, 7, 2015, p.47.

The Economist, "Half mast," December 10, 2016, p.88.

The Economist, "Pardon me," November 7, 2015, p.37.

The Economist ,"Pointillist power," March 16, 2019, p.72.

The Economist, "The Reluctant European," October 19, 2015, p.8.

The Economist, "You say raki, I say ouzo," April 23, 2016, p.41.

The Economist, "Why storms are getting worse," September 22, 2018, pp.54-56.

The Economist, In special section 'The World If' "The $78 trillion free lunch," July 15, 2017, pp.6-8.

The Economist, "The isle is full of noises," April 7, 2018, p.43.

The Economist, "2020 Division," January 5, 2019, pp. 23-4.

The Economist, "A clean sweep," June 2, 2018, pp.30, 32.

The Economist, "A Curse in Disguise," July 26, 2018, p.28.

The Economist, "A darker shade of blue," November 28, 2015, p.70.

The Economist, "A despot's guide to foreign aid," April 16, 2016, p.40.

The Economist, "A dismal dynast" and "Ex Factor," January 21, 2017, pp.8, 37-8.

The Economist, "A flicker of light in a prison state," August 4, 2018, pp.39-40.

The Economist, "A Green Revolution," March 12, 2016, pp.21-3.

The Economist, "A home in the country," September 29, 2018, pp.56-8.

The Economist, "A thousand golden stars," July 22, 2017, p.35.

The Economist, "Agreeing to Agree," September 3, 2016, p.38.

The Economist, "Asia is at last waking up to the threat of a trade war," June 30, 2018, p.25.

The Economist, "Australasia Feels the Heat," September 12, 2015, p.39.

The Economist, "Back into the Fold," September 20, 2014, p.30.

The Economist, "Bello No Brussels here," July 9, 2016, p.28.

The Economist, "Big fish," April 9, 2016, p.53.

The Economist, "Britain, a lonely domino," September 29, 2018, p.52.

The Economist, "Bunyan Back from the dead," May 6, 2017, p.40.

The Economist, "Buttonwood, Loathe thy Neighbour," January 9, 2016.

The Economist, "Charlemagne Please Mr. Erdogan," January 28, 2017, p.47.

The Economist, "Charlemagne: High wall, Narrow Sea," November 14, 2015.

The Economist, "China's investments around the Indian Ocean are indebting recipients and antagonizing India," March 10, 2018, p.44.

The Economist, "Climate change, What goes up" and "Nothing so concentrates the mind," September 21, 2019, pp.26-30, 66-7.

The Economist, "Corking the Genie," January 23, 2016, p.35.

The Economist, "Coups, I did it again," July 26, 2018, p.35.

The Economist, "Daggers out at the chicken's neck," July 29, 2017, p.34.

The Economist, "Debt relief for Dolphins," September 9, 2017, pp.44-5.

The Economist, "Dorian's Wrath," September 7, 2019, pp.34-5.

The Economist, "The day after tomorrow," September 28, 2019, pp. 56-7.

The Economist, "Sea changes," September 28, 2019, p.57.

The Economist, "The best offence is a good defence," Septemer 28, 2019, p.81.

The Economist, "The Vatican's Finances," October 12, 2019, p.78.

The Economist, "China in the Pacific," October 26, 2019.

The Economist, "Elections in Bhutan," October 13, 2018, pp.40-1.

The Economist, "Foam Flecked," September 3, 2016, p.34.

The Economist, "Forget whiter than white," July 8, 2017, pp.10-11.

The Economist, "Gateway to the Globe," July 26, 2017, pp.13-16, 35.

The Economist, "Coups, I did it again," July 26, 2018, p.35.

The Economist, "Global Power," July 25, 2015, p.67.

The Economist, "Half of Belize, please," April 21, 2018, p.29.

The Economist, "How to Make Eritria Less Horrible," August 4, 2018, pp.10-11.

The Economist, "Indian foreign policy," December 23, 2017, pp.47-8.

The Economist, "Liberty and disintegration," December 8, 2018, p.14.

The Economist, "London or local," November 3, 2018, p.36.

The Economist, "Lord of the Isles," December 22, 2018, p.75.

The Economist, "Making waves," March 12, 2016, p.38.

The Economist, "Mark Three," December 17, 2014, p.30.

The Economist, "Master and commander, for now," December 15, 2018, p.38.

The Economist, "Most failed state," September 10, 2016, p.41.

The Economist, "No enchanted evening," March 21, 2015, p.33.

The Economist, "No Jemmah Tommor," January 28, 2017, p.61.

The Economist, "Nothing to see here," August 13, 2016, p.49.

The Economist, "Off the map," July 14, 2018, p.72.

The Economist, "Opening the gates," May 19, 2018, pp.SP 3-12.

The Economist, "Pacific gyre," November 3, 2018, pp.40, 42.

The Economist, "Palace in the jungle," March 12, 2016, pp.46-7.

The Economist, "Palm-fringed pandemonium," August 19, 2017, p.30.

The Economist, "Paradise Lost," September 10, 2017, pp.29-30.

The Economist, "Piqued in Papua New Guinea," November 24, 2018, p.33.

The Economist, "Plaguing Paradise," August 29, 2015, p.30.

The Economist, "Ports in the Horn," July 26, 2018, pp.33-4.

The Economist, "Presidential pardon," June 13, 2015, pp.35-6.

The Economist, "Repair Job," November 18, 2017, p.63.

The Economist, "Seeds of Suspicion," April 27, 2019, pp. 38-9.

The Economist, "Sibling Rivalry," October 29, 2016, p.34.

The Economist, "Small but not too beautiful," August 13, 2016, p.38.

The Economist, "Small is Beautiful, "June 1, 2019, p.31.

The Economist, "Solomon's Choice," May 4, 2019, p.34.

The Economist, "Stongman vs. Strongman," November 17, 2016, p.46.

The Economist, "Tear down these walls," February 27, 2016, pp.37-8.

The Economist, "The 1.2 billion opportunity," Special report Business in Africa, April 16, 2016.

The Economist, "The corrections" and "Lots not to like," August 12, 2017, p.22, 36.

The Economist, "The Great Divergence," March 12, 2016, pp.25-6.

The Economist, "The Great Wharf," April 21, 2018, pp.33-4.

The Economist, "The holdout", September 10, 2016, p.41.

The Economist, "The march of democracy slows," August 20, 2016, p.37.

The Economist, "The Mother of Invention," December 24, 2016, p.54.

The Economist, "The risk of relying on Chinese cash," July 26, 2018, p.30.

The Economist, "The superpowers' playground," April 9, 2016, pp.49-50.

The Economist, "The World Economy, An Open and Shut Case," October 1, 2016, p.33.

The Economist, "Tourism in South-East Asia," April 14, 2018, pp.32-3.

The Economist, "Trading Places," January 20, 2018, p.35.

The Economist, "Vladmer's Choice," December 23, 2017, pp.47-8, 53-5.

The Economist, "Where does the aid go?" June 11, 2016, p.62.

The Economist, "Worth celebrating?" June 11, 2016, p.51.

The Economist, "Fisticuffs at dawn," August 26, 2017, p.31, 34.

The Economist, "The rising seas," August 17, 2019, pp.15-6.

The Economist, "Nearly sweet nothing," December 16, 2017, p.32.

The Economist, "Going to Bits," January 14, 2017, p.48.

The Economist, "In the lurch" and "Murder in Paradise," October 21, 2017, pp.21-6, 52.

The Economist, "Special report Emerging Markets Out of the traps," October 7, 2017, pp.1-20.

Edwards, John. "How the Spaniards Got There First." The New York Review of Books, January 14, 2016, pp.60-62.

Ewing, Jack. "Dimitris Christofias, 72, Cyprus Who Served One Disaster-Filled Term." The New York Times, Monday, June 24, 2019, p.A21.

French, Howard W. "Africa's Lost Kingdoms." The New York Review of Books, June 27, 2019, pp.47-49.

Freytas-Tamura, Kimono De. "Tourists Flood Iceland after Two Catastrophes." The New York Times, Thursday, November 17, 2016.

Friedman, Lisa. "As Another Storm Churns On, Islands Already Hurting Seek Aid at U.N." The New York Times, Wednesday, September 20, 2017, p.A6.

Gettleman, Jeffrey. "Book Culture Rises in Bhutan." The New York Times, Monday, August 20, 2018, p.C1-2.

Gillis, Justin. "Our Climate Change Future Has Arrived." The New York Times, Sunday, October 13, 2019, p.SR 3.

Goldstone, Heather and Elsa Parten. "Here's What Happens if Your Island Nation Goes Underwater." The New York Times, Wednesday, December 10, 2015.

Goodman, Peter S. and Liz Alderman. "Iceland's Tourism Industry Goes Dry." The New York Times, Monday August 26, 2019, pp.1, 4-5.

Gross, Matt. "Devastation Everywhere, but Still a Destination." The New York Times, Sunday, March 25, 2018, p.TR 10.

Grynberg, Roman and Roy Mickey Joy. "The Accession of Vanuatu to the WTO." The Journal of World Trade, 2000, Kluwer Law International, 2000, p.695.

Gutierrez, Jason and Hannah Beech. "Philippines Demands Chinese Inquiry Into Sinking of Boat." The New York Times, Friday, June 14, 2019, p.A4.

Higgins, Andrew. "Cypriots Fear Russian Meddling in Settlement Talks." The New York Times, Monday, February 6, 2017, p.A4, A6.

Hockenos, Paul. "Europe is Falling." The New York Times, Sunday, August 21, 2016, p.16.

Holmes, Stephen. "The Identity Illusion." The New York Review of Books, January 17, 2019, pp.44-5, 48.

Horn, Sebastian, Carmen Reinhart, and Christ of Trebesch. "China's Overseas Lending." Kiel Working Paper 2132, Kiel Institute for World Economics, 2019.

Horowitz, Jasdon and Steven Eblangerr. "Italy Welcomes China's Leader, and Vast Infrastructure Project." The New York Times, Saturday, March 23, 2019, p.A4.

Horton, Chris. "Island Nation Said to Weigh Cutting Ties With Taiwan." The New York Times, Friday, September 6, 2019, p.A9.

Hughes, Helen. "Industrializing Small Countries." Industry and Development, 1984, pp.89-9.

International Monetary Fund. "World Economic Outlook Database." April 12, 2017.

Ives, Mike. "A Vision for Floating Cities To Fend Off Rising Seas." The New York Times, January 28, 2017, p.A4.

Ives, Mike. "Promised Billions for Climate Change. Poor Countries Are Still Waiting." The New York Times, September 10, 2018, p.A10.

Ives, Mike. "Remote Pacific Nation, Threatened by Rising Seas." The New York Times, July 3, 2016, p.10.

Jacobs, Andrew and Jane Perlez. "U.S. Wary of a Chinese Base rising as Its Neighbor in Africa." The New York Times, Sunday, February 28, 2017, pp.A1, A9.

Jacobs, Andrew. "Joyous Africans Take to Rails, With China's Help." The New York Times, Wednesday, February 8, 2017, pp.A1, A3.

Jakobsdottir, Katrin. "An Ice-Free Iceland Is Not a Joke." The New York Times, Sunday, August 18, 2019, p.SR 2.

June 23, 2018, pp.B1, B7.

Karasz, Palko. "No Free Ride? Luxembourg Disagrees." The New York Times, Friday, December 7, 2017, p.7.

Kay, John. "The Economics of Small States." Hume Occasional Paper no. 81 2, October 1, 2008.

Keating, Joshua. "The U.S. Likes the World Map the Way It Is." The New York Times, Sunday, September 24, 2017, p.SR 10.

Kekic, Laza. "The Economist Intelligence Unit's Index of Democracy." in The World in 2007, The Economist, 2007.

Kitti, Michael. The New York Times, Saturday, March 29, 2015.

Knowles, Rachel and Kirk Semple. "Pummeled islands in Bahamas Escape a 2nd Blow." The New York Times, Sunday, September 15, 2019, p.4.

Knowles, Rachel, Elisabeth Malkin, and Frances Robles. "'Dystopian Mess' as Storm Cripples the Bahamas." The New York Times, Wednesday, September 4, 2019.

Kolbertt, Elizabeth. "Global Warming." The New Yorker, October 22, 2018, pp.23-4

Krauss, Clifford. "Guyana's Coalition Government Falls After No-Confidence Vote." The New York Times, Sunday, December 23, 2018, p.A8.

Krauss, Clifford. "Political Rifts Return After Guyana Changed Course." The New York Times, Saturday, August 11, 2018, p.A5.

Krugman, Paul. "Earth, Wind, and Liars." The New York Times, Tuesday, April 17, 2018.

Kulp, Scott A. and Benjamin H. Strauss. "New elevation data triples estimates of global vulnerability to sea level rise and coastal flooding." Nature Communications 10, 2019.

Lanchester, John. "World on Fire." The New York Times Book Review, April 28, 2019, pp.1, 20.

Landler, Mark and Edward Wong. "Bolton Views Bigger U.S. Role in Africa, but Goal Is Countering China's Sway." The New York Times, Friday, December 14, 2018.

Lewis, W. Arthur. "Economic Development with Unlimited Supplies of Labour." Manchester School of Economic and Social Studies, 1954, pp.139-91.

Leys, Colin. "What is the Problem about Corruption?" Journal of Modern African Studies, III, No. 2, pp.215-230.

Lipset, Seymour Martin. "Some Social Requisited of Democracy: Economic Development and Political Legitimacy." American Political Science Review 3, 1959.

Loeak, Christopher. "A Clarion Call from the Climate Change Front Line." The Huffington Post, January 23, 2016.

Lu, Denise and Christophe Flauelle. "Erased by Rising Seas by 2050." The New York Times, Wednesday, October 30, 2019, p.A6.

Magra, Iliana. "Barbaric Penalties in Brunei: Stoning for Gay Sex and Amputations for Theft." The New York Times, April 4, 2019, p.A6.

Malkin, Elisabeth. "Bahamians Hunker Down as Dorian Strikes as 'Catastrophic' Storm." The New York Times, Monday, September 2, 2019, p.A14.

McKibben, Bill. "World at War." New Republic, August 16, 2016.

McKibben, Bill. "How Extreme Weather is Shrinking the Planet." The New Yorker, November 26, 2018.

McNeil, Donald G., Jr., "A Buzzing Thing in the Sky' Delivers Vaccines to Vanuatu." The New York Times, Tuesday, December 18, 2018, p.A10.

Millard, James A. "Is China a Colonial Power?" The New York Times, Sunday, May 6, 2018, p.SR 7.

Moose, Hassan and Jeffrey Gentleman. "President of Maldives Besieges Supreme Court Over Imprisoned Foes." The New York Times, February 6, 2018, p.A6.

Myers, Steven Lee, and Chris Horton. "2 Pacific Nations Sever ties With Taiwan, Boltersing Chinese Influence in Region." The New York Times, Saturday, September 21, 2019.

Myers, Steven Lee. "Squeezed by an Indian-China Standoff, Bhutan Holds Its Breath." The New York Times, August 16, 2007, pp.A1, A9.

Nairin, Suchi. "Tiny atoll in Pacific cries out for help." The Times of India, June 6, 2008.

Najar, Nida. "Ex-President Is Arrested In Maldives." The New York Times, February 23, 2015, p.A6.

Neuman, William and Jeffrey C. Mays. "Plan for Rising Seas: not Retreat, but Advance." p.A27.

Nijhuis, Michelle. "Early Warnings." The New York Review of Books, June 27, 2019, pp.37-39.

Osnos, Evan. "Making China Great Again." The New Yorker, January 8, 2018, pp.36-45.

Plumer, Brad. "Warming Poses Grave Danger to World's Oceans." The New York Times, Thusday, September 26, 2019, pp.Ai, A9.

Popper, Nathaniel. "On Money: We know plenty about the losers in global trade. . . . winners?" The New York Times Magazine, September 11, 2016, pp.20-4.

Porter, Eduardo and Karl Russel. "Migrants Are on the Rise Around the World, And Myths About Them Are Shaping Attitudes." The New York Times, Wednesday, June 20, 2018, pp. 1,13.

Radu, Daniela Julianna. "Tax Havens Impact on the World Economy." Proceedings – Social and Behavioral Sciences, Elsevier, Oxford, 2013.

Rasheed, Zaheena and Geeta Anand. "Ex-Vice President of Maldives Convicted of Trying to Kill President." The New York Times, Friday, June 10, 2016, p.A8.

Roberts, Sam. "Bribery and Corruption, or 'Honest Graft.'" The New York Times, Sunday, June 19, 2016, p.SR5.

Robies, France, Kirk Semple, and Richard Perez-Pena. "Vicious Storm, Roaring West. Hits Caribbean." The New York Times, Thursday, September 7, 2017, pp.A1, A14.

Rogers, Katie. "Eyeing Greenland, Trump Again Mixes Real Estate With Diplomacy." The New York Times, Saturday, August 17, 2019, p.A9.

Rubin, Alissa A. "New Wave of Popular Fury May Crash Down in 2017." The New York Times, Tuesday, December 6, 2016, p.A11.

Rytz, Mathieu. "Sinking Islands, Floating Nation." The New York Times, February 7, 2018, p.A3.

Schendler, Auden and Andrew P. Jones. "Stopping Climate Change Is Hopeless. Let's Do It." The New York Times, Sunday, October 7, 2018, p.SR 10.

Schultz, Kai. "A Law School in a Kingdom of Buddhism." The New York Times, Sunday, October 9, 2016.

Schultz, Kai. "From Bhutan to New York's Dairyland Heartland." The New York Times, January 27, 2016, pp.A20-22.

Schultz, Kai. "In Bhutan, Happiness Index as Guage for Social Ills." The New York Times, Wednesday, January 18, 2017, p.A6.

Scobbi, Iein. "Case concerning Certain Phosphate Lands in Nauru (Nauru v. Australia) Preliminary Objections Against Judgement." The International.

Semple, Kirk and Azam Ahmed. "Hurricane Maria Does 'Mind-Boggling Damage to Dominica, Leader Says." The New York Times, Wednesday, September 20, 2017.

Semple, Kirk, Elisabeth Malkin, and Frances Robles. "Deaths Reported As Brutal Storm Pounds Bahamas." The New York Times, Tuesday, September 3, 2019.

September 22, 2019, p.A7.

Semple, Kirk. "Battling Over Where, or Whether, to Rebuild After Storms." The New York Times, Tuesday, October 8, 2019, p.A8.

Starr, Paul. "The Battle for the Suburbs." The New York Review of Books, September 26, 2019, pp.38-40.

Stevenson, Alexandra and Motoko Rich. "Trans-Pacific Trade Allies Move On without the U.S." The New York Times, Sunday, November 12, 2017, p.A8.

Stiglitz, Joseph E. "The Euro, How the Common Currency Threatens the Future of Europe." The New York Times Book Review, Sunday, August 21, 2016, p.17.

Swanson, Ana and Jim Tan Kersley. "Trade Deals Still in Motion, U.S. or No U.S." The New York Times, Thursday, January 25, 2018, pp.B1, B3.

Tarabay, Jamie. "Australia's Military Ties in the Pacific, Balancing U.S. and Chinese Alliances." The New York Times, Wednesday, June 12, 2019, p.A9.

Thrush, Glenn. "China's Weight Fuels Reversal By Trump On Foreign Aid." The New York Times, Monday, October 15, 2018, pp.B1, B3.

Warne, Kennedy. "Against the Tide." National Geographic, November 2015, pp.125-135.

Webb, A.P. and P. S. Kench. "The dynamic response of reef islands to sea-level rise." Global and Planetary Change, Elsevier Retrieved 2010, July 2013.

Wolfe, Matthew. "Without a Trace." Harper's Magazine, February 2019, pp.25-36.

Wong, Edward. "China's Edge Over U.S. Contractors in Africa: Truckloads of Loans." The New York Times, Sunday, January 13, 2019, pp.1-2.

Worrell, DeLisle. "Policies for Stabilization and Growth in Small Very Open Economies." Occasional Paper No. 85, Group of Thirty, Washington, D.C., 2017.

Worrell, DeLisle. "The Barbados Economy, The Road to Prosperity." unpublished paper, 2017.

INTERVIEWS

Adedeji, Adebayo, Director, Institute of Public Administration, University of Ife, Nigeria (later Director General of the African Union) Multiple times 1967-1970.

Antoine, Timothy, Governor, Eastern Caribbean Central Bank (ECCB), February 6, 2016.

Bakija, Jon, Professor of Economics, Williams College, October 14, 2014.

Baptiste, Michael, Macroeconomic Advisor, Ministry of Finance and Energy, February 10, 2016.

Bernhardsson, Magnus, Professor, Williams College, September 19, 2014.

Cagilaba, Lielani Monoci Seruwaia, Reserve Bank of Fiji, multiple times, 2016-2017.

Davis, James C., Professor Emeritus, University of Pennsylvania, multiple times 2013-2015.

Delgado, Rosa Brito Delgado, Central Bank of Cabo Verde, multiple times September- December 2018.

Edwards Q. C., Celia, Old Fort, St. George's, Grenada, West Indies, February 13, 2016.

Gudjonsson, Hiynar, Iceland Consul and Trade Commissioner in New York, interview with Hans Humes 18 Sep

Gunnlaugsson, Sigmundur, Prime Minister of Iceland, September 24, 2014.

Humes, Hans, Chief Executive Officer, Greylock Capital Management, LLC, multiple times 2013-2019.

Hyde, John, Professor Emeritus, Williams College, November 21, 2014, multiple (2014-2018).

Jones, Kimbell, Monday, September 10, 2018.

Kimbi, Tchenga Wanchi Michael, Ministry of Economy, Planning, and Regional Development, Cameroons, September 19, 2014, March 29, 2015 and June 6, 2015.

Localio, Liam, Greylock Capital investment broker, January 17, 2018.

Mahon, James, Woodrow Wilson Professor, Williams College, multiple, 2011-2018.

Marshall, Joachim, police officer, Dominica, November 25, 2018.

Mediretta, A. J., Partner, Greylock Capital Management, multiple times, 2010-2018.

Moriarty, Francis, columnist, November 11, 2017.

Murray, David J., professor emeritus and provost, Open University, United Kingdom, and Visiting Professor, University of Fiji, multiple times 2001-2018.

Nyanzi, Suleiman, Bank of Uganda executive, interviews September 21, 2015 and May 14, 2016.

Pettis, Michael, Professor, University of Beijing Interview September 25, 2015.

Peyrefitte, Michael, as Speaker of the Belize House of Representatives, (December 26 and 30, 2013), and as Attorney-General, February 17 and 20, 2017.

Sareer, Ahmed, Permanent Representative of Maldives to the UN (and, Ambassador to the United States and Chairman of AOSIS), November 2, 2015.

Simati, Aunese Makoi, Permanent Representative of Tuvalu to the United Nations, November 2, 2015.

Sinnott, Brendan, retired Director General, the European Union, multiple interviews, 2014-2019.

Teasdale, Anthony, Director of Research, European Union, and co-author of The Penguin Companion to European Union, multiple times, 2014-2018.

Worrell, DeLilse, Governor, Central Bank of Barbados (2007-2018), January 22, 2018 and April 20, 2018, and email June 3, 2018 and July 1, 2018.

Worrell, DeLilse, Governor, Central Bank of Barbados (2007-2018), interview with Hans Humes September 14, 2020.

Yoder, Douglas, Deputy Director, Miami-Dade County Water and Sewer Department, member Florida Water Resources Advisory Commission, multiple times, 2018-2019.

The author is also indebted to *The World Almanac and Book of Facts*.

INDEX

OTHER BOOKS BY SAMUEL HUMES IV

The Structure of Local Government Throughout the World (with Eileen Martin)

The Structure of Local Government—A Comparative Survey of 81 Countries (with Eileen Martin)

Local Governance and National Power, A Worldwide Comparison of Tradition and Change in Local Government, 1991

Managing the Multinational, Confronting the Global-Local Dilemma, 1993

Government and Local Development in Western Nigeria (with Robert F. Ola) 1995

Belgium: Long United, Long Divided, 2014